The Systematics of

Lasiopogon

(Diptera: Asilidae)

Robert A. Cannings

Victoria, Canada

Published by the Royal British Columbia Museum, 675 Belleville Street, Victoria, British Columbia, V8W 9W2, Canada.

All drawings by Robert A. Cannings, except for the drawing on the cover and figure 1 by Rolof Idema (reprinted with permission).
All photographs by Robert A. Cannings, except for figure 22 by S.G. Cannings (reprinted with permission).

Edited, designed and typeset by Gerry Truscott, RBCM.
Set in Times New Roman PS (body text 10/12).
Cover design by Chris Tyrrell, RBCM.

Printed and bound in Canada by Hignell Book Printing.

National Library of Canada Cataloguing in Publication Data
Cannings, Robert A., 1948-
  The systematics of Lasiopogon (Diptera: Asilidae)

  This book is the published form of the author's doctoral thesis
defended in October 1999. Cf. Preface.
  Includes bibliographical references: p.
  ISBN 0-7726-4636-8

  1. Robber flies. I. Royal British Columbia Museum.

QL537.A85C36 2001   595.77'3   C2001-960264-2

CONTENTS

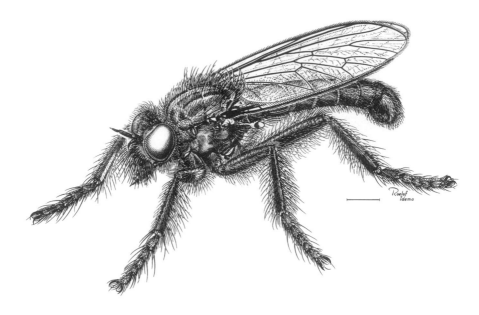

Figure 1. *Lasiopogon hinei* Cole and Wilcox, male; this is the only *Lasiopogon* species living in both the Old and New worlds. Scale line = 1 mm.

PREFACE

This book is the published form of my doctoral thesis defended in October 1999 in the Department of Environmental Biology at the University of Guelph, Guelph, Ontario.

The genus *Lasiopogon* is a speciose and widespread group of robber flies (Diptera: Asilidae) inhabiting the north temperate parts of the Earth. It is the most northerly ranging genus of asilid flies and the relationships of the species on either side of the Bering Strait offer excellent opportunities for biogeographical research. No study has ever examined the genus as a complete entity and no clearly defined intrageneric relationships have been proposed, although Bezzi (1921) suggested some tentative and general ones for the European fauna. Initially, my goal was to revise the world fauna and subject it to a phylogenetic and biogeographic analysis. Basic to this was the description of new species, the redescription of existing ones stressing genitalic characters, the creation of workable identification keys and the documentation of detailed species distributions.

At the outset, treatments of the Asilidae in the most recent Nearctic (Martin and Wilcox 1965) and Palaearctic (Lehr 1988) catalogues listed 37 and 14 species respectively, for a total of 51. But after I assembled large amounts of material, I determined that *Lasiopogon* contained 73 valid species names; I reduced this to 69 after detailed study. Further examination of the material uncovered 43 undescribed species (not including 6 proposed but not yet published by Kovár and Hradský (1996)), and it is clear that more species await discovery, especially in central Asia. Appendix 1 contains a preliminary checklist of the species.

A comprehensive phylogenetic revision of the whole genus in the time available was not practicable. Although I examined and identified all the material assembled, and documented many new species, I analysed only a fraction of the species in detail. The study includes a cladistic overview of some defined species groups and a detailed taxonomic and phylogenetic analysis of the seven species groups and 29 species in the monophyletic *opaculus* section, comprising about 25 per cent of *Lasiopogon* species and largely Nearctic in distribution. I describe 14 new species and redescribe the others.

The phylogenetic hypothesis of the genus as a whole is treated only by employing putative species groups defined by exemplar species. Thus, the phylogeny of these groups is considered tentative and will not be resolved more

1

explicitly until all known *Lasiopogon* species are described or redescribed. Despite the incomplete nature of this revision, I am convinced that it is worth publishing in its present form so that it can be used as a published basis for future species-group revisions and phylogenetic analyses.

The identification keys cover all known North American species and those in Asia east of 60°E longitude. Europe contains many undescribed species, and the bulk of the fauna there was not studied in enough detail to allow the production of keys. The three keys in this book contain undescribed species as well as described ones; although this emphasizes the incomplete nature of the publication, it permits the accurate identification of all known species, which more conservative keys would not allow.

I use the evolutionary relationships developed in the phylogenetic hypotheses to explain the present biogeographic patterns and possible origins of the species discussed. The most derived clades of the *opaculus* section are Asiatic, and the biogeographic analysis includes discussion of links between the Nearctic and Palaearctic *Lasiopogon* faunas throughout the Tertiary and Quaternary. In addition, I examine the place of *Lasiopogon* among its close relatives. My study accepts, with reservations, the status of the Stichopogoninae (Artigas and Papavero 1990) as presently conceived, and I outline a phylogenetic hypothesis of the subfamily.

INTRODUCTION

GENERAL BIOLOGY OF THE ASILIDAE

The robber fly family (Diptera: Asilidae) contains more than 6700 described species worldwide (Geller-Grimm 2000). Robber flies are predators that as adults pursue other insects (usually flying ones), seize them and kill them with paralysing saliva injected through the hypopharynx (tongue). They then suck up the liquefied contents of the prey through the proboscis (Whitfield 1925, Wood 1981). The morphology of the adult fly, especially the prominent eyes, the mouthparts and the raptorial legs, reflects this mode of prey capture and feeding. Robber flies usually hunt in open, well-lit areas and are most active in the warmest parts of the day; overcast skies greatly curtail their activity. Different genera, and often different species within a genus, have different hunting behaviour and preferences for perching sites.

There is usually little obvious difference between the sexes, except for the genitalia, although females tend to be larger than males and often have broader abdomens. Colour patterns sometimes differ between males and females. In *Lasiopogon* these differences are minor; often the tomentum on the abdomen is more extensive (but less dense) in the female. But in some other genera, the differences are marked. Males of some species of *Cyrtopogon*, for example, have prominent, dark marks on the wings. Other secondary sexual characteristics occur in males, such as the expanded silver abdominal apex in *Nicocles*, the striking white abdomens of *Efferia* and the tarsal ornamentation of some *Cyrtopogon* species.

Records of prey taken by Asilidae indicate that they are mostly opportunistic predators, feeding upon any insect that they can subdue and kill. But some species show a strong preference for prey from one or two insect orders (Wood 1981); in many instances this may simply reflect the availability of prey. *Lasiopogon* is known to attack several orders of insects, but is most commonly found with Diptera as prey (Hobby 1931, Lavigne 1972, Lavigne and Holland 1969, Melin 1923, Poulton 1906, Weinberg 1978).

Detailed life-history studies of robber flies are rare. Melin (1923), studying Asilidae in Sweden, showed that in northern species, at least, the larva is the overwintering stage and the pupal stage lasts two to six weeks. He estimated that

3

the life cycle of *Laphria* species was at least three years and that of *Lasiopogon cinctus* (Fab.) was at least two. It is likely that larval growth is faster in warmer regions and many species probably live only one year (Theodor 1980).

Robber fly larvae prey on the eggs, larvae and pupae of other insects in the soil or in rotting wood, although the immature larvae of a few species are ectoparasitic on their hosts (Knutson 1972, Wood 1981).

Hull (1962) treated the world genera of Asilidae, and Wood (1981) keyed the North American genera. Engel 1930 is long out of date but is the only publication that has dealt with the entire family at the species level in the Palaearctic, although Lehr has produced a significant body of systematic and ecological publications for various groups of the Palaearctic fauna (e.g., Lehr 1962, 1984a, b, 1996). Majer (1997) keyed the European genera. Various publications examined parts of the Holarctic fauna from a regional geographic perspective (e.g., Adisoemarto 1967, Baker and Fischer 1975, Cole 1969, Bromley 1934 and 1946, Hine 1909, James 1941, Nelson 1987, Oldroyd 1969b and 1970a, Weinberg and Bächli 1995) or a taxonomic one (e.g., Lehr 1984a, Martin 1975, Wilcox 1966, Wilcox and Martin 1936). Wood (1981) gave a summary of the morphology, biology and classification of the North American fauna. Oldroyd (1970b, 1974) and Londt (e.g., 1985, 1994) treated parts of the Afrotropical fauna. Lavigne (1999) and Lavigne et al. (1978) produced bibliographies for asilid literature subsequent to Hull's review (1962), and Geller-Grimm (2000) compiled a comprehensive bibliography on the internet.

INTRODUCTION TO *LASIOPOGON*

Members of the robber fly genus *Lasiopogon* live around the temperate regions of the Northern Hemisphere from Britain to far-eastern Russia, from Alaska to New England (map 1). *Lasiopogon* is perhaps the most northerly ranging genus of the Asilidae; species reach the Arctic Ocean in North America and along the lower reaches of the great Siberian rivers. Its diversity is greatest in the Cordillera of western North America and in Europe, although it may also be diverse in central Asia, especially in China, where its true status is unclear. This study finds that *Lasiopogon* comprises 69 described and at least 49 undescribed species.

The flies are small to medium-sized, rather setose asilids averaging 8 to 10 mm long (figure 1), usually brown or black with varying amounts of grey or brown tomentum colouring the body. In some species the genitalia and parts of the legs are ferruginous. The facial tubercle and mystax are prominent. The adults perch on the ground or on logs and rocks (less frequently on low vegetation) and are opportunistic predators of small insects, especially Diptera. The larvae also are predators and live in sandy soil, especially along sea beaches and streambeds, but also in subalpine meadows, dry forests, grasslands and heaths. Except for taxonomic work or mentions in faunistic studies, little has been

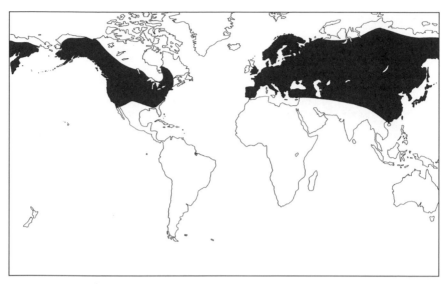

Map 1. *Lasiopogon* world distribution. The southern boundary of Palaearctic fauna from the Caspian Sea to the South China Sea is estimated.

written about *Lasiopogon*. The exceptions are the ecological and behavioural discussions by Lehr (1984b) and Lavigne (1972), Lavigne and Holland (1969), and a paper on cytology (Metz and Nonidez 1924).

Oldenberg (1924) wryly noted: "The genus *Lasiopogon* has long belonged to the problem children of the Asilidae." At first glance, species of the genus are similar morphologically, and some of the earlier attempts to record species and their ranges resulted in problems that more detailed work has since failed to eliminate. The identity of many species of *Lasiopogon*, especially in Europe, has long been obscured by confused taxonomy, erroneous identification keys, and incorrect identifications in collections. For example, the species limits of *Lasiopogon montanus* Schiner, the most widespread montane *Lasiopogon* species of Europe, and its closest relatives have been misunderstood ever since 1862 when *L. montanus* was named (Shiner 1862). Most of this confusion was the result of a vague understanding of species limits, inadequate descriptions and inattention to the original material. As a result, the validity of *L. bellardii* Jaennicke has been unrecognized for decades (Cannings 1996). The same confusion has perpetuated the mistake that *L. pilosellus* Loew and *L. macquarti* (Perris) are members of the Alpine fauna (Weinberg and Bächli 1995) even though the former is known only from Turkey and the latter from southwestern France.

Dissected male genitalia offer stable and unequivocal characters for setting species limits in *Lasiopogon*. This is especially helpful, because many of the characters traditionally used in *Lasiopogon* descriptions, such as colour and chaetotaxy, are variable and taxonomically unreliable. But no species were originally described using characters from dissected genitalia and, not surprisingly, the systematics of the genus has suffered. The relationships of *Lasiopogon* to other genera have never been adequately defined and no phylogenetic analysis has ever been published for the genus.

The flawed taxonomy, the plethora of undescribed species and absence of a study on the Palaeartic and Nearctic faunas as a whole indicate that a comprehensive generic revision of *Lasiopogon* is overdue. This widespread and speciose genus also offers excellent opportunities for biogeographic analysis.

TAXONOMIC HISTORY

Loew (1847) divided the genus *Dasypogon* Meigen into many groups. He proposed the name *Lasiopogon* as a group (subgenus) name for *Dasypogon cinctus* (Fabricius), the common species of northern Europe, and *D. pilosellus* and *D. tarsalis*, both of which he described from Turkey. He also synonymized *D. hirtellus* Fallén and *D. cinctellus* Meigen with *D. cinctus*, and tried to address the confusing taxonomy that already plagued the group at that early date (see pp. 40-41).

The last of the *Lasiopogon* species to be placed originally in *Dasypogon, D. macquarti*, was named by Perris (1852). By 1862 Schiner had named *Lasiopogon montanus* and in 1867 Jaennicke described *L. bellardii*; *Lasiopogon* was now being used as a full generic name. Loew himself published descriptions of the first North American species – *L. bivittatus* (1866), *L. tetragrammus* (1874), and *L. opaculus* (1874) – but by the latter date he changed the name to *Daulopogon* because he had learned that *Lasiopogon* was in use for a genus of plants in the Asteraceae. Osten Sacken described *Daulopogon arenicola* (1877) and adopted the new generic name in his catalogue (1878). This name change was unwarranted, however, and Williston restored *Lasiopogon* in his influential manual on North American Diptera (1908). Johnson (1900) was last to use the name *Daulopogon* when he named *D. terricola*. Back (1909) assembled the type material of these five known Nearctic species (he mentioned, but did not include, the recently described *L. quadrivittatus* (Jones 1907)) and summarized the fauna with a key and descriptions.

By the time Bezzi (1921) described eight species and revised the genus in Eurasia, there were 16 known species in the Old World. Bezzi was the first to try to make sense of the relationships among the Eurasian fauna. He divided the species into three loosely characterized groups: a montane (Alps/Apennine) group with large, robust and setose species; a second type represented by *L. cinctus* from lowland northern and central Europe; and a third group of smaller, more delicate species, much less setose, with the abdomen largely pollinose. This last group he considered more-or-less Mediterranean in distribution. Bezzi also keyed the six known Nearctic species based on descriptions in the literature. His is the only publication in which species from both the New and Old worlds are discussed, even if only briefly.

In his major treatment of the Asilidae in *Die Fliegen der palaearktischen Region*, Engel (1930) described *L. bezzii* and included the three species (*L. intermedius, L. lichtwardti* and *L. nanus*) named by Oldenberg (1924) in his insightful contribution to the study of the genus. Engel's treatment today is badly dated, but remains the latest synthesis of the European *Lasiopogon* fauna. Timon-David (1950) described *L. fourcatensis* from the Pyrenees, the sole new European species to be recognized between 1930 and Hradský's initial work in the 1960s.

Lindner (1966) tried to clarify the relationships among the species Bezzi described from the Alps. However, because he did not dissect male genitalia he was unable to recognize the species accurately; in fact, he made matters worse by synonymizing several valid names. In 1966, Castellani and Crivaro added new records to the literature on European *Lasiopogon* species. Unfortunately, their paper was oddly formatted, indenting some species accounts under others (*L. cinctus, L. immaculatus* and *L. montanus*), leading Lehr (1988) to believe many species had been synonymized therein. Because of this misunderstanding, Lehr's (1988) treatment of *Lasiopogon* in the *Catalogue of Palaearctic Diptera* synonymized a number of valid species.

Hradský and Moucha (1964) and Hradský (1982) described *L. soffneri* and *L. peusi* from Bulgaria and Greece, respectively. Hradský (1981) named both the Japanese species (*L. akaishii* and *L. rokuroi*) and *L. eichingeri*, the only known Chinese species at the time, while Richter (1962, 1977) described *L. avetianae* from the Caucasus and *L. tuvensis* from the Altai. Lehr (1984a) recorded 7 new Eurasian species, ranging from Ukraine and Uzbekistan to Primorskiy Kray on the Pacific Ocean. He also added some insights, especially ecological ones, into the phylogeny of various genera of the Stichopogonini (Lehr 1984b). Lehr's work brought the total of valid Palaearctic species to 34; this was reduced to 33 when I synonymized *L. sibiricus* with *L. hinei* (Cannings 1997; see below). Kovár and Hradský (1996) recognized six more species from Europe, but the descriptions are still in manuscript.

In the New World, after Back's 1909 summary of the genus, taxonomists in the western United States began collecting asilids in earnest, taking up where Osten Sacken left off. Cole (1916) named *L. drabicola* and the widespread *L. cinereus*. Melander (1923) added 7 new western species to the fauna and keyed the 15 known American species. Cole (1924) described *L. littoris*, bringing the total of Nearctic species to 16 by the late 1920s, the same total that was known in the Palaearctic at that time. In 1934, Curran split *Alexiopogon* from *Lasiopogon* to hold the aberrant *L. terricola* (Johnson), with its unusual coloration and lack of dorsocentral and scutellar bristles.

The most significant single publication on *Lasiopogon* was the revision of the North American species by Cole and Wilcox (1938). The authors examined most of the specimens in American collections at the time, and described 21 new species from localities as far apart as Alaska and New England, New Mexico and

Georgia. Like Melander (1923), they stressed the importance of the form of the epandrium in defining species, but still did not investigate the detailed morphology of the male and female genitalia that is revealed by dissection, and so failed to recognize some valid species. In his 1923 key, Melander included a species, *Lasiopogon arizonensis* Schaeffer (1916), that Cole and Wilcox (1938) noted as belonging in *Cophura* Osten Sacken. Pritchard (1943) confirmed the placement of the species in *Cophura*.

Since Cole and Wilcox 1938, only 2 other species – *L. prima* (Adisoemarto 1967) and *L. polensis* (Lavigne 1969) – have been named in North America, bringing the number of species (including *L. terricola*) to 39 at the beginning of my study. I have reduced the number of valid names to 36 but added new ones. In an earlier publication (Cannings 1997), I discussed four species having Beringian affinities and synonymized *L. sibiricus* Lehr with *L. hinei* Cole and Wilcox, recognizing *L. hinei* as having populations in both the Palaearctic and Nearctic.

The higher classification of the Asilidae is in flux; it suffers from a serious lack of phylogenetic analysis (Yeates 1994) and there is no satisfactory cladogram at the subfamily and tribal levels. Phylogenies for tribes have been proposed in only a few studies (e.g., Adisoemarto and Wood 1975, Fisher 1986). Papavero (1973a) outlined the history of robber fly classifications. Macquart (1838), using wing venation and antennal structure, first divided the Asilidae into Dasypogonites, Laphrites and Asilites, corresponding to the three subfamilies, Dasypogoninae, Laphriinae and Asilinae, that were subsequently maintained by most 19th-century entomologists. The Dasypogoninae was characterized by an open marginal cell (cell r1). *Leptogaster* and its relatives were soon separated into a fourth subfamily, the Leptogastrinae, and later, Martin (1968) raised it to a family, a change that was refuted by Oldroyd (1969).

Problems in this simplified, four-family classification of the Asilidae were rife, not the least of which was that the Dasypogoninae were clearly a residual plesiomorphic group that included a wide diversity of genera. *Lasiopogon*, in this classification, belongs to the Dasypogoninae (Hull 1962, Martin and Wilcox 1965, Wood 1981) and has traditionally been placed in the Tribe Stichopogonini. Hardy (1930) erected this tribe to include dasypogonine species with acanthophorites in the female terminalia. Hardy and later Karl (1959) reduced the number of subfamilies to two: the Dasypogoninae and the Asilinae.

Karl, in a pioneering cladistic work largely dealing with the male genitalia, included the Leptogastrini in the Asilinae because of the apomorphic one-segmented maxilliary palp, and linked the Laphriini and relatives with the Dasypogonini, Damalini and others in an enlarged Dasypogoninae. He placed *Lasiopogon* and the Stichopogonini in this latter group but gave them separate lineages. Hull (1962), less thoughtfully, restricted the tribe to species characterized by a dorsally divergent frons, but this included *Willistonina* Back, a taxon that clearly belonged elsewhere with *Cyrtopogon, Stenopogon* and others.

Later authors have attempted various modifications to the classification. Papavero (1973a) placed the Stichopogonini in a new subfamily, Stenopogoninae, based on Hull's Tribe Stenopogonini (Hull 1962). This detail was followed by Lehr (1988) in the *Catalogue of Palaearctic Diptera*. Artigas and Papavero (1990) raised the tribe to subfamily rank in the Stichopogoninae; Fisher and Wilcox (1997) adopted this classification in their catalogue of the robber flies of the Nearctic Region.

Several authors now recognize 10 genera of Stichopogoninae (Artigas and Papavero 1990, Lehr 1988): *Afghanopogon* Hradský (Afghanistan), *Argyropogon* Artigas and Papavero (Argentina), *Clinopogon* Bezzi (Africa, Australia, Fiji, Indonesia), *Eremodromus* Zimin (Central Asia, Middle East, Egypt, Sudan), *Lasiopogon* Loew (widespread in North America and Eurasia), *Lissoteles* Bezzi (Mexico south to Peru on Pacific coast, except for one species on the Atlantic coast of Central America); *Rhadinus* Loew (Central Asia, Middle East, North Africa, Sudan, South Yemen), *Stackelberginia* Lehr (Central Asia), *Stichopogon* Loew (worldwide), *Townsendia* Williston (United States to South America). *Psilinus* Wulp was included in the subfamily by Oldroyd (1974) and Lehr (1988), but Londt (1993) assigned it to *Rhabdogaster* Loew and confirmed that it belongs in the Stenopogoninae.

Lasiopogon, although long classified with the above genera, shows significant differences in the form of the facial tubercle and mystax, the form and density of setae, and especially in the structure of the male genitalia.

MATERIALS AND METHODS

SOURCES OF MATERIAL

This study is based on the examination of approximately 18,000 specimens of *Lasiopogon* and other asilids borrowed, with the help of the curators and collections managers listed in parentheses, from the 85 institutions or individuals listed below. Collection codes in the text indicate the location of type material and other specimens; they are based, with some additions, on those assigned to the insect collections of the world by Arnett, Samuelson and Nishida (1993).

AGSC   A.G. Scarbrough Collection (private), Baltimore, Maryland, U.S.A.

AMNH   American Museum of Natural History, New York, New York, U.S.A. (D. Grimaldi).

ANSP   Academy of Natural Sciences of Philadelphia, Philadelphia, Pennsylvania, U.S.A. (J. Gelhaus, D. Azuma).

BCPM   Royal British Columbia Museum, Victoria, British Columbia, Canada (R.A. Cannings, J.A. Cosgrove).

BMNH   Natural History Museum, London, United Kingdom (J. Chainey).

BPBM   Entomology, Bernice P. Bishop Museum, Honolulu, Hawaii, U.S.A. (N.L. Evenhuis).

BYUC   Monte L. Bean Life Science Museum, Brigham Young University, Provo, Utah, U.S.A. (R. Baumann).

CAES   Connecticut Agricultural Experiment Station, New Haven, Connecticut, U.S.A. (C.T. Maier).

CASC   Department of Entomology, California Academy of Sciences, San Francisco, California, U.S.A. (P.H. Arnaud, Jr, N.D. Penny, K. Ribardo).

CDAE   California Department of Food and Agriculture, Sacramento, California, U.S.A. (E.M. Fisher).

CNCI   Canadian National Collection of Insects, Agriculture Canada, Ottawa, Ontario, Canada (J.M. Cumming, D.M. Wood).

CRNC   R. Nelson Collection (private), Provo, Utah, U.S.A.

CSUC   Department of Entomology, Colorado State University, Fort Collins, Colorado, U.S.A. (B.C. Kondratieff).

CUCC   Department of Entomology, Clemson University, Clemson, South Carolina, U.S.A. (B. Robinson, K.M. Hoffman).

CUIC   Department of Entomology, Comstock Hall, Cornell University, Ithaca, New York, U.S.A. (J.K. Liebherr, E.R. Hoebeke).

DEBU   Department of Environmental Biology, University of Guelph, Ontario, Canada (S.A. Marshall).

DEIC   Deutsches Entomologisches Institut, Eberswalde-Fonow, Germany (F. Menzel).

DENH   Department of Entomology, College of Life Sciences and Agriculture, University of New Hampshire, Durham, New Hampshire, U.S.A. (D.S. Chandler).

EDNC   North Carolina Department of Agriculture, Raleigh, North Carolina, U.S.A. (K.R. Ahlstrom).

EDUM   J.B. Wallis Museum, Department of Entomology, University of Manitoba, Winnipeg, Manitoba, Canada (R.E. Roughley).

EMFC   E.M. Fisher Collection (private), Sacramento, California, U.S.A.

ENAM   Ecole Nationale Superieure Agronomique de Montpellier, Montpellier, France (F. Leclant).

EMEC   Essig Museum of Entomology, Department of Entomology, University of California, Berkeley, California, U.S.A. (J.A. Chemsak).

EMUS   Entomology Museum, Utah State University, Logan, Utah, U.S.A. (W.J. Hanson).

ESUW   Department of Plant, Soil and Insect Sciences, University of Wyoming, Laramie, Wyoming, U.S.A. (R.J. Lavigne).

ETHZ   Eidgenössische Technische Hochschule Zürich, Zürich, Switzerland (B. Merz).

FMNH   Field Museum of Natural History, Chicago, Illinois, U.S.A. (A.F. Newton, P.P. Parillo).

FSCA   Florida State Collection of Arthropods, Florida Department of Agriculture, Gainesville, Florida, U.S.A. (G. Steck).

FGGC   F. Geller-Grimm Collection (private), Frankfurt, Germany.

HLDH   Hessisches Landesmuseum, Darmstadt, Germany (W. Schneider).

IBPV   Entomology Department, Institute of Biology and Pedology, Russian Academy of Sciences, Vladivostok, Russia (P. Lehr, A. Lelej).

INHS   Department of Entomology, Illinois Natural History Survey, Champaign, Illinois, U.S.A. (D.W. Webb; K.R. Methven).

IZAS   Institute of Zoology, Academia Sinica, Beijing, China (Y. Shi).

KSUC   Department of Entomology, Kansas State University, Manhattan, Kansas, U.S.A. (H.D. Blocker).

KUIC   Entomological Laboratory, Kagoshima University, Kagoshima, Japan (A. Nagatomi, D. Yang).

LACM   Los Angeles County Museum of Natural History, Los Angeles, California, U.S.A. (B.V. Brown).

LEMQ Lyman Entomological Museum, McGill University, Ste.-Anne-de-Bellevue, Quebec, Canada (T. Wheeler, C.-C. Hsiung).

LGBC L.G. Bezark Collection (private), Sacramento, California, U.S.A.

MCNV Museo Civico di Storia Naturale, Venice, Italy (L. Munari).

MCZC Department of Entomology, Museum of Comparative Zoology, Harvard University, Cambridge, Massachusetts, U.S.A. (P. Perkins, M.S. Kelley, C.T. Graham).

MDBC M.D. Baker Collection (private), Ames, Iowa, U.S.A.

MHNN Musée d'Histoire Naturelle, Neuchatel, Switzerland (J.-P. Haenni).

MHRC M. Hradský Collection (private), Zasmuký, Czech Republic.

MNHN Entomologie, Muséum National D'Histoire Naturelle, Paris, France (L. Tsacas).

MSNM Museo Civico di Storia Naturale, Milano, Italy (C. Leonardi).

MSUC Department of Entomology, Michigan State University, East Lansing, Michigan, U.S.A. (F.W. Stehr).

MTEC Department of Entomology, Montana State University, Bozeman, Montana, U.S.A. (M.A. Ivie).

MZLU Universitets Zoologiska Institut, Lund University, Lund, Sweden (R. Danielsson).

MZUF Museo Zoologico "La Specola", Florence, Italy (L. Bartolozzi).

NHMW Naturhistorisches Museum, Vienna, Austria (R. Contreras-Lichtenberg).

NHRS Naturhistoriska Riksmuseet, Stockholm, Sweden (B. Viklund).

NMBA Naturhistorische Museum des Stifts Admont, Admont, Austria (B. Hubl, E. Krasser, J. Goetze).

NMBS Naturhistorisches Museum, Bern, Switzerland (E. Obrecht).

NMCC Naturmuseum Chur, Switzerland (via G. Baechli).

NYSM New York State Museum, Albany, New York, U.S.A. (J.K. Barnes).

ODAC Oregon Department of Agriculture, Salem, Oregon, U.S.A. (R. Westcott).

OSUC Department of Entomology, Ohio State University, Columbus, Ohio, U.S.A. (C.A. Triplehorn).

OSUO Department of Entomology, Oregon State University, Corvallis, Oregon, U.S.A. (A. Brower, J. Lattin, G. Brenner).

QMOR Collection entomologique Ouellet-Robert, University of Montreal, Montreal, Quebec, Canada (P. P. Harper, M. Coulledon).

ROME Royal Ontario Museum, Toronto, Ontario, Canada (D.C. Darling).

RMNH Nationaal Natuurhistorische Museum, Leiden, Netherlands (J. van Tol).

RUIC Department of Entomology and Economic Zoology, Rutgers State University, New Brunswick, New Jersey, U.S.A. (M.M. May).

SEMC Snow Museum of Entomology, Department of Entomology, University of Kansas, Lawrence, Kansas, U.S.A. (R.W. Brooks).

SMDV Spencer Entomological Museum, Department of Zoology, University

	of British Columbia, Vancouver, British Columbia, Canada (G.G.E. Scudder, K. Needham).
SMFD	Forschungsinstitut Senckenberg, Frankfurt, Germany (W. Tobias).
SMNH	Royal Saskatchewan Museum, Regina, Saskatchewan, Canada, (K. Loney, R. Hooper).
SMNS	Staatliches Museum für Naturkunde, Stuttgart, Germany (H.-P. Tschorsnig).
SMTD	Staatliches Museum für Tierkunde, Dresden, Germany (U. Kallweit).
TAMU	Department of Entomology, Texas A & M University, College Station, Texas, U.S.A. (J. Oswald, E.G. Riley).
TAUI	Department of Zoology, University of Tel Aviv, Tel Aviv, Israel (A. Freidberg).
UAIC	Department of Entomology, University of Arizona, Tucson, Arizona, U.S.A. (D.R. Maddison).
UASM	Strickland Entomological Museum, Department of Biological Sciences, University of Alberta, Edmonton, Alberta, Canada (G. Ball, D. Shpeley).
UCDC	Department of Entomology, University of California, Davis, California, U.S.A. (R.O. Schuster).
UCRC	Department of Entomology, University of California, Riverside, California, U.S.A. (S.I. Frommer).
UGCA	Department of Entomology, University of Georgia, Athens, Georgia, U.S.A. (C.L. Smith).
UMHF	Universitetets Zoologiska Museum, Helsinki, Finland (P. Vilkmaa).
UMMZ	Museum of Zoology, University of Michigan, Ann Arbor, Michigan, U.S.A. (B.M. O'Connor, M. O'Brien).
UNSM	University of Nebraska State Museum, Lincoln, Nebraska (B.C. Ratcliffe, W.E. Hall).
USNM	United States National Museum of Natural History, Smithsonian Institution, Washington DC, U.S.A. (F.C. Thompson, G.F. Hevel).
USUC	Department of Biology, Utah State University, Logan, Utah, U.S.A. (W.J. Hanson).
WSUC	James Entomological Museum, Department of Entomology, Washington State University, Pullman, Washington, U.S.A. (R.S. Zack).
ZISP	Zoological Institute, Russian Academy of Sciences, St. Petersburg, Russia (V. Richter).
ZMAN	Zoologische Museum, Universiteit van Amsterdam, Amsterdam, Netherlands (B. Brugge).
ZMHB	Museum für Naturkunde der Humboldt-Universität zu Berlin, Berlin, Germany (H. Schumann).
ZMPA	Institute of Zoology, Polish Academy of Sciences, Warsaw, Poland (P. Trojan).
ZSMC	Zoologisches Staatsammlung, Munich, Germany (W. Schacht).

SPECIMEN PREPARATION

Specimens collected in the field were killed in vials containing ethyl acetate fumes and prepared for drying within 12 hours of capture. Specimens were pinned through the right side of the scutum with a 00 insect pin, then pinned into a Schmitt box for drying at room temperature. During mounting, the specimens were arranged to reveal critical characters: the wings were spread at about 45° to show the dorsum of the abdomen, the abdomen was propped so that the terminalia (in lateral view) were in a line with the head and thorax, and the legs were drawn down so as not to obscure the thoracic pleura.

Terminalia were clipped off the abdomen with microscissors. In males the cut was made at the base of segment 7, in females at the base of segment 6 (the spermathecae extend well back into the abdomen).

Male genitalia were placed in a vial of 10% KOH and allowed to soak for 12 hours. This treatment resulted in a specimen excellent for dissection. Alternatively, the vial containing the KOH and genitalia can be heated in a beaker of water on a hotplate for 10 minutes, but this method is somewhat more damaging to the membranes and does not produce as consistent a result as the cold-soaking treatment. The specimens were removed from the vials and washed in glacial acetic acid and then distilled water (15 minutes in each solution). They were then placed in a deep well slide containing glycerine and dissected under a dissecting microscope (Wild M8).

Dissection of the male genitalia consisted of three steps. First, using fine dissecting probes made from insect pins, the genital capsule was separated from the undifferentiated abdominal segments by cutting the intersegmental membrane between segment 8 and the hypandrium and epandrium. The genital capsule was then split into dorsal and ventral halves by severing the lateral connections between the hypandrium and epandrium and by cutting the subepandrial membrane at its base. The final task was to remove the phallus complex from its anchors in the bowl formed by the fused hypandrium/gonocoxae) by severing the lateral connections of the parameral sheath with the gonocoxal apodemes. The resulting four pieces could then be examined from all angles: (1) tergite and sternite 8, hidden from view in the undissected insect; (2) the epandrium/proctiger complex with its ventral subepandrial sclerite; (3) the fused hypandrium/gonocoxite complex without the obscuring phallus, and with the important gonostyli; (4) the phallus complex.

The male *Lasiopogon* genital capsule is rather heavily sclerotized, and these dissections are considerably more difficult if the KOH treatment is replaced by the more gentle lactic acid one (Sinclair et al. 1994).

Female genitalia were macerated in a similar fashion to those of the males, although they were left in the KOH for 18 hours instead of 12. This is counterintuitive, because the spermathecae, with their significant characters, are extremely delicate and potentially easily damaged by excess maceration. But the destruction of surrounding tissues tangled among the spermathecal tubes is more important to a successful dissection than any additional weakening of the spermathecae that might occur. Actually, the spermathecae stand up well to the additional maceration. The lateral abdominal membranes were cut to the apex of segment 8 and the genitalia placed in Evan's Blue (0.5%) stain for five minutes. This lightly colours the fine, transparent spermathecae, making them visible among the tracheae and other material. The tergites, back to tergite 8, were then removed, exposing the furca (sternite 9) and the spermathecal ducts. The remaining body contents, such as the tracheae, were separated from the integument and the were spermathecae disentangled. Care was taken not to tear the long, delicate ducts, although this is difficult. The furca was separated from sternite 8 and was mounted, together with the attached spermathecae, in glycerine on a microscope slide.

The dissected material was stored in glycerine in microvials pinned below the specimen from which the material came.

MEASUREMENTS AND DRAWINGS

Most measurements were made through a Wild M8 stereoscopic dissecting microscope; the maximum magnification used was 100X. Most of the measurements of components of the genitalia were obtained with a Leitz Laborlux S compound microscope and a micrometer slide with a minimum scale interval of 0.01 mm. The maximum magnification used was 250X.

Measurements listed below are given in each species description. These measurements (range and mean) are based, where possible, on a sample of 10 specimens (5 males, 5 females).

Intraspecific body size can be extremely variable in the Asilidae, and is presumably dependent on natural genetic variation among individuals and larval nutrition. Body length may vary depending on the variable amount of abdominal expansion at the time of drying. I attempted to include the complete range of size shown by the specimens examined when selecting specimens for measurement. In addition, I selected samples from as wide a geographical range as possible. Nevertheless, the measurements should be taken as a guide only; examination of longer series of specimens often results in values outside the ranges given. Several measurements are incorporated into ratios to compensate for differences correlated with body size. Measurements are expressed as follows:

Body length   The linear distance from the anterior extremity of the face (gibbosity) to the posterior extremity of the abdomen (terminalia). When using the general terms "small", "medium-sized" and "large" to refer to the size of a specimen, I define small as less than 6mm long, medium-sized as 6-12 mm long and large as over 12 mm long.

Head (figure 2a-c).

> **Head width (HW)**   In anterior view, the greatest distance between the lateral margins of the eyes.
>
> **Face width (FW)**   The distance between the eyes at the level of the antennae (least width).
>
> **Vertex width (VW)**   In dorsal view, the distance between the eyes along a line touching the anterior margins of the posterior ocelli. The ratio FW/VW is a measure of the dorsal divergence of the eyes.
>
> **Vertex depth (VD)**   In anterior view, the distance between the lowest point of the vertex and dorsal margin of the eye, a measure of the depth of the interocular depression.
>
> **Gibbosity height (GH)**   In lateral view, the greatest distance between the anterior margin of the eye and the apex of gibbosity.
>
> **Gibbosity length (GL)**   In lateral view, the distance between the dorsal margin of gibbosity and the ventral margin of the eye. The ratio GH/GL is a relative measure of the height of the gibbosity.

Antenna (figure 2d).

> **Length flagellomere 1 (LF1)**   In lateral view, the distance from the base of the flagellomere to the apex.
>
> **Width flagellomere 1 (WF1)**   In lateral view, the greatest width of the flagellomere. The ratio WF1/LF1 gives a measure of the elongation of the flagellomere.
>
> **Length flagellomere 2+3 (LF2+3)**   In lateral view, the distance from the base of flagellomere 2 to the apex of flagellomere 3. The ratio LF2+3/LF1 is used in the species descriptions.

Wing (figure 2e)

> **Position of r-m crossvein (DC1)**   The base of the discal cell (forking of M_1 and M_3) to the r-m crossvein.
>
> **Length of discal cell (DC2)**   The base of the cell (see DC1 above) to the centre of the junction between the m-m crossvein and M_2. The ratio DC1/DC2 is the discal cell index (DCI).

The drawings are mostly semi-diagrammatic representations of structures. They are not necessarily intended to depict all the morphological details, but to illustrate important features. For example, illustrations of the hypandrium and gonocoxites do not include setae. Notes on the illustrations of particular structures are given in the discussion of those structures in the section on *Lasiopogon* morphology.

I made the drawings using a Wild M8 dissecting microscope and camera lucida, or a Leitz Laborlux S compound microscope equipped with a drawing tube, depending on the magnification required.

Consistent orientation and stability of genitalic components is essential during the measuring and drawing of these structures under the compound microscope, so I temporarily embedded the dissected genitalia in glycerine gel. I made the

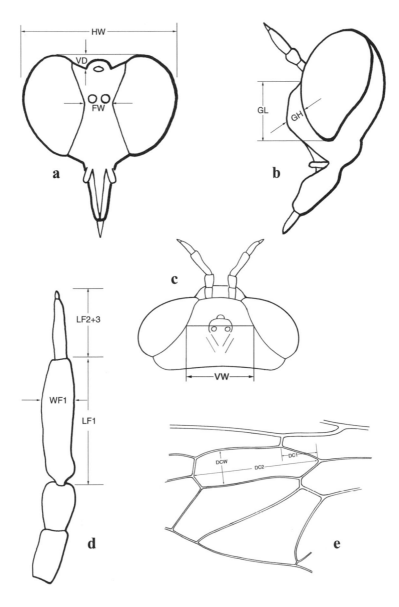

Figure 2. Measurements of *Lasiopogon*: **a**, anterior view of head (FW = face width, HW = head width, VD = vertex depth); **b**, lateral view of head (GL = length of gibbosity, GH = height of gibbosity); **c**, dorsal view of head (VW = vertex width); **d**, antenna (LF1 = length of first antennal flagellomere, LF2+3 = combined length of second and third antennal flagellomere, WF1 = greatest width of first antennal flagellomere); **e**, discal cell of wing (DC1 = length from base of discal cell to r-m crossvein. DC2 = length of discal cell. DCW = greatest width of discal cell).

glycerine gel by mixing equal amounts of boiling water and powdered gelatine, then stored the resulting solid in a covered Petri dish. When ready, I placed a small amount of the gel on a depression slide, then heated it on a hotplate until it became fluid. Placing the genitalic structure to be drawn in the liquid, I oriented it in the desired position as the liquid solidified. Fixed in the solid gel, the structure could be microscopically examined without it drifting or turning in the field. To draw wings, antennae and spermathecae, I mounted them on flat slides in glycerine.

PRESENTATION OF SPECIES DESCRIPTIONS

Species descriptions (beginning on page 71) are presented in alphabetical order. Each begins with a list of synonymies; for previously described species or subspecies, reference to the original publication of the valid name and each junior synonym is presented chronologically.

"Diagnosis" is a summary of critical identifying features. Species are described in a standard fashion based on the discussion of morphology (pp. 21-39) and the measurements taken. Where significant variation in colour or setation occurs between males and females, there is a separate section for females; e.g., under Thorax or Abdomen. A determination made with the keys should be confirmed first by comparison with the diagnosis, then by using the description.

Allotypes are not designated for new species but are included where designated by other authors. Data for primary types of both new and previously described species are cited in full, under the heading "Type Material", following the system described by O'Hara (1983). Labels are listed from the top downward, with data from each label enclosed in quotation marks and the lines of each label delimited by oblique slash marks. Data are recorded exactly as found on the labels, with additional information, including my interpretations of data, given in square brackets. If the genitalia have been dissected, this is indicated and the name of the person who did the dissection is indicated in square brackets. The repository of the primary types is indicated. Label data for paralectotypes and paratypes other than allotypes are simply listed alphabetically by location. Other specimens studied are listed in the same manner under "Other Material Examined", and their total number is given. Countries are arranged alphabetically; localities are listed alphabetically by country (and provinces and states in North America). All collection dates are presented in the order of day, month (lower-case Roman numerals) and year. Repositories for all specimens are in parentheses after the collector's name. Distribution is further depicted on range maps for the species.

Information on the type locality is presented under the heading "Type Locality"; the label data on the primary type listed verbatim under "Type Material" are not always sufficient to characterize the location. Where necessary, comments concerning taxonomic history and synonymies are given under

"Taxonomic Notes". The derivation of the specific epithet is documented under "Etymology". A summary of the species' range is presented under "Distribution". "Phylogenetic Relationships" summarizes sister species relationships and membership in other monophyletic groups. Under the general heading "Natural History" is included information on habitat, seasonal occurrence, prey and other biological notes; this is taken from label data, the literature and my own observations.

Names of species recognized as new but not described have been designated with the first three letters of the species group to which they belong and a number; e.g., *Lasiopogon* biv-1 sp. nov.

In this study, I have distinguished a number of monophyletic species groups in the genus *Lasiopogon*. In naming these groups and those lineages I have informally called clades, sections and so on, I use the specific epithet of the first-named species included in the group (e.g., the *opaculus* group) rather than the species' binomial (e.g., the *Lasiopogon opaculus* group). The names of these lineages are descriptive only and should not be considered an incorrect use of the species name (Wheeler 1991).

SPECIES CONCEPT FOR *LASIOPOGON*

The delimitation of *Lasiopogon* species is based on the morphospecies concept. Consistent, definable gaps in morphological characters, especially in the male genitalia, represent the boundaries of species. At present, adult morphology is the most comprehensive source of characters for convenient comparison in these flies.

I examined the male genitalia (especially the epandrium, gonocoxite, phallus and, most particularly, the gonostylus) of specimens from one location to ascertain uniformity of structure and compared these structures with those of specimens from other localities. I assumed that groups of specimens with the same genitalic character states belonged to the same, widely distributed, reproductively continuous species, and that such similarity represented genetic continuity. I inferred, therefore, that a group with distinctive genitalic structure is reproductively isolated from other groups, in the absence of evidence to the contrary (Askevold 1991). Then, by finding other morphological characters that correlated with these genitalic characters, I constructed species identities.

PHYLOGENETIC ANALYSIS

The approach I used to reconstruct the phylogenetic relationships of *Lasiopogon* follows the cladistic method originally proposed by Hennig (1965, 1966). Elaborations and refinements of this method have been discussed by many authors (e.g., Maddison et al. 1984, Watrous and Wheeler 1981, Wheeler 1986, Wiley 1981).

Only morphological characters, mostly from the external and genitalia of both sexes, were used in the analyses. I defined states for each character and inferred that they are ancestral (plesiomorphic) or derived (apomorphic) through outgroup analysis (Maddison et al. 1984 and others). Outgroups are any groups used in an analysis that are not included in the taxon under study (the ingroup). A number of genera of Stenopogoninae and Stichopogoninae were considered as outgroups in character assessment. I paid special attention to *Bathypogon* (Stenopogoninae: Bathypogonini), whose male genitalic structure corresponds most closely to the unusual genitalia of *Lasiopogon*. I then recognized monophyletic groups (Hennig 1966) on the basis of shared derived states (synapomorphies) arranged into a succession of nested hierarchies, producing trees showing common ancestry (cladograms). Groups of related taxa (clades) must be supported by synapomorphies; shared ancestral character states are uninformative. Autapomorphies, those derived states unique to taxa terminal on branches of the cladogram, are also uninformative in phylogeny and were not used in the three analyses. But computer analysis does include these on the cladogram and they have been retained as useful in providing information on distribution of character states.

I weighted characters using the system described by Marshall (1985, 1987). Three categories of characters are weighted in order of their apparent value in elucidating relationships. Those coded "+++" are characters uniquely derived in the ingroup, thus providing strong support of monophyly. Characters coded "++" are apparently independently derived in the ingroup but occur in outgroup(s); they may provide strong evidence for monophyly, but because they are subject to homoplasy, misinterpretation of homologies may result. The characters coded "+" are widespread in both the ingroup and outgroups; homoplasy is common. Character weights were not incorporated in the Hennig86 analyses, but reflect the confidence that I have in the character's utility to define robust clades. They are used in discussions concerning the strength certain characters provide in support of monophyletic groups.

The phylogenetic reconstruction of *Lasiopogon* carried out here is based on three analyses: (1) genera of Stichopogoninae; (2) *Lasiopogon* species groups; and (3) species of the *opaculus* section. The three character matrices (tables 1-3, pp. 268, 274 and 296) were analysed using Hennig86 (version 1.5) (Farris 1988), a computer program for analysis of phylogenetic relationships using parsimony. Hennig86 options "bb" and "ie" were used. The cladograms (figures 250-60) were produced using CLADOS (version 1.2) (Nixon 1992) from Hennig86 output and were subsequently redrafted to improve readability.

MORPHOLOGY OF *LASIOPOGON* AND STRUCTURES USED IN CLASSIFICATION

I have followed McAlpine 1981, with few exceptions, for general terminology of dipteran morphology, and Wood 1991 and Sinclair et al. 1994 for additional clarification of male genitalia structure. I prefer this terminology over that used by Hull (1962) and Theodor (1980) in their works on Asilidae, by Cole and Wilcox (1938) in their revision of Nearctic *Lasiopogon*, and the genitalic terminology of the Asilidae proposed by Karl (1959). Terms given in this section are used throughout the text and especially in the species descriptions and diagnoses.

Where colour is used in descriptions it refers to the condition in mature adults where possible. Tenerals or young specimens can be significantly paler than mature ones. But considerable variation can occur in mature specimens either in the same population or, more commonly, in geographically distant parts of a species' range. When it was encountered, such variation is described briefly. Description of colour is not based on any standardized colour code and is kept as simple as possible. Where two colours are mixed, the names are hyphenated with the dominant colour second (e.g., gold-grey). A solidus (forward slash) indicates variation between two colours (e.g., brown/black means that the colour varies from brown to black).

Ground colour is the colour of the cuticle, often obscured by tomentum, the microscopically superficial extensions of the cuticle that are usually called pollinosity in the Asilidae (Cole and Wilcox 1938, Hull 1962) or pruinosity (McAlpine 1981). As Fisher (1977) noted, the term "pollen" implies a dust-like or granular nature, whereas under high magnification tomentum appears as minute recumbent curved microtrichia. Areas without tomentum are usually shiny.

Setation is well developed but variable. Because of its greater variability, chaetotaxy in the Asilidae has not been applied taxonomically as successfully as it has in the Cyclorrhapha. I have designated some groups of setae, especially on the head (e.g., frontal setae, orbital setae), that are not usually separately discussed in asilid descriptions. Because of their variable numbers and position, these setae are not as precisely defined as those in many other taxa, especially in higher ones. Setae include setulae (very short, stubble-like setae), hairs (delicate, flexible, soft setae), bristles (strong, stiff setae), and spines (short, thick setae).

21

Lasiopogon species are small robber flies 5 to 12 mm long, compact but slender, with wings extending more-or-less to the apex of the abdomen (figure 1). The body is moderately setose, grey or brown tomentose, at least on the head and thorax, and usually on the abdomen. The abdomen usually has bands of grey or light brown tomentum on the apices of the tergites contrasting with darker tomentum (glabrous, shining cuticle) basally, but frequently is completely covered in grey tomentum. The cuticle is black, dark brown or sometimes, in part, ferruginous, especially on the antennae, the lateral angle of postpronotal lobe, the apices of posterior abdominal segments, the male and female terminalia, and the legs. The colour of the setae, either white/yellow or brown/black, is useful in species identification. This is especially true for setae on the mystax, scutum, katatergites, scutellum, legs and tergite 1. Females have fewer setae than males do, especially on scutum, legs and abdomen.

HEAD (figure 3)

In lateral view the face has a protuberant gibbosity, rather more emphatic dorsally, and bearing a mystax of many long, often strong, setae. The gibbosity is usually rather abruptly reduced to the eye level just ventral to the antennal bases. The head is covered with tomentum in some shade of silver, grey, gold or brown. The eye is long, strongly convex anteriorly and receding anteroventrally. The anterior eye facets are only slightly enlarged. The occiput is well developed, convex behind the eyes, more-or-less flat posteriorly. Occipital setae mainly occur on the dorsal half of the head, are variable in number and form, and are often bristle-like, short or long, straight or curved anteriorly and laterally. Hairs are normally white, long, fine and dense on the ventral half of the occiput, especially adjacent to the eye margin and in the genal and postgenal areas. The proboscis is usually directed obliquely forward, in line with, or up to an angle of about 45° to the hind margin of the eye; the labium is basally expanded and apically obtuse and rounded, bearing fine, pale, moderately long hairs ventrally. The palp is small, cylindrical and two-segmented; the second segment bears a large apical pore and only a few bristles.

In anterior view the face at the antennae is 0.30 to 0.15 times the width of the head, diverging strongly to the vertex. The frons is excavated with gently sloping sides; the vertex is saddle-shaped, the excavation variable, infrequently negligible. The relative widths of the face and vertex, and the depth of the vertex are sometimes useful in characterizing species. The expansion of the vertex led entomologists to place *Lasiopogon* in the Tribe Stichopogonini (Hull 1962, Lehr 1988, Papavero 1973b) or Subfamily Stichopogoninae (Artigas and Papavero 1990) along with *Stichopogon, Rhadinus, Townsendia* and others. The setae of the frons and vertex are variable, but conveniently separated into three main groups: frontal setae in a small patch posterior to each antenna; orbitals in an

Figure 3. *Lasiopogon* head
setation: lateral (top left),
anterior (top right) and
dorsal (bottom).
ant s = antennal setae.
F1 s = setae of first
flagellomere.
fr s = frontal setae.
gen s = genal setae.
lab s = labial (proboscis)
setae. mys = mystax.
occ s = occipital setae.
ocel s = ocellar setae.
orb s = orbital setae.

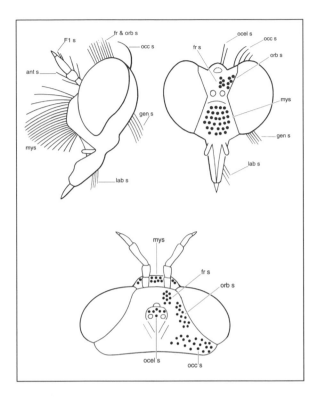

elongate area adjacent to the eye margin, extending from the level of the antennae posteriorly to the level of the lateral ocelli, where the setal field usually widens medially; ocellar setae usually consisting of two or three pairs of long bristles and several pairs of weaker hairs on the prominent ocellar tubercle.

The antennae (see also figure 2d) are attached just above the middle of the head, the scape and pedicel subequal in length, the former cylindrical, the latter globular. The first flagellomere (third antennal joint of Cole and Wilcox 1938) is less than twice the combined length of scape and pedicel. It is flattened but not wider than the first two segments. The shape of this flagellomere varies from rectilinear to elongate oval. The second flagellomere (arista of Cole and Wilcox 1938) is shorter than the first; it is narrow and attenuate and bears a minute third flagellomere (apical spine of Hull 1962), which is glabrous and nipple-like. In some specimens, the apex of the second flagellomere appears bifid because an apical projection extends parallel to the third flagellomere. All segments except the third flagellomere are densely covered with short (0.01-0.02 mm), curved, recumbent setulae (tomentum). The scape and pedicel have numerous robust setae; the first flagellomere is bare (except for tomentum) or sometimes with one or more somewhat appressed bristles on the posterior edge.

THORAX AND APPENDAGES (figures 4-7)

The prothorax is tomentose and, in addition, has fine, moderately long, white hair on most sclerites; the postpronotal lobe (humeral lobe) bears fine hair, sometimes bristles. The lateral angle of the postpronotal lobe is raised and glabrous, usually shining brown or ferruginous. Behind this runs the narrow ridge of the paratergite. A fully sclerotized bridge links the prosternum and proepisternum (see Hardy 1948).

The scutum is moderately arched, mostly finely tomentose and usually with dark stripes of three sorts. Dorsocentral stripes are usually present and are some shade of brown. I consider the narrow, paired stripes on the scutum midline to be the acrostichal stripes (median stripes (Cole and Wilcox 1938)); they are normally the same colour as the basal tomentum (grey, brown-grey, gold-brown), but are somewhat darker. In some species the narrow area between these stripes is more-or-less the same colour as the dorsocentral stripes, producing a medial stripe. The presence, colour and pattern of these stripes is useful in species identification. Large, often diffuse intermediate spots (Fisher 1986) frequently mark the scutum laterally, anterior and posterior to the transverse suture.

Setation is strikingly variable, typically black/brown, white/yellow or a mixture of both: dorsocentrals vary from strong bristles to fine hairs, or may be absent; acrostichals, if present, are usually short and fine; short setae sometimes cover much of the rest of the scutum, at least the presutural part (notal setae). The scutellum is flat and tomentose; apical scutellar setae and associated hairs are almost always present. Other groups of strong setae usually present on the scutum in the postsutural area are the postalars and supra-alars (in my interpretation, the latter are restricted to the postsutural area). Laterally in the presutural area, two groups of strong setae occur. Because these setae do not fall clearly into the categories listed by McAlpine (1981), for convenience here I follow the terminology of Cole and Wilcox (1938): bristles adjacent to the suture are presuturals and those more anterior (less frequently present) are posthumerals (figure 5).

Laterally, the mesothorax is normally densely tomentose with little setation. The most prominent bristles are those of the katatergite (metapleuron), which form a distinct line. Finer, shorter hairs may be present among the bristles. Setae occur in only four other locations: in two patches on the posteriodorsal region of the anepisternum (mesopleuron), on the posterior part of the katepisternum (sternopleuron), and sometimes on the anepimeron (pteropleuron). The hairs on the latter two sclerites are almost always fine and usually white.

The postmetacoxal area is membranous.

The legs are usually moderately stout, but the femora or tibiae are never swollen or strongly club-shaped. Setation is usually abundant but variable within and among species, and is usually more sparse in females. Tomentum on the legs varies from heavy (e.g., *L. littoris, L. ripicola*) to absent (*L. yukonensis*) and becomes thinner from coxa to tarsus. The femora are usually coated with sparse

Figure 4. Thoracic sclerites: lateral. a bas = anterior basalare. ab tg 1 = abdominal tergite 1. ab tg 2 = abdominal tergite 2. anatg = anatergite. anepm = anepimeron. anepst = anepisternum. apn = antepronotum. a spr = anterior spiracle. cx1 = procoxa. cx2 = mesocoxa. cx3 = metacoxa. halt = halter.
kepm = katepimeron.
kepst = katepisternum.
ktg = katatergite.
medtg = mediotergite.
mnot = scutum of mesonotum.
mr = meron. mtanepst = meta-
nepisternum. mtepm = mete-
pimeron. mtkepst = metkate-
pisternum. pal cal = postalar
callosity. patg = paratergite.
p bas = posterior basalare.
pprnl = postpronotal lobe.
prepm = proepimeron.
proepst = proepisternum.
p spr = posterior spiracle.
trn sut = transverse suture.

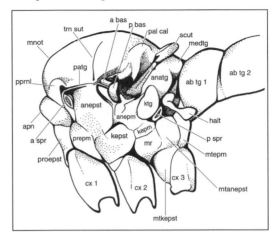

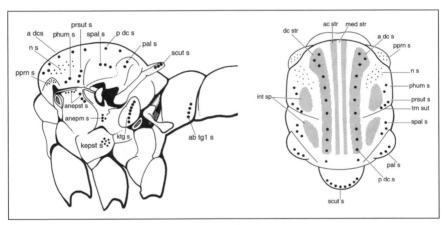

Figure 5. Thoracic setation: lateral (left) and dorsal (right). ab tg1 s = abdominal tergite 1 setae. ac str = acrostichal stripe. a dc s = anterior dorsocentral setae. anepm s = anepimeron setae. anepst s = anepisternum setae. dc str = dorsocentral stripe. int sp = intermediate spots. kepst s = katepisternum setae. ktg s = katatergite setae. med str = medial stripe. n s = notal setae. pal s = postalar setae. p dc s = posterior dorsocentral setae. phum setae = posthumeral setae. pprn s = postpronotal setae. prsut setae = presutural setae. scut s = scutellar setae. spal s = supraalar setae. trn sut = transverse suture.

tomentum, often with the venter bare; tomentum on the tibiae and tarsi is frequently less evident. The trochanters are normally bare.

Many species have a small, narrow, nipple-like protuberance on the anterior margin of the hind coxa (figure 6). I refer to this structure as the coxal peg, after Yeates 1994.

Coxal setae are most prominent on the anterior face of the procoxae, the anterior lateral angles of the mesocoxae, and the posterior lateral edges of the mesocoxae and metacoxae. They can also occur on the posterior lateral margins of the procoxae, the ventral edges of the mesocoxae and the anterior ventral margins of the metacoxae. The setae are bristles or hairs.

Setae on the femur take several forms.

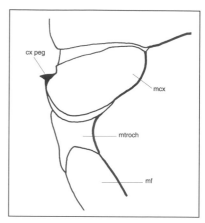

Figure 6. Coxal peg. cx peg = coxal peg. mcx = metacoxa. mf = metafemur. mtroch = metatrochanter.

The basic vestiture is of hairs covering much of the surface, often short and more-or-less appressed in the less hirsute forms, longer and more prominent in others, especially on the lateral and ventral faces. Straight hairs or weak bristles, usually pale but sometimes dark, often arise vertically from the ventral surface. These are usually long, sometimes up to twice as long as the femur width. Bristles, ranging from weak and long to strong and short, normally line the posterior dorsal face of the profemur and the anterior dorsal face of the mesofemora and metafemora. In some species irregular ranks of bristles replace the single line. The mesofemur bears the fewest bristles. Each femur has a variable number of short bristles dorsally and laterally at the apex.

Tibial hairs are variable. Often short and depressed, these hairs are often mixed, dominated or replaced by long hairs, especially laterally and ventrally. As in other asilids, there is a distinctive brush of short, dense hair on the apical half or two-thirds of the ventral and anteroventral faces of the foretibia. A similar, shorter brush can occur at the apices of the mesotibiae and metatibiae in some species. Tibial bristles form more-or-less regular rows: the foretibia has dorsal and posterolateral rows and one ventrolateral (posterior) row; the mesotibia has these as well as one anteriolateral and one ventrolateral (anterior) row; the metatibia has an anteriolateral and a ventrolateral (anterior) row. The apex of each tibia typically bears a ring of long, strong bristles. Tibial bristles can be short or long, on the foretibia from two to five times longer than the tibial width. These bristles can be much shorter on the mesotibia. There is no terminal sigmoid or bent spine.

Tarsal hairs are short and more-or-less decumbent. Bristles are apical, usually longer on the basal segments than on apical ones.

The tarsal claws in *Lasiopogon* are long with acute apices; they are reddish brown with black apices. Pulvilli are rectangular, about two to four times as long as their width. Claws, empodium and pulvilli extend approximately the same distance from the apex of the fifth tarsomere.

The wing (figure 7) is about a quarter to a third as wide as its length. Venation is a basic asilid pattern: R_{2+3} ending in the costa, cell R_{2+3} widely open; R_4 gently arched basally, ending just anterior to the wing apex; cells r_5, m_1, m_2 and cua_1 widely open, m_3 narrowed at the margin or sometimes closed; cup closed at, or just before, the wing margin; discal cell elongate, r-m crossvein at, or basal to, the mid-point of the cell. The wings are usually rather clear, but microsetulae covering the wing membranes sometimes colour them pale brown or even darker. Some species have vague brown spotting at certain vein junctions.

The halter knob is either immaculate or marked. In some species (*L. currani*, *L. fumipennis*, *L. lichtwardti*), the spot is always present, black and definite, and can be used confidently in species identifications. In a few, notably *L. cinctus*, the spot can be present or absent, definite or diffuse, and therefore unreliable in specific determinations. Despite these vagaries, its presence or absence can be especially useful in separating otherwise similar species; e.g., in *L. bellardii* the spot is usually absent and in *L. montanus* it is usually present (Cannings 1996).

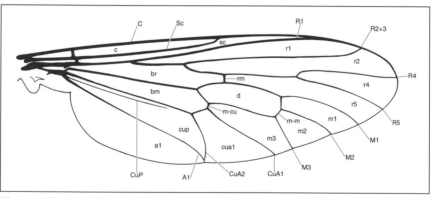

Figure 7. Wing venation. A1 = first anal vein. a1 = anal cell. br = basal radial cell. bm = basal medial cell. C = costa. c = costal cell. CuA1 = CuA2 = anterior branches of cubitus. cua1 = anterior cubital cell. CuP = posterior branch of cubitus. cup = posterior cubital cell. d = discal cell. M1, M2, M3 = posterior branches of media. m1, m2, m3 = medial cells. m-cu = medial-cubital crossvein. m-m = medial crossvein. R1 = anterior branch of radius. R2+3, R4, R5 = posterior branches of radius. r1, r2... = radial cells. r-m = radial-medial crossvein. Sc = subcosta. sc = subcostal cell.

ABDOMEN

At its base, the abdomen is about as wide as the scutum and is slightly tapered to the apex and flattened dorsoventrally, especially in females. In both sexes, seven segments are visible anterior to the modified segments of the genitalia. In the male, segment 8 is hidden under segment 7, and segment 9 makes up the bulk of the visible genitalia. The male terminalia are prominent, the arched epandrium halves usually as wide as or wider than the adjacent abdominal segment. Male genitalia are rotated 180°, at least after the first mating, and thus specimens almost always have the ventral hypandrium and gonocoxite complex situated dorsally and the laterally clasping epandrium ventrally. In the female, segment 8 makes up much of the visible terminalia; its sternite is a lobed, shovel-like structure used in oviposition. The heavy spines of the tergite of segment 10 (acanthophorite) are often visible posterior to tergite 8.

Segments 1 to 7 are variably patterned with tomentum. Although a few species have bare, shining tergites, most have bands of pale tomentum apically on the segments. The basal areas may be bare or covered with tomentum darker than the apical bands. The bands may be uniformly narrow dorsally and laterally; but in some species the bands are widened along the lateral edge of the tergite or along the dorsal midline, or expanded basally in these areas. In some species the entire tergite on all segments is covered with pale tomentum. Sternites are usually more-or-less covered with tomentum.

Tergite 1 is short and broad, and bears several rather long setae dorsolaterally. Associated dorsolateral and ventrolateral hairs are long, becoming shorter and more posteriorly directed on subsequent segments. The dorsal setae are usually short and more-or-less appressed, becoming longer laterally and merging with the ventrolateral hairs. In some species these setae are mere setulae. In some females especially, the posterior setae can be short and bristle-like. Setae on the sternites are sparsely distributed, vertically oriented hairs, and are usually pale.

MALE GENITALIC STRUCTURE (figures 8-14, 245-48)

In *Lasiopogon* tergite 9, the epandrium, is divided mid-dorsally into a pair of large, horizontally moving claspers (superior forceps). The two halves are completely separate, but are articulated basally in the membrane (figure 8a); in many species a small rectangular sclerite (basal epandrial sclerite) lies between the bases of the two epandrium halves in the dorsal membrane (figures 8a, 245i) immediately posterior to tergite 8 (prior to any rotation). In cross-section, the epandrium halves are concave medially, with this surface angled somewhat ventrally and covered with membrane, more-or-less sclerotized apically. The apical sclerotized part (apical shelf), indicated by a line in the drawings (medial view), usually bears setae (figure 25); these appear horizontally oriented in the drawings, but

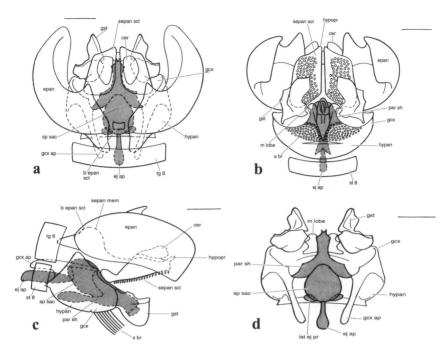

Figure 8. Male genitalia (*Lasiopogon monticola*): **a**, dorsal; **b**, apical; **c**, lateral; **d**, hypandrium/gonocoxite complex, dorsal.
b epan scl = basal epandrium sclerite. cer = cercus. epan = epandrium.
ej ap = ejaculatory apodeme. gcx = gonocoxite. gcx ap = gonocoxal apodeme.
gst = gonostylus. hypan = hypandrium. hypopr = hypoproct. lat ej proc = lateral ejaculatory
process. m lobe = medial lobe of gonocoxite. par sh = paramere sheath. s br = setal brush.
sepan mem = subpandrial membrane. sepan scl = subpandrial sclerite. sp sac = sperm sac.
st 8 = sternite 8. tg 8 = tergite 8. Scale lines = 0.3 mm.

are actually at a strong angle to the surface of the epandrium. The less sclerotized membrane basally may also bear setae, but these are less abundant. The drawings show the epandrium in lateral view with the dorsal side uppermost. Because most specimens have secondarily rotated genitalia (see below) with the ventral side uppermost, the orientation of the drawings might cause some initial confusion. The shape of the epandrium is more clearly seen in dissected specimens than in intact ones, although estimates of width to length are taken from undissected specimens and thus do not include the apodeme.

The divided epandrium is a derived condition relative to the ground plan of the Brachycera (Sinclair et al. 1994) and the presence of hinged epandrium halves is, in the Stichopogoninae, an autapomorphy of *Lasiopogon*. Karl (1959) believed that an epandrium split into basally attached halves was plesiomorphic in the Asilidae and that both completely divided halves or an entire sclerite were apomorphies.

In many species with genitalia rotated 180° (e.g., Laphriinae) the tergite and sternite of segment 8 are strongly reduced (Fisher 1986, Karl 1959). In *Lasiopogon*, such reduction is minor; the sclerites are relatively shorter in their midline length than those of segment 8 and the sternite is somewhat reduced in size relative to the tergite. Throughout all species, the reduction is rather uniform and I decided that no characters from segment 8 were useful for species identification or phylogenetic analysis.

Structures posterior to segment 9 are collectively called the proctiger (Wood 1990). This consists of paired, quadrilateral, pad-like cerci and the hypoproct with associated subepandrial membrane and its sclerite (Sinclair et al. 1994) (figure 8). According to Sinclair et al. (1994) the hypoproct is probably sternite 10, and I concur, but only using superficial evidence: in *Lasiopogon*, the proctiger in the male has a similar appearance and location relative to the cerci as the apex of sternite 10 has in the female. The subepandrial membrane forms the roof of the space between the dorsal epandrium/proctiger and the ventral hypandrium/gonopod/phallus complex. Imbedded in the membrane is a broad, roughly V- or U-shaped sclerite with a lightly sclerotized, bilobed apex (hypoproct) (figures 8, 13). This is the subepandrial sclerite (Sinclair et al. 1994), otherwise known as the ventral epandrial sclerite *sensu* Irwin and Lyneborg (1981), the median plate (Hardy 1934-35), the surstyli *sensu* Adisoemarto and Wood (1975), bacilliform sclerite *sensu* McAlpine (1981), or the ventral lamella of proctiger *sensu* Karl (1959).

The main part of the sclerite is armed with short, usually stout, ventrally projecting, spine-like setae. These spines are useful in phylogenetic analysis; they are variable in shape but fall into two major groups, those that are strongly tapered and apically attenuate and those that are peg-like or at least largely parallel-sided and only moderately acute. The surface of the spines is covered by parallel grooves (figure 14). Microsetae cover parts of the sclerite and adjacent membranes. The pattern of these setae is indicated by stippling on the drawings of the sclerite. The subepandrial sclerite varies in shape and is useful in distinguishing some species groups. Sclerites mostly heart-shaped and sclerotized along much of the midline (figure 247a-f) (bearing tapering, attenuate spines) are typical of predominantly Eurasian groups and of some in western North America; more rectangular sclerites largely divided medially (figure 247j-k) (with peg-like spines) are found in the monophyletic *opaculus* section + *bivittatus* group, a mainly Nearctic lineage with some members in eastern Asia.

The subepandrial sclerite is continuous with the hypoproct, which forms the posterior, bifid apex of the plate and bears longer, more delicate setae. In some groups, the hypoproct is shelf-like, lying in a more dorsal plane to the main subepandrial sclerite (figures 13, 247k). The subepandrial sclerite originates in the intersegmental membrane between segment 9 and the proctiger (Sinclair et al. 1994) and articulates at its anterolateral corners (basally) with the medial projections of the gonocoxal apodemes. The associated membrane joins the dorsal face of the phallus.

The hypandrium (sternite 9) and epandrium are separate in the ground plan of the Brachycera (Sinclair et al. 1994). In *Lasiopogon*, unlike in many other asilids, the hypandrium and the paired gonocoxites (basistylus, inferior forceps) do not retain their plesiomorphic separation, but rather are extensively fused to form a bowl-like complex that opposes the epandrium and its associated structures ventrally (figure 8). The line of fusion of the hypandrium and gonocoxites is evident on the convex ventral surface where a transverse slit forms the base of the medially opposed gonocoxite lobes. These lobes, when present, may be gently rounded or elongate and acute (figure 246), and in one case (*L. nitidicauda*), bear sclerotized teeth. There may be small basal processes medially that are associated with the articulation of the gonostyli; in the *opaculus* section these are expanded medial lobes, blunt or toothed and heavily sclerotized apically (figure 8d). The gonocoxal apodemes are prominent, arising basolaterally on the dorsal margins of the gonocoxites and angling basoventrally. The basal portions articulate with the arms of the parameral sheath, holding the phallus in place; they also articulate with the base of the subepandrial sclerite (figure 8). Apically (anteriorly), the apodemes take various forms; for example, they may be linear or spatulate and often are further anchored to the hypandrium by sclerotized webs. In the *cinctus* group they are long, spatulate (figure 246c) and joined ventrally by massive sheathing sclerotization. The strong exsertion of apodemes from the base of the fused hypandrium/gonocoxites is an apomorphy of *Lasiopogon*.

A gonostylus (claspers (Hardy 1934-35), dististyle) is inserted on the inner (dorsal) face of the base of each of the gonocoxite lobes (figure 8d). The basal position of this articulation is an apomorphy of the Asiloidea + Eremoneura (Cumming et al. 1995, Sinclair et al. 1994). The gonocoxite and gonostylus together are termed the gonopod (McAlpine 1981). In *Lasiopogon* the gonostyli exhibit a wide variety of forms, from complex, folded and flattened structures (figures 12, 246) to elongate hooks, and are of great importance in phylogenetic analysis and species diagnosis. They move in a dorsoventral direction, an apomorphy of the Muscomorpha (*sensu* Woodley 1989). In doing so, they apparently press the venter of female sternite 8 against the subepandrial sclerite.

The dorsal concavity of the hypandrium/gonopod complex contains the phallus (aedeagus + parameres) complex (figure 8d). The aedeagus is an elongate flask-shaped tube (figures 9, 248); most authors consider it a derivative of segment 10 (McAlpine 1981). Basally (anteriorly) it is expanded into a bulbous, transparent sperm sac (vesica (Karl 1959); pump chamber (Theodor 1976)) that bulges slightly anterodorsally. The ejaculatory apodeme is fused to the membrane on its anterior surface and acts as a piston, collapsing the base of the sac and forcing sperm through the aedeagal tube and out the gonopore at its tip. This apodeme takes a variety of forms, but it is usually elongate, sometimes shallowly V-shaped in lateral view, and variously shaped apically and in cross-section. Anterolaterally in the sperm-sac membrane are embedded two quadrilateral or triangular lateral ejaculatory processes (aedeagus apodeme or cross apodeme (Karl 1959); basal plates (Theodor 1976)).

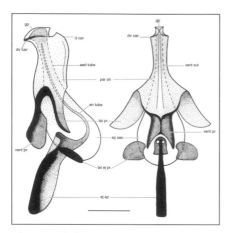

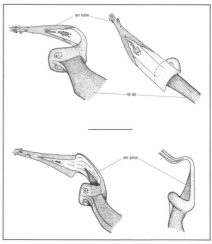

Figure 9. Phallus (*Lasipogon coconino*):
lateral (left) and ventral (right).
aed tube = aedeagal tube. d car = dorsal
carina. dv car = dorsoventral carina.
ej ap = ejaculatory apodeme.
en tube = endoaedeagal tube.
gp = genital pore. lat pr = lateral process.
lat ej ap = lateral ejaculatory process.
par sh = paramere sheath. sp sac = sperm sac.
vent pr = ventral process. vent sut = ventral
suture. Scale line = 0.3 mm.

Figure 10. Endoaedeagal tube
(*L. quadrivittatus* above, *L. cinctus* below).
ej ap = ejaculatory apodeme.
en proc = endoaedeagal process.
en tube = endoaedeagal tube.
Scale line = 0.2 mm.

Inside the sperm sac, attached to the base of the ejaculatory apodeme, the endoaedeagal tube (figures 9, 10) arches dorsally at its base (where it is ribbon-like and sclerotized only laterally) and runs posteriorly, becoming tube-like, to the gonopore. Its surface bears scattered, sclerotized granules. As the ejaculatory apodeme pushes the base of the sperm sac inwards to force sperm out of the gonopore, the endoaedeagal tube slides through the aedeagal tube and its apex perhaps even exits the gonopore (Theodor 1976; figure 11). The tube has the appearance of, and suggests the action of a fine, moveable cleaning structure, acting much like the sliding wire inside the hollow nib of a mechanical drafting pen. It is unknown what the real function of the tube is, but it might, through its motion, keep the aedeagal tube open and aid the smooth delivery of semen. Sinclair et al. (1994) postulated that the endoaedeagal tube is a potential apomorphy of the Asiloidea (Therevidae + Scenopinidae + Asilidae + Apioceridae + Mydidae).

The endoaedeagal process (figure 10) is a basal (posterior) elongation of the ejaculatory apodeme and is considered a synapomorphy for the Brachycera, although it is lost in many taxa (Sinclair et al. 1994). Sinclair et al. (1994) believed that it is not homologous with the endoaedeagal process because both structures are present in *L. cinctus*. But the process appears, at least in a superficial examination, to be the enlarged base of (or at least is fused to) the endoaedeagal tube in

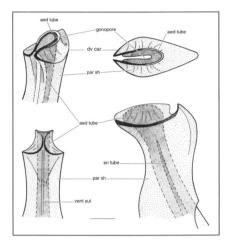

Figure 11. Phallus apex (*L. quadrivittatus*), clockwise from top left: ventrolateral, apical, lateral, ventral. aed tube = aedeagal tube. en tube = endoaedeagal tube. dv car = dorsoventral carina. par sh = paramere sheath. vent sut = ventral suture. Scale line = 0.1 mm

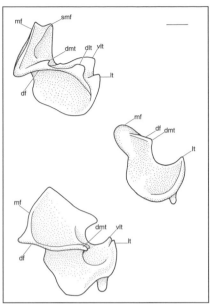

Figure 12. Gonostylus, dorsal view (top to bottom: *L. albidus*, *L. hinei* and *L. akaishii*). df = dorsal flange. dlt = dorsolateral tooth. dmt = dorsomedial tooth. lt = lateral tooth. mf = medial flange. smf = secondary medial flange. vlt = ventrolateral tooth. Scale line = 0.1 mm.

many species; Theodor (1976) illustrated this in a number of genera. The process is reduced or absent in many species of *Lasiopogon*, such as *L. quadrivattatus*, but forms a short pointed base to the endoaedeagal tube in others, such as *L. cinctus*.

Parameres are paired structures that link the base of the aedeagus and the dorsomedial margin (base) of each gonocoxal apodeme (McAlpine 1981, Sinclair et al. 1994). Karl (1959) did not recognize the existence of parameres in the Asilidae. In the ground plan of the Brachycera the parameres are fused medially to form a sheath around the aedeagus that anchors it to the gonocoxites (Sinclair et al. 1994, Wood 1991). In *Lasiopogon* the sheath surrounds the aedeagus in its apical half expanding posteriorly to surround the apical part of the sperm sac and connecting it to the gonocoxal apodeme by lateral wing-like processes. This connection can be heavily sclerotized or only slightly so, depending on the species. The ventral edge of the sheath is produced into an elongate lobe, which takes various forms; it is attached by membrane to the dorsal (inner) surface of the hypandrium/gonocoxites. The apex of the sheath, which is variable in its shape and elongation, closely surrounds the aedeagal tube at the gonopore, the opening at the apex of the phallus. The tube is fused along the length of the median ventral suture of the sheath. The margins of this suture are weakly joined or are not closed.

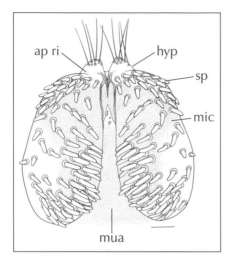

Figure 13. Subepandrial sclerite (*Lasiopogon quadrivittatus*): ventral.
ap ri = apical ridge. hyp = hypoproct.
mic = microsetae. m u a = medial unscero-
tized area. sp = spine. Scale line = 0.1 mm.

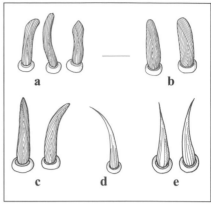

Figure 14. Subepandrial sclerite spines:
a, *L. cinereus*; **b**, *L. quadrivittatus*;
c, *L. monticola*; **d**, *L. canus*; **e**, *L. apenninus*.
Scale line = 0.05 mm.

FEMALE GENITALIC STRUCTURE (figures 15-17, 249, 250)

Abdominal segment 7 is the most posterior undifferentiated segment of the female *Lasiopogon*. The tergite (epigynium) and sternite (hypogynium) of segment
8 lack the usual pollinosity and setation and they form the bulk of the heavily
sclerotized terminalia. Sternite 8 is usually keel-shaped with a bifid, shovel-like
apex (hypogynial valves) separated from the main body of the sternite by membrane and lighter sclerotization (figure 15). The ventral surface of the valves can
bear carinae (figure 233) and setae of variable strength; the dorsal (inner) surface
is lined with a separate sclerotized plate attached to the ventral surface apically.
In the *akaishii* group a pair of additional medial processes are developed between
the valves (figure 29). The sternite may be heavily sclerotized and formed into a
permanent trough-shaped structure (figure 249a-c), or may have a lightly sclerotized midline (figures 45, 249d), which allows the sternite to fold and flatten
from side to side. Tergite 8 is arched, usually laterally compressed, and more-or-
less tapers posteriorly.

Tergite 10 is much smaller than tergite 8, and consists of a medially divided
sclerite (acanthophorites), each half bearing six to nine heavy, blunt spines (figure 15). The number of spines is usually, but not always, equal on either side of
the acanthophorites; frequently the anterodorsal spine is much smaller than the
others. Usually, spines can be counted accurately only in dissected material.
Mediodorsally the two acanthophorites are linked by a small sclerotized remnant
of tergite 9. Tergite 10 is entire and separate from tergite 9 in the most plesiomor-

Figure 15.
Female genitalia
(*L. monticola*): lateral,
undissected (top) and
dissected (bottom); sternite
8, ventral (middle).
acc gl = accessory gland.
ac sp = acanthophorite
spines. cer = cercus.
hypg v = hypogynial valve.
lat lobe = lateral lobe of
sternite 8.
lat lobe s = lateral lobe
setae. sp = spermatheca.
st 8 = sternite 8.
st 9 = sternite 9.
tg 8, 9, 10 = tergites.
Scale lines:
top and middle = 0.3 mm;
bottom = 0.2 mm.

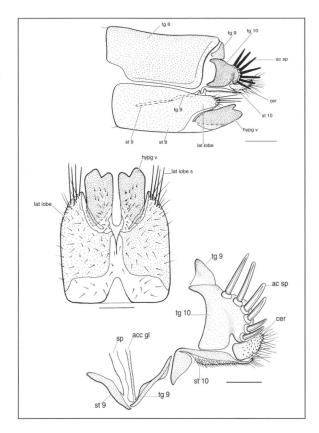

phic taxa of the lower Brachycera (e.g., *Phara*, a primitive tabanid) (Irwin 1976). Yeates (1994) considered these tergites fused in some bombyliids,. He also indicated this fusion in the asilid genera *Dioctria* and *Leptarthrus*, contradicting the interpretation of Adisoemarto and Wood (1975), which I follow here.

Tergite 9 is reduced to three small sclerites (Adisoemarto and Wood 1975, Irwin 1976, Lavigne and Bullington 1981). As noted, one links the two halves of tergite 10 dorsomedially (figure 15). The lateral pair are embedded in the intersegmental membrane near the basal margin of tergite 10 and approximate the apical arms of sternite 9 (furca) (figures 15, 17). This condition, where the lateral portions of the tergite are separate, surrounded by membrane and ventrally situated, is similar to that in *Cyrtopogon sabroskyi* Lavigne and Bullington (Lavigne and Bullington 1981) and other asilid species examined. Adisoemarto and Wood (1975) interpreted the origin in Dioctriini and Echthodopini in the same manner, unlike Reichardt (1929) and Crampton (1942), who considered them part of sternite 9. As noted by Lavigne and Bullington (1981), this is an apomorphic condition assumed to have been derived from a situation similar to that seen in some Therevidae where tergite 9 is a thin sclerite partly fused with tergite 10, and with ventrolateral margins connected to the posterolateral edges of sternite 9 (Irwin

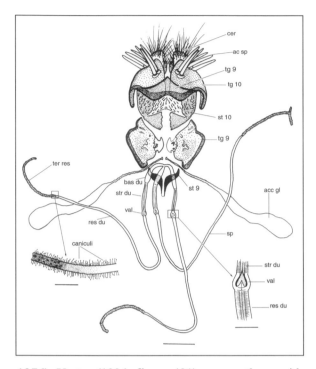

Figure 16. Spermathecae (*Lasiopogon cinereus*): dorsal view. acc gl = accessory gland. ac sp = acanthophorite spines. bas du = basal duct. cer = cercus. res du = reservoir duct. sp = spermatheca. str du = striated duct. st 9 = sternite 9 (furca). st 10 = sternite 10. tg 9 = tergite 9. tg 10 = tergite 10. ter res = terminal reservoir. val = valve. Scale line = 0.2 mm; enlargements = 0.05 mm.

1976). Yeates (1994, figure 421) apparently considered these sclerites part of sternite 8 in the asilid *Dioctria atricapilla*. These two tergite 9 sclerites lie adjacent to the dorsal surface of the lateral lobes of sternite 8. In some species, an additional pair of sclerites has developed in the connecting membrane between tergite 9 sclerites and the lateral lobes of sternite 8 (figure 249a-b).

Sternite 9, the genital fork (furca), is a Y- or V-shaped sclerite that lies in the dorsal wall of the genital chamber (vagina) (figures 15-17). The genital fork is divided medially in some species. As in most Asilidae, there are three spermathecae, opening in the membrane between the arms of the furca (figure 16).

Spermathecae take various forms in the Asilidae. Theodor (1976) examined the spermathecae of many asilid genera and described their internal and external structure in detail. He states that except in a few genera such as *Leptogaster* in which the middle spermatheca is strikingly different from the lateral ones, the three spermathecae are identical in form. But Theodor often used only one exemplar from each genus in his study. In some specimens of *Lasiopogon cinereus* (figure 16), at least, the central spermatheca has a forked apex, rather than the straight one found on the lateral spermathecae. In most other *Lasiopogon* species examined, the middle terminal reservoir is longer and thicker than the lateral ones. The significant variation in gross form and structural detail documented by Theodor is not correlated to the presently accepted classification of the family.

The spermathecae are usually contained in the apical half of the abdomen; often they do not reach past the base of the eighth segment. In *Lasiopogon* they

Figure 17. Female sternite 9 and associated structures (*L. monticola*): ventral view. st 9 = sternite 9 (furca). st 10 = sternite 10 with lobes. tg 9 = tergite 9. Scale line = 0.1 mm.

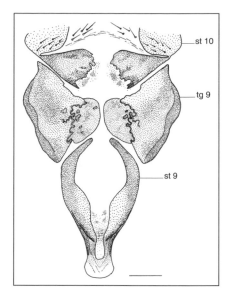

usually reach at least the apex of the seventh segment, looping back on themselves. In other Asilidae they can be extraordinarily long; in *Heteropogon*, for example, they are coiled back on themselves three or four times, circumscribing almost the entire abdominal cavity (Artigas and Papavero 1991b). The terminal reservoir (apical tube) in which the semen is stored is more-or-less sclerotized in most species and is lined with glandular cells that empty into the spermathecal lumen through tiny sclerotized tubes that Theodor (1976) called canaliculi (figure 16). In dried, macerated material the sheathing glandular cells are usually gone but their sclerotized caniculi remain; they are scattered over the outer surfaces of the spermathecae in varying densities and lengths. They are more prevalent distally and usually also occur on the accessory glands.

Throughout the family, the reservoir can be a bulb, sac or tube of variable thickness and length. Tube-shaped spermathecae are often coiled in various ways. The reservoir usually empties into a reservoir duct, which in turn (in most Asilidae), opens through a sclerotized, funnel-shaped valve into differentiated ducts that may also be sclerotized or striated (figure 16). Muscles attach above the valve and extend to the base of the duct. This whole complex is an ejection apparatus; the amount of sperm used during oviposition is evidently regulated by the valve (Theodor 1976). Paired accessory glands empty into the genital chamber immediately posterior to the spermathecal openings (figure 16). These glands produce material that coats the eggs during egg-laying (McAlpine 1981).

In most *Lasiopogon* species the reservoir is a long, slender, delicate tube of variable morphology (figures 16, 250). The terminal reservoir is the sclerotized apex of this tube and, in the *opaculus* section, can be coiled, hooked, forked or straight and rod-like. Canaliculi are usually scattered over the surface of the reservoir and, to variable degrees, proximally along the thin-walled reservoir ducts. A short basal section has thick, apparently striate walls (figure 31). The reservoir can make up much of the spermathecal length; e.g., in *L. hinei* it is about 85 per cent of the total length (figure 97). In most Eurasian and some western North American species, the reservoir duct widens into a bulb at about mid-length (figure 250a). Alternatively, the duct can be shorter than usual, and the terminal reservoir apically broad, as in *L. fumipennis* (figure 250b). The valve in *Lasiopogon* is only weakly sclerotized and empties into a short, striated duct (figure 16). At its base, this duct empties into a thin-walled basal duct, which can be relatively narrow or a more substantial, elongate, wrinkled sac; the lumen is indicated by twisted and folded ribbon-like structures which are star-shaped in cross-section. In *L. quadrivittatus* this sac is large, a full 20 per cent of the total spermathecal length (figure 187), and the lumen in cross-section is shaped like a six-pointed star. The three spermathecae end separately immediately adjacent to each other in a triangular pattern or in a common duct (figure 250a-b) of variable length that empties into the roof of the genital chamber. The paired accessory glands are elongate, narrow sacs that can be almost as long as the spermathecae (figure 16). They also may bear canaliculi.

The cerci are lightly sclerotized lobes embedded in membrane posterior to tergite 10; they bear a terminal brush of setae (figure 15). Lying in the ventral membrane of this terminal complex is the lightly sclerotized plate of sternite 10, covered in varying densities with hairs or short setae. It is usually vase-shaped, the narrow end directed posteriorly, and is often divided medially. A pair of small sclerites are usually attached to the base of the sclerite, between it and the paired remnants of tergite 9. These sclerites can be triangular (figure 17), thin and oval or reduced to only a slight darkening of the membrane; they are bent ventrally roughly at right angles to the main part of sternite 10 (figure 15). Sometimes they are absent. Whether these are extensions of sternite 10 (they are sometimes fused with the anterior edge of this sclerite) or derive from the intersegmental membrane is unknown. I assume that they are part of sternite 10 and call them sternite 10 lobes. The anus opens on the midline between sternite 10 and the cerci.

GENITALIC INVERSION

Rotation of the male genitalia is common in Diptera (McAlpine 1981). This change allows the male to remain upright during tail-to-tail copulation and reduces the pair's vulnerability to predation. Many Asilidae copulate tail-to-tail, and the male rotates its terminalia 180°, the epandrium ending up in a ventral position. This is the case in *Lasiopogon*, first recognized by Lundbeck (1908). *Lasiopogon* specimens are often captured with genitalia in various stages of rotation, the amount depending on the time elapsed since emergence and whether the insect has mated or not. All freshly teneral specimens that I have seen – which, presumably, are unmated – have unrotated genitalia. This observation is supported by Weinberg (1978), who collected a population of recently emerged specimens of *Lasiopogon montanus* in the Carpathian Mountains of Romania. For a short time after emergence, specimens show no rotation. As sclerotization progresses, however, the genitalia rotates up to 90°, and completely to 180° after the initial copulation. Almost all male specimens examined in insect collections show complete 180° rotation, indicating that initial copulation occurs relatively soon after emergence. Rotation is to the right or left and, according to Weinberg, occurs about equally in either direction in *L. montanus*. She also documented the location of the torsion: segment 7 was rotated 45°; segment 8, 90°; terminalia, 180°. In females captured during copulation, genitalia had rotated 80°, twisting the abdomen from segment 5 posteriorly (Weinberg 1978). This rotation disappears after copulation is complete.

COPULATION

During copulation the male genitalia articulate mainly with the ventral parts of the female terminalia. During the virgin male's copulation, the male genitalia are upside down with respect to the female; genitalic rotation (inversion) is then completed, allowing the male to perch or fly upright tail-to-tail with the female. Reichardt (1929) described this in detailed morphological terms in the Asilini, in which inversion does not occur. After the initial copulation, the male genitalia of *Lasiopogon* are permanently inverted. In *Lasiopogon*, the epandrium halves clasp sternite 8 laterally, their apices extending to the apex of sternite 7. The apex of the female genitalia slides between the gonocoxite lobes, forcing the brush of bristles on the gonocoxites vertically. The bristles intermingle with the setae of the cerci and the spines of the acanthophorites of the female. The phallus slides through the notch formed by the apical lobes (hypogynial valves) of sternite 8 and enters the genital chamber and the spermathecal duct. The subepandrial sclerite with its (presumably) sensory spines runs along the ventral surface of sternite 8 and the hypogynial valves. The gonostyli apparently press down on the dorsolateral surface of female sternite 8, helping to open the gonopore for the insertion of the phallus.

TAXONOMY OF *LASIOPOGON*

THE GENUS *LASIOPOGON* LOEW

Synonymy

Lasiopogon Loew, 1847. Linn. Ent. 2: 508 (*Dasypogon* subgenus).

Type species *Dasypogon pilosellus* Loew 1847, des. Rondani 1856 (as *Dasypogon hirtellus* Meigen, not Fallén) Dipt. Ital. Prodr. 1: 156; Coquillett, 1910. Proc. U.S. Nat. Mus. 37: 558.

Daulopogon Loew, 1874. Berlin Ent. Zeitschrift 18: 377. Unjustified new name for *Lasiopogon* Loew. Type species *Dasypogon pilosellus* Loew.

Alexiopogon Curran, 1934. North American Diptera, p.183. Type species *Daulopogon terricola* Johnson by monotypy.

Back (1909) noted that Loew, after establishing the name *Lasiopogon*, discovered that it was already in use in botanical nomenclature (a Palaearctic and Afrotropical genus of Asteraceae), and so changed the name to *Daulopogon*. This name was adopted by Osten Sacken (1878) in his catalogue of North American Diptera. But Williston (1908) was justified in restoring the older name *Lasiopogon*, because as the zoological and botanical systems are separate, there is no rule demanding such a change.

Type Species Designation

According to Lehr (1988), the type species of *Lasiopogon* is *Asilus cinctus* Fabricius (first described species); but Rondani (1856) designated *Dasypogon pilosellus* Loew, 1847 (as *Dasypogon hirtellus* Meigen 1820, not Fallén). Coquillett (1910, p. 558) followed this designation.

There is an unresolved taxonomic problem with the first-described species, *Lasiopogon cinctus*. This is the common species of northern Europe and the best known species in the genus, appearing in the literature hundreds of times. But the Fabricius type of *Asilus cinctus* in Copenhagen is a specimen of *Choerades fuliginosa* (Panzer) (Laphriinae) (W. Michelson, pers. comm.). Fabricius's original description (1781), is so brief ("... Abdomen shining black with margins of segments white...") that it could describe either the *Lasiopogon* or the *Choerades* species. Loew (1847) was aware of the situation and established the continuing use of *Asilus cinctus* for the *Lasiopogon* species. He believed, for some unexplained reason, that the use of the name *Asilus cinctus* for the *Choerades* (= *Laphria*, in part) species was invalid. He explained that Meigen had properly distinguished the species when he referred to it as *Dasypogon cinctus* in 1804, but was not entitled to change the name to *D. cinctellus* in 1820 (Meigen 1851, reissue). In addition, Fallén (1814) had already used the name *D. hirtellus*. Nevertheless, Fabricius (1805) himself seemed to be clear on the matter; when Meigen created the genus *Laphria*, Fabricius referred his *Asilus cinctus* to the new genus, placing *Asilus cinctus* and *Asilus fuliginosus* Panzer in synonymy. There seems no reason to suspect the present type specimen is not Fabricius's specimen. Despite all this confusion the names *Lasiopogon cinctus* and *Choerades/Laphria fuliginosa* have been in constant use for 150 years and should be conserved by a proposal to the International Commission on Zoological Nomenclature before the *cinctus* group is revised. During this revision, the question of types would have to be resolved.

Generic Diagnosis

Lasiopogon species are those with the characteristics of the subfamily Dasypogoninae *sensu lat.* or Stichopogoninae (Artigas and Papavero 1990) with the following combination of characters:

1. Grey or brown tomentum on the head, the thorax and, usually, the abdomen; the abdomen is usually banded with heavier tomentum on the apices of segments, but sometimes completely grey tomentose. The cuticle is black, dark brown, or sometimes, in part, ferruginous.

2. The frons diverging laterally above the antennal bases.

3. A prominent gibbosity on the face, bearing a well developed, brush-like mystax over most of its surface.

4. The scape and pedicel of the antennae more-or-less equal in length; flagellomere 1 flattened and elongate, but not wider than the pedicel; flagellomere 2 longer than flagellomere 3; 2 and 3 together shorter than flagellomere 1.

5. Fused prosternum and proepisternum.

6. Metacoxal pegs, when present, slender, weak and acute (except in the aberrant *L.* unc-7sp. nov.).

7. The anal cell of the wing closed near the margin; M3 normally open.

8. The male genitalia laterally clasping epandrium halves completely separate and basally hinged in membrane; a basal epandrial sclerite often present.

9. The epandrium halves bearing basal apodemes.

10. The subepandrial sclerite V- or U-shaped, its apex directed posteriorly, and armed with peg-like or spine-like setae.

11. Bowl-shaped hypandrium. The gonocoxae apically lobe-like, broadly fused posterolaterally to the hypandrium, the junction indicated by a transverse slit that, continuing at a right angle medially to the apex, separates the gonocoxite lobes. A brush of long, stiff setae arming the hypandrium posteromedially and/or the gonocoxites medially. Prominent gonocoxal apodemes.

12. The phallus complex vase-shaped, the sperm sac elliptical and swollen, and the parameral sheath more-or-less elongate; the lateral processes of the parameral sheath prominent, wing-like or fused ventrally into a skirt-like structure; the lateral ejaculatory processes present; the ejaculatory apodeme prominent, linear or spatulate in cross-section; the endoaedeagal tube present.

13. Tergite 10 of the female genitalia developed into spine-bearing acanthophorites; no spines developed on the anteroventral areas of tergite 10.

14. Three spermathecae, each a looped tube, usually long, with the apical portion (terminal reservoir) irregularly coiled, curled, hooked or rod-like and more-or-less sclerotized; the genital fork (furca = sternite 9) Y- or V-shaped, often medially divided, sometimes keeled.

15. Sternite 8 of the female prominent and keel- or trough-shaped with a bifid, shovel-like apex bearing strong sclerites (hypogynial valves).

16. Tergite 9 of the female reduced to three small sclerites, the medial one linking the two halves of tergite10 mediodorsally, the lateral ones embedded in the membrane along the basal margin of tergite 10, approximating the arms of sternite 9 basally.

Similar Genera

In this study, the Stichopogoninae is restricted to genera sharing a dorsally expanded frons, fused hypandrium/gonocoxites, and fused prosternum and proepisternum: *Lasiopogon, Stichopogon, Townsendia, Lissoteles, Rhadinus* and others. The other genera are usually less setose flies than *Lasiopogon*, distinguished by a flat face and, most importantly, their unrotated male genitalia with an undivided, dorsoventrally clasping epandrium. This is not a trivial difference – see pages 272-3 for more discussion of this matter.

Hull (1961) linked *Willistonina* Back to *Lasiopogon* and *Stichopogon*, but in *Willistonina*, the hypandrium and gonocoxites are not fused and the prosternum is separate from the proepisternum; Artigas and Papavero (1993) properly placed it in the Stenopogoninae and raised a new tribe, Willistonini, to hold it.

Other than *Lasiopogon*, only a very few genera of asilids outside the Asilinae, Ommatiinae and Apocleinae have the epandrium halves freely articulating in membrane, dorsoventrally oriented and laterally clasping; most are a single unit, fused at least at a point on the midline at the base or are a pair of horizontal flaps. Evidently, the Australian *Bathypogon* Loew is the closest stenopogoninine to *Lasiopogon*. The gonocoxites and hypandrium are fused in a fashion reminiscent of *Lasiopogon*, but the gonocoxite lobes have the typical stenopogoninine processes and the prosternum is reduced. The epandrium halves are completely separate and articulated in membrane as in *Lasiopogon*; even a basal sclerite is present in some species (this is the only genus, other than *Lasiopogon,* that I have seen with this structure).

Artigas and Papavero (1991a) placed *Bathypogon* with the South American *Carebaricus* Artigas and Papavero in the tribe Bathypogonini of the Stenopogoninae; I have not seen *Carebaricus* and cannot vouch for any similarities in epandrium structure between it and *Bathypogon*; Artigas and Papavero (1991a) indicated that the epandrium is fully divided but that the hypandrium and gonocoxites are completely separate. This suggests that the two genera are not closely related. The South American *Tillobroma* Hull has free epandrium halves but it also belongs in the Stenopogoninae; it was split from its close relative *Hypenetes* Loew from South Africa. The two genera are probably Gondwanian sisters.

It is difficult to establish the presence of completely separated epandrium halves in undissected specimens. There are several genera that have split but not separated epandria and that are similar in appearance to *Lasiopogon*. The most obvious are *Cyrtopogon* Loew and *Lestomyia* Williston. Both genera contain flies with similar shape and habits to *Lasiopogon*; they are mostly setose flies with a gibbous face and strong mystax. Both have species sympatric with *Lasiopogon* species. *Cyrtopogon*, a large Holarctic genus especially common in woodlands, belongs in the Stenopogoninae, and thus, unlike *Lasiopogon*, has the prosternum disassociated from the proepisternum. *Lestomyia* is restricted to the Nearctic and is particularly common in grasslands and other semiarid habitats. With its bent foretibial spine, it belongs in the Dasypogoninae. Neither of these genera have an expanded frons.

Immature Stages

Larvae and pupae play no part in this systematic treatment of *Lasiopogon*. I searched for larvae and pupae of *Lasiopogon cinereus* in the sand along the banks of the Coquihalla River in southern British Columbia. Although this is the densest population of *Lasiopogon* I have observed, I did not find any immatures.

Lasiopogon cinctus, the common species of northern Europe, is the only species whose immature stages have been described. Lundbeck (1908) was the first to record them. Melin (1923) gave the best treatment of the immatures and their habitats; his information has been reproduced in other works (Ionescu and Weinberg 1971, Séguy 1927).

Figure 18. Ocean beach dune habitat of *Lasiopogon littoris* and *L. arenicola*. Oso Flaco Lakes, San Luis Obispo County, California, U.S.A.

Figure 19. River cobble and boulder habitat of *L. cinereus*. Coquihalla River, Hope, British Columbia, Canada.

Natural History

Adults of *Lasiopogon* species live mainly in sandy or rocky habitats, especially near water. They sit on the ground, stones, logs or sometimes on low vegetation and fly up after prey. Presumably, the larvae burrow in the soil and pupate there. Along the Pacific coast of North America, *L. actius, L. arenicola, L. bivittatus, L. littoris* and others hunt on the ocean beaches and dunes (figure 18). Many species inhabit stream and river banks. *L. cinereus* is restricted to substrates in or immediately adjacent to stony mountain streams in western North America (figure 19); *L. bezzii* flies along rivers in lowland Italy; and *L. septentrionalis* lives on the sandy edges of Siberian rivers (figure 20). But water is not a prerequisite. Many species are common in dry mountain forests and subalpine meadows and a few live in sandy grasslands. For example, in Alaska and Yukon, *L. canus* and *L. hinei* live on dry, grassy hillsides and tundra (figure 21), as well as along streams; *L. cinctus* is common in heaths, dry meadows and pinewoods in northern Europe; *L. montanus* and *L. bellardii* live in meadows high in the Alps (figure 22), and *L. monticola* and *L. delicatulus* inhabit similar places in the Rocky and Cascade mountains of North America; and *L. albidus* and *L. chaetosus* hunt in the dry sagebrush steppes of the American Great Basin, far from any water (figure 23).

Species of *Lasiopogon* are adapted to mesic conditions and shun high temperatures. In lowland and mid-elevation environments, adults are typically active in

Figure 20. Sand and gravel riparian habitat of *L. septentrionalis*. Kolyma River drainage, Magadanskaya Oblast, Russia.

Figure 21. Dry montane habitat of *L. canus, L. prima* and *L. hinei*. Sheep Mountain, Kluane, Yukon Territory, Canada.

Figure 22. Alpine meadow habitat of *L. bellardii*. Flumserberg, Flums, Switzerland.

Figure 23. Sandy grassland habitat of *L. chaetosus*. Abert Lake, Lake County, Oregon, U.S.A.

spring and early summer, usually from late March to June. Some, living in places that become very hot, even in spring, fly in the winter. *L. gabrieli* is active from December to March in southern California and in Taiwan, *L. solox* has been collected only in January. *L.* unc-7 sp. nov. lives on the arid plateau of western Nevada and flies in the cool fall rather than in the spring as its congeners do. On the other hand, species living high in the mountains or at high latitudes are active mostly from June to August.

The best accounts of adult behaviour were given by Lavigne and Holland (1969) for *L. cinereus* and *L. polensis*, Lavigne (1972) for *L. quadrivittatus*, and Melin (1923) for *L. cinctus*.

Little is known about nocturnal resting behaviour. In Wyoming, Lavigne and Holland (1969) recorded *L. cinereus* at dawn perched one to two metres high in willow bushes three to five metres away from the stream bank. Apparently, the flies had spent the night there.

All the *Lasiopogon* species studied (Lavigne 1972, Lavigne and Holland 1969, Melin 1923), captured their prey in flight. After spotting potential prey, the hunter quickly turned its body to face the flying insect. It usually chose a small, slow-flying prey; but it frequently went after prey too large to handle, in which case it returned to the same perch without making contact. *L. cinereus* made attack flights every two or three minutes until one was successful. Even females *in copula* often flew out from the perch and captured prey, dragging the male behind.

These *Lasiopogon* killed their prey while still in flight and usually brought them back to the take-off site to feed. Sometimes, as Lavigne and Holland (1969) observed for *L. polensis*, the hunter could not readily subdue larger prey and fell to the ground with its victim before killing it. Some flies manipulated their prey during feeding, inserting the proboscis several times in different parts of the body. These flies usually used the fore tarsi or the fore and hind tarsi to reposition the corpse; *L. quadrivittatus* often rolled onto its side or back and manipulated the prey with all tarsi (Lavigne 1972). *Lasiopogon* can inflate and deflate soft-bodied prey by pumping its basal abdominal segments; presumably, this progressively enlarges the surface area of soft tissue for digestive action (Lavigne and Holland 1969). When feeding was complete, the fly withdrew its proboscis by pushing off the with the fore tarsi. The recorded feeding times of *L. polensis* ranged from five to twenty minutes.

Most prey species recorded are small Diptera. For *L. cinctus* in Sweden, Melin (1923) listed flies from many families, from tipulids and chironomids to lonchaeids and anthomyiids. Poulton (1906) recorded a tipulid, *Pachyrrhinia lineata* Scop., in Britain. Weinberg (1978) noted feeding habits of *Lasiopogon montanus* on Vîrful cu Dor in the southern Carpathian Mountains. It preyed on Diptera of the families Therevidae, Muscidae and Empididae (mostly the last); it attacked only these flies, even though Hymenoptera and other insects of various orders were common. Almost all the *Lasiopogon* specimens captured with prey were females, and several examples of cannibalism were noted; for example, the Wyoming material of *L. quadrivittatus* included three females with male prey. In Colorado, 16 of 18 prey specimens taken by *L. quadrivittatus* were flies; the other two were Homoptera (Cicadellidae and Psyllidae; see species account) (Lavigne 1972). Small Homoptera were also second to Diptera in a Wyoming study of *L. polensis* (Lavigne and Holland 1969). Of 63 specimens killed, there were 40 Diptera, 10 Cicadellidae, eight aphids, 4 psyllids and a single hymenopteran. The same study showed that the availability of suitable prey will often influence prey selection: *L. cinereus*, which hunts from rocks on the edge or in the middle of, streams, preyed mostly on aquatic insects (40 Ephemeroptera, 16 Diptera and 1 trichopteran; see species account).

Marshall (pers. comm.) described a wing-spreading display performed by *L. marshalli* sp. nov. When a fly, perched on a riverside stone, was approached by a second fly, the initial occupant of the perch faced the newcomer and spread its wings horizontally. I have seen a similar display in *L. cinereus*. I assume this behaviour is associated with feeding territoriality.

Lasiopogon has no recognizable courtship display. Weinberg (1978) recorded the mating of *L. montanus* Schiner in alpine meadows in Romania. A male would fly briefly around a female perched on the low grass and then dive, striking the female and causing her to fly; coupling then occurred. *L. polensis* males usually intercepted females in the air and, falling to the ground, linked genitalia (Lavigne and Holland 1969). Copulation lasted from 13 to 20 minutes, about the same duration as in *L. quadrivittatus* (Lavigne 1972). Males of *L. cinereus* normally mounted females while they perched; incomplete matings in this species lasted 47 to 71 minutes (Lavigne and Holland 1969). Copulation in *Lasiopogon* is tail to tail.

Lavigne (1972) observed *L. quadrivittatus* laying eggs in soil at the base of plants. It inserted the abdomen to a depth of three millimetres and removed it 30 to 45 seconds later. The egg of *L. cinctus* is narrowly oval (about 1 x 0.55 mm) with a shiny whitish-yellow shell (Melin 1923).

Melin (1923) found *L. cinctus* larvae "in sandy pastures, wooded pastures, fallows and clover-fields; in gravel pits; in piles of sand and banks of sand along roads and railways; in sandy hillocks thrown up by moles or voles in different places; in dune-sand amongst the roots of grass and *Salix repens*, and in sandy tussocks. But I have also found the larva in a grassy sward with soil rich in mould." In dry sand or sandy soil, Melin often found these larvae at depths of up to half a metre.

Melin documented two pupation and emergence dates for *L. cinctus* in Sweden. A female pupated on 12 April and emerged as an adult on 29 April; dates for a male were 10 May and 26 May, respectively. Thus the duration of the pupal stage is about 17 days. Evidence suggests that this species takes at least two years to develop (Melin 1923).

Distribution

Lasiopogon is a Holarctic genus (map 1, p. 5). In North America it ranges north to the Arctic Ocean in the Alaska/Yukon/Mackenzie region of the extreme northwest (the most northerly record is at 68°26'N in northern Alaska) and south in the Cordillera to northern Baja California (about 31°N) and the mountains and plateaus of central Arizona and New Mexico. There is no Nearctic transboreal fauna. The genus occurs transcontinentally on the Great Plains south to Oklahoma. In eastern North America the Appalachian fauna extends north to James

Bay and south along the mountains and eastern Piedmont plateaus to the edge of the lowlands in Georgia and Mississippi and onto the gulf lowlands in Alabama (about 31°25'N).

In Eurasia, *Lasiopogon* species range from Britain to Far-eastern Russia, north at least to the treeline and into the tundra along rivers (Lehr 1984a). It is known as far north as 70°05'N on the Lena River (*L. hinei*). The genus ranges south to the Mediterranean Sea, Turkey, the Caucasus, Kazakhstan, the highlands of western China, Korea and Japan. All these areas are north of 30°N. The one species that lives in Taiwan (on the Tropic of Cancer), *L. solox* Enderlein, should not be considered a member of the Oriental fauna because it inhabits medium elevations and is part of the Palaearctic fauna that occurs in the island's central highlands.

IDENTIFICATION KEYS TO THE *LASIOPOGON* SPECIES OF THE NEARCTIC AND EAST PALAEARCTIC REGIONS

The three identification keys below include all *Lasiopogon* species known in North America and eastern Asia. Species from Europe and Asia west of 60°E longitude are not included because many of these species were not analysed in detail during the study. Undescribed species are included so that the keys function accurately.

Specimens are keyed, for the most part, without using genitalic characters visible only in dissection; establishing the presence of the basal sclerite usually requires dissection. Some species are keyed more than once, owing to variation in their colour, setation and other characters, and the key is often more successful if male specimens, rather than females, are used. Colour characters work only on mature specimens. For example, the legs and genitalia of recently emerged specimens are normally yellow/red-brown but turn brown or black in most species when the fly matures.

The dorsum of the epandrium is oriented ventrally in specimens that have undergone genitalic rotation; this includes almost all mature specimens.

The medial stripe is the same colour as the dorsocentral stripes; a pale space between acrostichal stripes is not a medial stripe.

Species outside the *opaculus* section and not treated in detail in this book are indicated with an asterisk (*).

Identification key to Nearctic species of *Lasiopogon* occurring east of the Rocky Mountains

1a Scutellum bare, at most with a few, fine apical hairs; dorsocentral bristles absent; abdomen mostly dark basally, red apically, the pale tomentum on the tergites restricted to the apicolateral angles; legs mostly red-yellow. (Alberta east on the Great Plains to the southern margins of the Great Lakes and through New England, south to Virginia, Ohio and North Dakota.) . . . .*L. terricola* (Johnson)*

1b Scutellum with several strong bristles; dorsocentral bristles present; abdomen brown or black, in a few species ferruginous apically on apical tergites, the tergites with apical grey or grey-brown/gold bands of tomentum or, in some species, with grey and brown tomentum covering entire tergite; legs brown/black or, in some species, ferruginous with various amounts of pale tomentum 2

2a Halter knob with dorsal black or brown spot, often diffuse 3

2b Halter unmarked; dorsum of knob white to yellowish brown 5

3a Antennal segment F1 mostly ferruginous; mystax setae brown; tarsi and tibiae often ferruginous; halter spot dark brown or black and covering all of dorsum of knob; abdominal tergites with brown tomentum basally, the grey apical bands more than 0.5 times the tergite length (southeastern Ontario south to Georgia in, and east of, the Appalachian Mountains). . . . .*L. currani* Cole and Wilcox*

3b Antennal segment F1 mostly brown/black; mystax black or mixed black and white; legs brown/black; halter spot brown/black, sometimes light and irregular and covering only part of dorsum of knob; abdominal tergites shining black basally or with the black cuticle visible under thin tomentum 4

4a Mystax mostly white or white with black setae dorsally; abdominal tergites shining black with apical tomentum bands 0.25-0.50 times the length of the tergite. Male with dorsum of epandrium strongly emarginate on medial margins in apical half, these emarginations bearing disc-like lobes (figure 245g); female with sternite 8 black (Alaska and Yukon south through northern British Columbia to central Alberta). . . . .*L. prima* Adisoemarto*

4b Mystax always mostly black, sometimes a few white setae ventrally; male abdominal tergite with apical tomentum bands very narrow, usually covering only the intersegmental membrane. Male with dorsum of epandrium strongly emarginate on medial margins in apical half but without disc-like lobe (figure 91); female with sternite 8 ferruginous; genitalia as in figures 91-97 (northern Eurasia; Alaska and Yukon south through northeastern British Columbia to central Alberta). . . . .*L. hinei* Cole and Wilcox

5a   Mystax setae white or pale yellow **6**

5b   Mystax setae dark brown or black **9**

6a   Tarsi, tibiae and bases of femora ferruginous; genitalia ferruginous, as in
     figures 78-81 (female unknown) (mountains of North Carolina).
     . . . .*L. flammeus* **sp. nov.**

6b   Legs brown/black with varying amounts of pale tomentum; genitalia
     brown/black **7**

7a   Scutellar and notal bristles white; genitalia as in figures 181-87 (north-
     eastern British Columbia and northern Alberta and Saskatchewan south to
     Colorado and Nebraska). . . . .*L. quadrivittatus* **Jones (part)**

7b   Scutellar and notal bristles black **8**

8a   Anterior margin of metacoxa with a coxal peg (figure 6); antennal scape
     usually with white setae; male with dorsum of epandrium strongly
     emarginate on apical half of medial margins; female with hypogynial
     valves of abdominal sternite 8 ferruginous, and with abundant pale hairs
     and dark bristles (plains and mountains of Alberta and Montana).
     . . . .*L. trivittatus* **Melander***

8b   Metacoxa without a coxal peg; antennal scape with brown setae; male
     with dorsum of epandrium strongly emarginate on basal half of medial
     margins (figure 144); female with hypogynal valves of abdominal sternite
     8 brown/black (sometimes with ferruginous tips), a few hairs visible
     microscopically at base (figure 148); genitalia as in figures 144-50
     (Oklahoma). . . . .*L. oklahomensis* **Cole and Wilcox**

9a   Legs and genitalia ferruginous; genitalia as in figures 205-11 (southern
     Appalachian Mountains from North Carolina to Georgia).
     . . . .*L. shermani* **Cole and Wilcox**

9b   Legs and genitalia brown/black (except for pale tomentum) **10**

10a Males **11**

10b Females (female of *L. chrysotus* unknown) **23**

11a Bristles on katatergite mostly white or pale yellow; bristles on basolateral
     corners of abdominal tergite 1 pale **12**

11b Bristles on katatergite mostly dark brown/black; bristles on basolateral
     corners of abdominal tergite 1 dark brown/black, pale or mixed dark and
     pale **16**

12a Epandrium in lateral view long and narrow, the greatest width about
     0.35-0.4 times the length, the apex bent dorsally (figure 213); genitalia as
     in figures 213-16 (Quebec and Michigan south to Virginia, West Virginia
     and Ohio) . . . .*L. slossonae* **Cole and Wilcox**

12b Epandrium in lateral view with the greatest width about 0.5-0.6 times the
     length, the apex rounded (figure 41) **13**

13a   Short appressed hairs on femora and tibiae almost all white; grey apical
       bands of tomentum on abdominal tergites very narrow, about 0.1-0.2
       times the length of the tergite; genitalia as in figures 197-200 (southern
       Appalachian Mountains from North Carolina to Mississippi; Piedmont
       (rarely in coastal areas) from South Carolina to Georgia)
       ....*L. schizopygus* **sp. nov.**
13b   Short, appressed hairs on tibiae and apical half of femur almost all
       brown/black; apical bands of tomentum on abdominal tergites 0.1-0.5
       times the length of the tergite. **14**

14a   Apex of gonostylus in lateral view strongly flattened apicobasally, the
       medial flange shallow, the height less than 0.25 times the dorsoventral
       width (figure 166); dorsal horizontal margin of gonostylus with a weak
       central tooth (figure 166); genitalia as in figures 165-68 (lowlands of
       southern Alabama) ....*L. piestolophus* **sp. nov.**
14b   Apex of gonostylus in lateral view arched and rounded, the medial flange
       deep, the height 0.4 or more times the dorsoventral width (figures 42,
       163); dorsal horizontal margin of gonostylus with a strong tooth or
       without a tooth (figures 145, 129) **15**

15a   Grey apical bands of tomentum on abdominal tergites about 0.2-0.5 times
       the length of the tergite; dorsal horizontal margin of gonostylus with
       strong central tooth (figure 153); genitalia as in figures 152-55 (southern
       Ontario and central Michigan west to Iowa, south to Mississippi and
       Georgia) ....*L. opaculus* **Loew**
15b   Grey apical bands of tomentum on abdominal tergites about 0.1-0.25
       times the length of the tergite; dorsal horizontal margin of gonostylus
       without central tooth (figure 42); genitalia as in figures 41-44 (central
       Appalachian Mountains: Kentucky, Tennessee, West Virginia)
       ....*L. appalachensis* **sp. nov. (part)**

16a   Anterior margin of metacoxa with coxal peg (figure 6); abdomen shining
       black, the width of the pale apical bands at most about 0.2 times the
       length of tergite, covering only slightly more than the intersegmental
       membrane; bristles on basolateral corners of abdominal tergites black or
       mostly black (central Alberta) ....*L. can-1* **sp. nov.***
16b   Anterior margin of metacoxa without a coxal peg; abdomen tomentum
       variable; bristles on basolateral corners of abdominal tergite 1 brown,
       black or white/pale-yellow **17**

17a   Bristles on basolateral corners of abdominal tergite 1 white/pale-yellow
       ....**18**
17b   Some bristles on basolateral corners of abdominal tergite 1 dark-
       brown/black **21**

18a   Apical bands of pale tomentum on abdominal tergites broad, their width
        0.5 or more times the length of the tergite; dorsal setulae on abdominal
        tergites mostly white (usually dark on areas of brown tomentum);
        epandrium in dorsal view with medial margins strongly concave in basal
        half (figure 237) **19**
18b   Apical bands of pale tomentum on abdominal tergites narrow, their width
        0.1-0.25 times the length of tergite; dorsal setulae on abdominal tergites
        all black; epandrium dorsal view with medial margins only moderately
        concave in basal half (figure 41) **20**

19a   Apical bands of pale tomentum on abdominal tergites grey, basal areas of
        tergites with light to medium brown tomentum. Genitalia dark brown,
        ferruginous basally; in lateral view the epandrium with greatest width
        about 0.45 times the length (figure 237). Face grey or gold-grey; genitalia
        as in figures 237-40 (Indiana and Ohio east to Virginia and Delaware)
        . . . .*L. woodorum* **sp. nov.**
19b   Apical bands of pale tomentum on abdominal tergites gold-grey, basal
        areas of tergites with dark brown tomentum. Genitalia dark brown/black;
        in lateral view the epandrium with greatest width about 0.55 times the
        length (figure 57). Face gold-silver; genitalia as in figures 57-60 (central
        Appalachian Mountains to eastern Tennessee) . . . . . .*L. chrysotus* **sp. nov.**

20a   Gonostylus with medial flange low, but not unusually so, about 0.4 times
        as high as the dorsal length; in dorsal view, face of flange not expanded
        medially (figure 42); ventrolateral tooth moderately developed (figure
        42); genitalia as in figures 41-44; difficult to separate from
        L. marshalli (central Appalachian Mountains: Kentucky, Tennessee,
        West Virginia) . . . .*L. appalachensis* **sp. nov. (part)**
20b   Gonostylus with medial flange unusually low, about 0.3 times as high as
        the dorsal length; in dorsal view, face of flange expanded medially (figure
        129); ventrolateral tooth strongly developed (figure 129); genitalia as in
        figures 128-31; difficult to separate from *L. appalachensis* (Appalachian
        Mountains in Virginia) . . . .*L. marshalli* **sp. nov. (part)**

21a   Tomentum of head and thorax pale grey; epandrium in dorsal view with
        medial margins only weakly concave, in lateral view with shallow tooth
        on apical 0.25 of the ventral margin (figure 62); genitalia as in figures 62-
        65 (northern British Columbia south through western Alberta to central
        California and southern Colorado) . . . .*L. cinereus* **Cole**
21b   Tomentum of head, thorax largely brown/gold-grey; epandrium in dorsal
        view with medial margins moderately or strongly concave, in lateral view
        no subapical shallow tooth (figures 128, 229) **22**

22a   Apical bands of pale tomentum on abdominal tergites 0.25-0.50 times the
        length of tergite; lateral setae on tergites 5-7 white, dorsal abdominal
        setulae mostly white; epandrium in dorsal view with medial margins

strongly concave (figure 229) antennal ratio LF 2-3/LF1 less than 0.55; anepimeron without fine setae; genitalia as in figures 229-32 (James Bay south to Michigan, southern Ontario, southern Quebec and Connecticut) ....*L. tetragrammus* Loew

22b Apical bands of pale tomentum on abdominal tergites about 0.2 times the length of tergite; lateral setae on tergites 5 to 7 dark; dorsal abdominal setulae dark; epandrium in dorsal view with medial margins moderately concave (figure 128); antennal ratio LF2+3/LF1 about 0.8; anepimeron with fine setae; genitalia as in figures 128-31 (Appalachian Mountains in Virginia) ....*L. marshalli* sp. nov. (part)

23a Bristles on katatergite mostly white or pale yellow; bristles on basolateral corners of abdominal tergite 1 mostly white or pale yellow; hairs on tergite 8 brown/black; hypogynial valves of sternite 8 with strong bristles 24

23b Bristles on katatergite dark brown/black; bristles on basolateral corners of abdominal tergite 1 mostly either dark brown/black or white or pale yellow; hairs on tergite 8 brown/black or white; hypogynial valves of sternite 8 bare, or with fine hairs or strong bristles 28

24a Short appressed hairs on metafemur mostly dark brown/black, a few white basally; genitalia as in figures 217-19 (Quebec and Michigan south to Virginia, West Virginia and Ohio) ....*L. slossonae* Cole and Wilcox

24b Short appressed hairs on metafemur mostly white, some dark apically .25

25a Grey or gold-grey tomentum on abdominal tergites forming definite apical bands, the basal areas of tergites with brown tomentum 26

25b Pattern of gold-grey apical tomentum and basal brown or grey-brown tomentum vague, the apical tomentum not forming a definite band and often extending dorsally to tergite base, reducing basal brownish tomentum to faint basolateral patches; genitalia as in figures 201-3 (southern Appalachian Mountains from North Carolina to Mississippi; Piedmont (rarely in coastal areas) from South Carolina to Georgia) ....*L. schizopygus* sp. nov.

26a Apical bands of grey tomentum extending basally along midline on dorsum of abdominal tergites, forming a grey triangle partly dividing basal brown tomentum into basolateral patches; genitalia as in figures 156-58 (southern Ontario and central Michigan west to Iowa, south to Mississippi and Georgia) ....*L. opaculus* Loew

26b Apical bands of grey tomentum not extending basally along midline of tergite 27

27a Spermatheca with basal half of the apical striate tube strongly sclerotized and not striate; with a dense patch of caniculi immediately apical to the striate portion of this tube (figure 171); genitalia as in figures 169-71 (lowlands of southern Alabama) ....*L. piestolophus* sp. nov.

27b Spermatheca with apical striate tube normal, without strongly sclerotized basal portion; caniculi immediately apical to striate tube more-or-less similar in density to those on rest of tube (figure 47); genitalia as in figures 45-47 (central Appalachian Mountains: Kentucky, Tennessee, West Virginia) ....*L. appalachensis* **sp. nov.**

28a Hairs on tergite 8 brown or black; hypogynial valves on sternite 8 bearing short spines; antennal ratio LF2+3/LF1 greater than 0.75; genitalia as in figures 132-34 (Appalachian Mountains in Virginia) ....*L. marshalli* **sp. nov.**

28b Hairs on tergite 8 white; hypogynial valves on sternite 8 bare, at most with fine hairs; LF2+3/LF1 less than 0.65 ....**29**

29a Bristles on basolateral corners of abdominal tergite 1 white or pale yellow; genitalia as in figures 241-43 (female of *L. chrysotus* sp. nov. is unknown, but almost certainly will key here) (Indiana and Ohio east to Virginia and Delaware) ....*L. woodorum* **sp. nov.**

29b Bristles on basolateral corners of abdominal tergite 1 dark-brown/black **30**

30a Anterior margin of metacoxa with a coxal peg (figure 6); hypogynial valves of sternite 8 ferruginous, bearing fine hairs (central Alberta) ....*L. can-1* **sp. nov.**

30b Anterior margin of metacoxa without a coxal peg; hypogynial valves of sternite 8 brown/black, without fine hairs **31**

31a Tomentum of head, thorax and legs pale grey; acrostichal stripes usually absent; usually a thin brown medial stripe present; a few white hairs on anepimeron; genitalia as in figures 66-68 (northern British Columbia south through western Alberta to central California and southern Colorado) ....*L. cinereus* **Cole**

31b Tomentum of head and thorax brown/gold-grey; acrostichal stripes present; anepimeron bare; genitalia as in figures 233-35 (James Bay south to Michigan, southern Ontario, southern Quebec and Connecticut) ....*L. tetragrammus* **Loew**

Identification key to the Nearctic species of *Lasiopogon* occurring in and west of the Rocky Mountains

1a Anterior margin of metacoxa with a coxal peg (figure 6) **2**
1b Anterior margin of metacoxa without a coxal peg **6**

2a Mystax white or mostly white **3**
2b Mystax black or brown, sometimes with a few white setae ventrally **5**

3a Anepisternal and coxal setae long, stout and white, the anepisternal setae about twice as long as the antennae; leg bristles white; one pair of black orbital setae; metacoxal peg stout and rounded apically (species active in autumn; Nevada plateau) . . . .*L.* **unc-7 sp. nov.***
3b Anepisternal setae absent or short, less than the length of the antennal scape and brown/black; coxal setae fine, white; leg bristles black/brown; more than one pair of black orbital setae; metacoxal peg slender, acute apically (figure 6) **4**

4a Abdominal tergites with basal areas lacking tomentum, the black or brown cuticle shining; in females the bare areas are often restricted to basolateral areas; antenna with F1 width at least 0.4 the length; mystax of female often dark dorsally (subalpine habitats, Cascade Mountains from Washington to northern California) . . . . *L. delicatulus* **Melander***
4b Abdominal tergites covered with tomentum, apically grey, basally brown; antenna with F1 width 0.25 to 0.3 the length (stream banks on east side of Rocky Mountains and adjacent plains in Alberta and Montana) . . . .*L. trivittatus* **Melander***

5a Base of abdominal tergites with brown tomentum; vertex with brown tomentum; notal pattern strong, with wide brown dorsocentral and medial stripes and brown lateral areas (coastal streams, Washington to northern Oregon) . . . .*L. pugeti* **Cole and Wilcox***
5b Base of abdominal tergites mostly shining black; vertex with grey tomentum; notal pattern obscure, usually dark grey with faint brown dorsocentral and medial stripes (Alaska, Yukon and extreme northwestern Northwest Territories) . . . .*L. canus* **Cole and Wilcox***

6a Dorsum of halter knob black or brown, or partially covered with black or brown markings **7**
6b Dorsum of halter knob plain white or yellow **12**

7a Mystax brown/black, sometimes a few white setae ventrally; abdominal tergites with varying amounts of tomentum; in dorsal view the epandrium halves without disc-like processes on the medial margins **8**
7b Mystax mostly white or white with black setae dorsally; abdominal tergites shining black with apical bands of grey tomentum covering 0.25-0.5 the length of the tergite; in dorsal view the epandrium halves with

medial margins strongly emarginate in apical half, these emarginations bearing disc-like processes (figure 245g); female with sternite 8 black (Alaska and Yukon south across northern British Columbia to central Alberta) . . . .*L. prima* Adisoemarto*

8a Some of the bristles on basolateral corners of abdominal tergite 1 black/brown 9

8b Bristles on basolateral corners of abdominal tergite 1 white or pale yellow 10

9a Tomentum of head and thorax pale grey; dorsocentral stripes brown, thin brown medial stripe present or absent; halter markings usually not covering entire dorsum of knob; legs and male genitalia black with grey tomentum (female genitalia shining); genitalia as in figures 62-68 (northern British Columbia and western Alberta south to central California and Colorado) . . . .*L. cinereus* Cole (part)

9b Tomentum of head and thorax grey/gold-brown; dorsocentral stripes strong and a medial stripe normally present; black or brown pigment strongly covering entire dorsum of halter knob; tibiae and genitalia (including both sternite 8 and tergite 8 in female) often ferruginous/yellow-red, the male genitalia without tomentum (southern British Columbia south to California and Colorado

. . . .*L. fumipennis* Melander* (part)

10a Dark pigment usually not completely covering dorsum of halter knob; katatergite setae black; in male, abdominal tergites with apical bands of tomentum very narrow, usually covering only the intersegmental membranes, the tergites shining black; female with brown hairs on tergite 8 and black bristles on apical lobes of sternite 8; genitalia as in figures 91-97) (Alaska and Yukon south across northeastern British Columbia to central Alberta) . . . .*L. hinei* Cole and Wilcox

10b Black or brown pigment strongly covering entire dorsum of halter knob; at least some katatergite setae white, golden or pale brown; in male, abdominal tergites covered with tomentum, not shining black; female with pale hairs on tergite 8; apical lobes of sternite 8 without bristles 11

11a Anterior scutum with short setulae, usually fewer than two anterodorsocentral setae longer than antennal scape + pedicel; mesonotal tomentum grey, usually suffused with some brown or gold, the dorsocentral stripes strong and a medial stripe normally present; tibiae and genitalia (including both sternite 8 and tergite 8 in female) often ferruginous/yellow-red (southern British Columbia south to California and Colorado)

. . . . *L. fumipennis* Melander* (part)

11b At least three anterodorsocentral setae as long as antennal scape + pedicel; mesonotal tomentum grey, stripes usually obscure; tibiae and genitalia brown (sternite 8 in female ferruginous/yellow-red) (mountains of south-eastern Wyoming) . . . .*L. polensis* **Lavigne***

12a Mystax mostly white or golden **13**
12b Mystax brown or black, sometimes with a few white setae ventrally . . .**25**

13a Scutellar bristles black/brown **14**
13b Scutellar bristles white **18**

14a Tibiae and tarsi ferruginous or yellowish, contrasting with brown/grey of femora (in *L.* tes-5 sp. nov., femora are often largely ferruginous) **15**
14b Tibiae and tarsi brown/black with pale tomentum **16**

15a Notal bristles strong and abundant, usually more than 4 anterior dorso-centrals and 3 or 4 strong postalars; femora brown with grey tomentum; epandrium ferruginous to brown with thin grey tomentum laterally, in dorsal view the medial margins strongly concave; female tergite 8 dark brown/black, with bristles on lateral lobes; genitalia as in figures 49-55) (dry grasslands and shrub-steppes of the Great Basin; Washington south to Nevada and Utah) . . . .*L. chaetosus* **Cole and Wilcox (part)**
15b Anterior dorsocentral bristles weak, usually 1 or 2; only 1 strong postalar; at least bases and apices of femora ferruginous; epandrium ferruginous, swollen, shining, and without tomentum, in dorsal view the medial mar-gins more-or-less parallel and bearing a small dark central tooth (cf. fig-ure 245f); female tergite 8 ferruginous and without bristles on lateral lobes (mountains of eastern Arizona) . . . .*L.* **tes-5 sp nov.***

16a Bristles on katatergite and basolateral corners of abdominal tergite 1 white/yellow; almost all setae on anterior part of scutum are minute setulae; epandrium in lateral view much wider apically than basally, the greatest width about 0.6 times the length; in dorsal view the epandrium halves with medial margins more-or-less parallel, each with a shallow tooth centrally (figure 181); genitalia as in figures 181-87) (mountains of northern New Mexico) . . . .*L. quadrivittatus* **Jones (part)**
16b Katatergite bristles all or mostly brown/black, rarely in *L. ripicola* all pale; at least some bristles on basolateral corners of abdominal tergite 1 brown/black; scutum with well-developed dorsocentral bristles; epan-drium in lateral view with dorsal and ventral margins more-or-less parallel, the greatest width about 0.5 times the length; apicoventral corner sharply angled (cf. figure 245i); in dorsal view the epandrium halves with medial margins moderately concave in apical half (cf. figure 245i) **17**

17a Large species, usually 11mm or longer; katatergite bristles mostly brown/black, some pale; dorsal setulae on abdominal tergites of female black/brown; apex of phallus with dorsal carina much longer than aedeagal tube; apex of dorsal carina expanded and oval in dorsal or ventral view (stream margins in the Columbia Basin of Washington and Idaho) . . . .*L. ripicola* **Cole and Wilcox***

17b   Small to medium-sized species, to 10mm long, but most 7-9mm; variable in coloration; katatergite bristles all brown/black or sometimes with a few pale; dorsal setulae on abdominal tergites of female black/brown, mixed white/brown/black or sometimes all white; apex of phallus with dorsal carina and aedeagal tube about equal in length; apex of dorsal carina linear, not expanded and oval in dorsal or ventral view (southern California) . . . .*L. drabicola* **Cole***

18a   At least lateral bristles on scutum brown/black; tarsal and some tibial bristles brown/black (coastal beaches, Queen Charlotte Islands south to Oregon) . . . .*L. actius* **Melander***

18b   All bristles white **19**

19a   Antenna with F2+3 longer than F1; scutum uniformly pale grey; small species (beaches on central coast of California) . . . .*L. littoris* **Cole***

19b   Antenna with F2+3 shorter than F1; scutum with dorsocentral stripes brown or grey, often obscure; body size variable **20**

20a   Males **21**

20b   Females **23**

21a   Epandrium in lateral view much wider apically than basally, the greatest width about 0.6 times the length (figure 181); in dorsal view the epandrium halves with medial margins more-or-less parallel, each with a shallow tooth centrally (figure 181); genitalia as in figures 181-84) (eastern Rocky Mountains and Great Plains from northeastern British Columbia south to Colorado) . . . .*L. quadrivittatus* **Johnson (part)**

21b   Epandrium in lateral view with dorsal and ventral margins more-or-less parallel, the greatest width about 0.5 times the length; apicoventral corner sharply angled (cf. figure 245i); in dorsal view the epandrium halves with medial margins moderately concave in apical half (cf. figure 245i) **22**

22a   Trochanters and joints of tarsi and tibiae brown/black; lateral abdominal setae long and dense, those on abdominal segment 2 about the same length as those on segment 1; sternite 8 mostly black/dark brown (beaches and dunes on central coast of California) . . .*L. arenicola* **Osten Sacken***

22b   Trochanters and joints of tarsi and tibiae ferruginous/yellow-red; lateral abdominal setae on segment 2 about half as long as those on segment 1; sternite 8 mostly ferruginous/yellow (dry grasslands and shrub-steppes east of Cascade Mountains) . . . .*L. albidus* **Cole and Wilcox***

23a   Eight or fewer major scutellar bristles arranged in one row; genitalia as in figures 185-87) (eastern Rocky Mountains and Great Plains from north-eastern British Columbia south to Colorado)
. . . .*L. quadrivittatus* **Jones (part)**

23b   Twelve or more major scutellar bristles arranged in several irregular rows
....**24**

24a   Trochanters and joints of tarsi and tibiae brown/black; lateral abdominal
setae long and dense, those on abdominal segment 2 about the same
length as those on segment 1 (beaches and dunes on central coast of
California) ....*L. arenicola* **Osten Sacken***

24b   Trochanters and joints of tarsi and tibiae ferruginous/yellow-red; lateral
abdominal setae on segment 2 about half the length of those on segment 1,
or even shorter (dry grasslands and shrub-steppes east of Cascade
Mountains) ....*L. albidus* **Cole and Wilcox***

25a   Males **26**
25b   Females **54**

26a   Genitalia ferruginous or yellow-red............................**27**
26b   Genitalia black, brown or dark red-brown, often with tomentum covering
the cuticle **35**

27a   In dorsal view, medial margins of epandrial halves with prominent spine
(figure 136) (Sierra Nevada) ....*L. monticola* **Melander (part)**
27b   In dorsal view, medial margins of epandrium halves without prominent
spine **28**

28a   Epandrium in lateral view with the greatest width at least 0.7 the length;
genitalia dark ferruginous; tibiae and tarsi brown/black; scutum with
medial stripe usually present; basal epandrial sclerite absent; phallus of
type shown in figure 248g **29**
28b   Epandrium in lateral view with the greatest width usually 0.6 or less the
length; genitalia bright ferruginous or yellow-red; tibiae and tarsi often
ferruginous or yellow-red; scutum without medial stripe; basal epandrial
sclerite present or absent; phallus one of types shown in figure 248h
and n..**30**

29a   Epandrium in lateral view twice as wide at apex as at base, apicoventral
corner a prominent angle; scutum usually with considerable brown
tomentum laterally and posteriorly (ocean beaches and coastal streams,
southwestern British Columbia to Oregon)
..........................*L. pacificus* **Cole and Wilcox* (part)**
29b   Epandrium in lateral view only slightly wider at apex than at base,
apicoventral corner blunt (figure 245e); scutum grey, the brown tomentum
more-or-less restricted to the dorsocentral and medial stripes (mountain
habitats from southern British Columbia and Alberta south to Colorado
and Utah) ....*L. aldrichii* **Cole and Wilcox* (part)**

30a   Epandrium clothed in long, strong setae, at least some as long as the
lateral width of epandrium; no basal umbo; in dorsal view medial margins

concave in apical half, the convex basal margins rather widely separated (figures 33, 49); apex of epandrium halves without small flange; basal epandrial sclerite present; phallus as in figures 35 and 51 . . . .**31**

30b Epandrium with fine, short setae, considerably shorter than lateral width of epandrium; basal umbo present; in dorsal view the medial margins touching or almost touching at umbo, only weakly concave in apical half and bearing a small dark central tooth, (cf. figure 245f); apex of epandrium halves often with small flange; basal epandrial sclerite absent; phallus as in figure 248h **32**

31a Mystax light to medium brown (most specimens have white mystax; these key out elsewhere); in lateral view epandrium width 0.4 the length, the setae white or pale brown (gonocoxite brush brown); femora dark brown with pale tomentum, tibiae yellow-brown or pale red-brown; white hairs laterally on basal tergites long, the longest on tergite 2 about 0.5 the length of the tergite; usually more than 5 anterodorsocentral bristles; genitalia as in figures 49-52 (dry grasslands and shrub-steppes, Washington south to Nevada and Utah)
. . . .*L. chaetosus* **Cole and Wilcox (part)**

31b Mystax dark brown/black; in lateral view epandrium width 0.5 the length, the setae brown; femora dark brown/black, ferruginous at base and apex, tomentum sparse; tibiae ferruginous; white hairs laterally on basal tergites very short, the longest on tergite 2 about 0.15 the length of the tergite; 3 or 4 anterodorsocentral bristles; genitalia as in figures 33-36 (mountains of eastern Arizona) . . . .*L. apache* **sp. nov.**

32a Setal brush on venter of hypandrium/gonocoxite complex brown **33**
32b Setal brush on venter of hypandrium/gonocoxite complex yellow/gold
. . . .**34**

33a In ventral view, lateral margins of epandrium halves curving evenly to midline apically; trochanters and bases of femora ferruginous, tibiae and tarsi ferruginous; tomentum of thorax usually gold-grey (Sierra Nevada and southern Cascade Mountains) . . . . . .*L. testaceus* **Cole and Wilcox***
33b In ventral view, lateral margins of epandrium halves bent abruptly inwards apically; femora, tibiae and tarsi brown/black; tomentum of thorax grey (San Bernardino Mountains of southern California) . . . .*L.* **tes-4 sp. nov.***

34a In dorsal view, width of epandrium half just basal to central tooth about 0.5 the length; most of the short setae laterally and dorsally on tergites 5 to 7 dark (Sierra Nevada) . . . .*L.* **tes-1 sp. nov.***
34b In dorsal view, width of epandrium half just basal to central tooth about 0.4 the length; most of the short setae laterally and dorsally on tergites 5 to 7 white (eastern slopes of Cascades in Oregon) . . . . .*L.* **tes-3 sp. nov.***

35a In dorsal view, medial margins of epandrial halves with prominent spine (figure 136); genitalia as in figures 136-39 (mountains of southern British

Columbia south to California and Utah) . .*L. monticola* **Melander (part)**

35b In dorsal view, medial margins of epandrium halves without prominent
spine **36**

36a Epandrium in dorsal view with medial margins bearing a small central
tooth (cf. figure 245f); basal epandrial sclerite absent; apex of epandrium
halves with weak flange; phallus of type shown in figure 248h (White
Mountains of central Sierra Nevada) . . . .*L.* **tes-2 sp. nov.***

36b Epandrium in dorsal view without small central tooth; basal epandrial
sclerite present or absent; shape of epandrium and phallus variable **37**

37a Greatest width of epandrium in lateral view at least 0.6 times the epandrium length; basal epandrial sclerite present or absent **38**

37b Greatest width of epandrium in lateral view 0.5 times the epandrium
length, or less; basal epandrial sclerite present **44**

38a Gonocoxite apical lobes elongate (figure 246a); epandrium with basal
umbos (figure 245e, f); scutum with medial stripe usually present; basal
epandrial sclerite absent; phallus of type shown in figure 248g **39**

38b Gonocoxite lobes absent (figures 71, 246h); epandrium without basal
umbos and shaped as in figures 70, 245i; scutum without medial stripe;
basal epandrial sclerite present; phallus as in figures 72 or 248k-m **42**

39a Apicoventral angle of epandrium produced into a strong tubercle **40**

39b Apicoventral angle of epandrium without a strong tubercle **41**

40a Mesonotal tomentum grey; dorsocentral stripes sometimes obscure,
medial stripe obscure or absent; femora shining black, without significant
tomentum (southern Yukon) . . . .*L. yukonensis* **Cole and Wilcox***

40b Mesonotal tomentum largely brown; dorsocentral and medial stripes
strong; femora black with brown tomentum (coastal streams in northern
California) . . . .*L.* **ald-1 sp. nov.***

41a Epandrium in lateral view hardly wider at apex than at base, apicoventral
corner blunt (figure 245e); scutum grey, the brown tomentum more-or-
less restricted to the dorsocentral and medial stripes (mountain habitats
from southern British Columbia and Alberta south to Colorado and Utah)
. . . .*L. aldrichii* **Cole and Wilcox* (part)**

41b Epandrium in lateral view much wider at apex than at base, apicoventral
corner a prominent angle; scutum usually with much brown tomentum
laterally and posteriorly (southwestern British Columbia south along the
coastal lowlands to northern Oregon)
. . . .*L. pacificus* **Cole and Wilcox* (part)**

42a Epandrium in dorsal view with medial margins strongly concave in apical
half (figure 70); katatergite setae black; genitalia as in figures 70-73
(mountains and plateaus of central Arizona and western New Mexico)
. . . .*L. coconino* **sp. nov.**

42b   Epandrium in dorsal view with medial margins almost parallel or only
       weakly concave (cf. figure 107); katatergite bristles pale or dark **43**

43a   Epandrium in lateral view as in figure 107, the width about 0.75-0.8 times
       the length; most katatergite setae white; lateral setae on tergites very
       short, the longest on tergite 3 only 0.25 times the length of the tergite, or
       less; genitalia as in figures 107-10 (mountains of Wyoming)
       . . . . *L. lavignei* **sp. nov.**
43b   Epandrium in lateral view with width about 0.6 times the length; kata-
       tergite setae black; lateral setae on tergites long, the longest on tergite 3
       more than 0.5 times the length of the tergite (stream sides in eastern
       Washington and Idaho) . . . . *L. martinorum* **Cole and Wilcox***

44a   Scutum usually with thin medial stripe; epandrium in lateral view with
       greatest width about 0.3-0.4 times the length and with a broad, shallow
       ventral tooth subapically (figure 62); phallus as in figure 64; gonostylus
       with broad lateral lobe (figure 63); mainly pale grey with large brown
       tergite patches in most of its montane range, dark brown in coastal
       Oregon, intermediate in coastal California; genitalia as in figures 62-65;
       species of stony stream banks (northern British Columbia and western
       Alberta south to California and Colorado) . . . . . . *L. cinereus* **Cole (part)**
44b   Scutum without medial stripe; epandrium in lateral view with greatest
       width about 0.4-0.5 times the length; phallus and gonostylus not as above;
       body colour variable **45**

45a   Epandrium shining black with a broad semicircular emargination apico-
       ventrally, producing a strong subapical ventral tooth; antenna with F1 long
       and narrow, the width 0.2-0.25 times the length, and F2+3/F1 about 0.6;
       anepimeron with prominent white hairs (ocean beaches from Washington
       south to northern California) . . . . *L. bivittatus* **Loew***
45b   Epandrium not as above; antennae and setation of epimeron variable . . **46**

46a   Some setae on basal segments of antennae white **47**
46b   Setae on basal segments of antennae brown/black **51**

47a   Epandrium in dorsal view with medial margins concave in apical half . **48**
47b   Epandrium in dorsal view with medial margins more-or-less straight
       (eastern Oregon and southern Idaho to southern California: may include
       more than one species) . . . . *L. zonatus* **Cole and Wilcox***

48a   Epandrium in dorsal view with medial margins strongly concave in apical
       half (cf. figure 70, 144); antennae with F2+3/F1 about 0.65 or more
       (southern California) . . . . *L. gabrieli* **Cole and Wilcox***
48b   Epandrium in dorsal view with medial margins moderately concave in
       apical half (cf. figures 41, 205, 245i); F2+3/F1 about 0.5 . . . . **49**

49a   Phallus with ventral margins of paramere sheath forming a strong, acute, subapical tooth (figure 248k); apical projection of dorsal carina narrow, the base hardly wider than the width at the midpoint; basal subepandrial sclerite present (valleys and hills of western California)
. . . . *L. californicus* **Cole and Wilcox***

49b   Phallus with ventral margins of paramere sheath more-or-less flat, not forming a strong, acute, subapical tooth; apical projection of dorsal carina with base much broader than the width at the midpoint; basal subepandrial sclerite present or absent **50**

50a   Mystax black/brown, sometimes a few white setae ventrally; basal epandrial sclerite present (central Sierra Nevada) . . . . . . *L.* **biv-2 sp. nov.***

50b   Ventral 0.25-0.5 of mystax white; basal epandrial sclerite only vaguely apparent (southern and central California lowlands)
. . . . *L. drabicola* **Cole* (part)**

51a   Epandrium in dorsal view with medial margins straight; dorsum of tergites shining brown/black, the apical band of grey tomentum dorso-lateral and lateral only (southern California, northern Baja California)
. . . . *L.* **biv-3 sp. nov.***

51b   Epandrium in dorsal view with medial margins moderately concave in apical half (cf. figures 205, 245i); dorsum of tergites with grey tomentum apically, brown basally, the grey extending basally along the midline (except in some specimens of *L. willametti*) **52**

52a   Scutum grey, without distinctive dorsocentral stripes, the stripes vaguely pale; phallus with apex of dorsal carina swollen but not flattened; phallus with ventral margins of paramere sheath expanded into a ventral carina (mountains of northern Baja California) . . . . *L.* **biv-3 sp. nov.***

52b   Scutum with dorsocentral stripes brown and distinct; phallus with apex of dorsal carina flattened, expanded and disc-like; phallus without ventral carina (cf. figure 248l, m) (species of ocean beaches and coastal rivers from Vancouver Island to northern California) **53**

53a   Longer hairs of abdomen mostly white; usually about 5 or 6 dark bristles basolaterally on tergite 1 (ocean beaches and coastal rivers from Vancouver Island to northern California; also east of Cascade range in Washington and Idaho) . . . . *L. willametti* **Cole and Wilcox***

53b   Longer hairs of abdomen mostly yellow/gold; usually more than 9 dark bristles basolaterally on tergite 1 (ocean beaches in Oregon)
. . . . *L. dimicki* **Cole and Wilcox***

54a   Tergite 8 largely ferruginous or yellow-red; tibiae, tarsi and sometimes part of femora ferruginous **55**

54b   Tergite 8 largely medium/dark brown or black; tibiae and tarsi brown/black **59**

55a Lateral lobes of sternite 8 with or without short hairs, but no bristles
(figure 249b); femora all brown/black or with extreme bases ferruginous
or red-brown (mountains and foothills from central Oregon south to
Arizona) **56**

55b Lateral lobes of sternite 8 with rather weak bristles (figure 37); femora
black/brown with both bases and apices ferruginous; genitalia as in
figures 37-39 (mountains of eastern Arizona) . . . .*L. apache* **sp. nov.**

56a Bases of femora ferruginous, more-or-less contrasting with the dark
brown/black of the rest of the femora (note: many female specimens of *L.
testaceus* and *L.* tes-3 sp. nov. are indistinguishable) **57**

56b Femora all dark brown/black **58**

57a Tomentum of thorax and apical bands of abdominal segments usually with
a definite gold/brown cast (Sierra Nevada and southern Cascade
Mountains of California) . . . .*L. testaceus* **Cole and Wilcox***

57b Tomentum of thorax and apical bands of abdominal segments mostly grey
with some faint gold (eastern slopes of Cascades in Oregon)
. . . .*L.* **tes-3 sp. nov.***

58a About 10mm long; tomentum of thorax and apical bands of abdominal
segments gold-grey; tibiae ferruginous, contrasting with dark brown
femora (Sierra Nevada) . . . .*L.* **tes-1 sp. nov.***

58b About 8-9mm long; tomentum of thorax and apical bands of abdominal
segments grey; tibiae light brown/brown, not contrasting strongly with
femora (San Bernardino Mountains of southern California)
. . . . *L.* **tes-4 sp. nov.***

59a Dorsum of abdominal tergites with apical grey tomentum extending
anteriorly on the midline in a broad triangle, separating two patches of
brown tomentum basally (this pattern vague in some species) **60**

59b Dorsum of abdominal tergites with apical grey tomentum forming a band
that does not extend anteriorly along the midline **66**

60a Short, appressed hairs on dorsum of metafemur black/brown, at least in
apical half; tergite tomentum pattern usually vague, the basal brown
patches weak (female of *L.* biv-1 sp. nov. [mountains of northern Baja
California] is unknown, but probably would key before couplet 62) . . .**61**

60b Short, appressed hairs on dorsum of metafemur white; tergite tomentum
pattern usually definite, the basal brown patches more-or-less clearly
defined (except in *L. martinorum*) **62**

61a Sternite 8 all ferruginous; apical lobes of sternite 8 without bristles;
lateral white setae on abdominal tergite 3 long, about 0.3 the length of the
lateral margin; medium-sized to small species, length usually 9mm or less
(White Mountains of central Sierra Nevada) . . . . .*L.* **tes-2 sp. nov.* (part)**

61b Sternite 8 dark brown/black; apical lobes of sternite 8 with bristles; lateral

white setae on abdominal tergite 3 sparse, minute; large species, length usually greater than 12mm; genitalia as in figures 74-76; (mountains and plateaus of central Arizona and western New Mexico)
. . . .*L. coconino* **sp. nov.**

62a   Setae on antennal scape brown/black **63**
62b   At least some setae on antennal scape white **64**

63a   Long lateral setae on abdominal segments 1 to 3 white; normally 4-7 black bristles laterally on abdominal tergite 1 (coastal beaches and river banks from southwestern British Columbia south to northern California)
. . . .*L. willametti* **Cole and Wilcox* (part)**
63b   Long lateral setae on abdominal segments 1 to 3 yellow/gold; normally 9-12 black bristles laterally on abdominal tergite 1 (coastal beaches in Oregon) . . . .*L. dimicki* **Cole and Wilcox***

64a   Tomentum largely pale silver-grey, the dorsocentral stripes and the basal brown patches on the abdominal tergites pale and often obscure; white anepimeron hairs negligible (stream sides in eastern Washington and Idaho) . . . .*L. martinorum* **Cole and Wilcox***
64b   Tomentum grey with brown and gold highlights, the dorsocentral stripes and basal brown patches on the abdominal tergites definite; white anepimeron hairs present and obvious (California) **65**

65a   Mystax with ventral third or half pale; setae ventrally on abdominal segments 6 to 7 white; small to medium-sized species, usually less than 9mm long (southern and central California lowlands)
. . . . *L. drabicola* **Cole and Wilcox* (part)**
65b   Mystax dark, or only a few ventral setae pale; setae ventrally on abdominal segments 6 to 7 brown/black; larger species, usually more than 9 mm long. . . . . .*L. californicus* **Cole and Wilcox* or L. biv-2 sp. nov.*** (Females of these species cannot be distinguished with any confidence. The latter species is, on average, larger and more hirsute than the former. *L. californicus* lives at low to moderate elevations in west-central California; *L.* biv-2 sp. nov. is restricted to the central Sierra Nevada.)

66a   Sternite 8, including hypogynial valves, ferruginous or amber **67**
66b   Sternite 8 variable in colour: whole sternite, including hypogynial valves, dark brown/black; or base of sternite dark brown/black with valves ferruginous; or only base of sternite ferruginous **68**

67a   Short setae on dorsum of metafemur brown/black; lateral setae on abdominal tergites 4 to 7 brown/black; apex of hypogynial valves extending past cerci and acanthophorite spines; apical lobes of sternite 8 without bristles (White Mountains of central Sierra Nevada) . . . .*L.* **tes-2 sp. nov.* (part)**
67b   Short setae on dorsum of metafemur white; lateral setae on abdominal tergites 4 to 7 white; apex of hypogynial valves just reaching tips of the

cerci and acanthophorite spines; apical lobes of sternite 8 with pale, rather weak bristles; (eastern Oregon and southern Idaho to southern California; may include more than one species) . . . . . .*L. zonatus* **Cole and Wilcox***

68a Hypogynial valves and lateral lobes of sternite 8 ferruginous; lateral lobes of sternite 8 without dark bristles (except in *L.* biv-3 sp. nov.); scutum normally with medial stripe, but sometimes obscure (absent in *L.* biv-3 sp. nov.) **69**

68b Hypogynial valves of sternite 8 brown/black; lateral lobes of sternite 8 with dark bristles; scutum without medial stripe, except in most specimens of *L. cinereus* **72**

69a Apical bands of pale tomentum on abdominal tergites emarginate dorso-centrally, the black, shining cuticle and dark brown tomentum reaching, or almost reaching, the posterior margin of the tergite; lateral lobes of sternite 8 with fine dark bristles; scutum without medial stripe (southern California and northern Baja California) . . . .*L.* **biv-3 sp. nov.***

69b Apical bands of pale tomentum on abdominal tergites not emarginate dorsocentrally, the basal margin of the band parallel to the apical margin of the tergite; lateral lobes of sternite 8 without dark bristles; scutum with medial stripe (sometimes faint) **70**

70a Metafemora with almost all dorsal short setae and the long, fine, ventral setae white (southern Yukon) . . . .*L. yukonensis* **Cole and Wilcox***

70b Metafemora with short dorsal setae brown/black (at least on the distal half) and the long, fine, ventral setae brown/black or white/gold **71**

71a Venter of metafemora with fine setae mostly white; setae on venter of abdominal sternite 7 mixed brown/black and white; scutum largely grey, the brown tomentum more-or-less restricted to the dorsocentral and medial stripes; scutellum with grey tomentum (mountain habitats from southern British Columbia and Alberta south to Colorado and Utah)
. . . .*L. aldrichii* **Cole and Wilcox***

71b Venter of metafemora with fine setae mostly gold/brown; setae on venter of abdominal sternite 7 golden or gold mixed with brown/black; usually considerable brown tomentum laterally and posteriorly on scutum, suffused with the grey basal tomentum; scutellum with brown tomentum.
. . . .*L. pacificus* **Cole and Wilcox*** and *L.* **ald-1 sp. nov.***
(Females of these species cannot be distinguished with any confidence. *L. pacificus* lives in southwestern British Columbia and south along the coastal lowlands to northern Oregon; *L.* ald-1 sp. nov. inhabits coastal streams in northern California.)

72a Short appressed setae on dorsum of metafemora brown/black; genitalia as in figures 140-41 . . . .*L. monticola* **Melander**

72b Short appressed setae on dorsum of metafemora white **73**

73a Setae on scutum around the anterior ends of dorsocentral stripes shorter
than antennal scape; tomentum on scutum grey with faint brown
dorsocentral stripes; genitalia as in figures 111-13 (mountains of
Wyoming) . . . . *L. lavignei* **sp. nov.**
73b Setae on scutum around the anterior ends of dorsocentral stripes as long
as, or longer than, antennal scape + pedicel; tomentum on scutum grey or
grey/brown with definite brown/gold-brown dorsocentral stripes **74**

74a Most fine setae on tibiae white; antennal ratio F2+3/F1 more than 0.6 and
normally greater than 0.7; body length normally less than 8mm (southern
California) . . . .*L. gabrieli* **Cole and Wilcox***
74b Fine setae on tibiae brown/black; antennal ratio F2+3/F1 variable; body
length greater than 8mm **75**

75a r-m crossvein at, or distal to, middle of discal cell; setae on ventrolateral
margins of abdominal tergites 1 to 7 white; setae on abdominal sternite 7
white/yellow (ocean beaches from Washington south to northern
California) . . . .*L. bivittatus* **Loew***
75b r-m crossvein proximal to middle of discal cell; setae on ventrolateral
margins of abdominal tergites 1 to 3 or 4 white, the rest brown/black;
setae on abdominal sternite 7 white or brown/black **76**

76a Setae on abdominal sternite 7 white; typically, the tomentum of the body
brown/grey, the brown/grey acrostical stripes often distinct; no medial
stripe; brown basal patches on the dorsum of abdominal tergites covering
0.4-0.5 times the length of tergites (ocean beaches and sandy banks of
coastal rivers from Vancouver Island to northern California; east of
Cascades along streams in northern Washington and Idaho)
. . . .*L. willametti* **Cole and Wilcox* (part)**
76b Setae on abdominal sternite 7 brown/black; typically, the whole body
covered with pale grey tomentum (extensive brown tomentum in
populations from southwestern Oregon and northwestern California);
acrostichal stripes faint or absent; a thin medial stripe usually present;
brown basal patches on the dorsum of abdominal tergites covering 0.5-
0.75 times the length of tergites (to 0.9 times the length in southwestern
Oregon and northwestern California); genitalia as in figures 66-68;
species of stony stream banks; (northern British Columbia and western
Alberta south to California and Colorado) . . . .*L. cinereus* **Cole**

Identification key to *Lasiopogon* species
of the East Palaearctic (east of 60° E longitude)

1a   Scutellar bristles white 2
1b   Scutellar bristles brown or black 3

2a   Scutum gold/brown-grey with dorsocentral and acrostichal stripes;
abdominal tergites with bands of .grey tomentum apically and thin brown
tomentum basally, the brown cuticle shining through, especially in males;
body length 7-10mm; genitalia as in figures 99-105 (Russia:
Buryatskaya/Baikal region) . . . .*L. kjachtensis* **Lehr**
2b   Scutum silver-grey, stripes obscure or absent; tergites covered with silver-
grey tomentum, which is somewhat more intense apically; very small
species, body length less than 6mm (Kazakhstan) . . . . . .*L. zaitzevi* **Lehr***

3a   Katatergite bristles white, sometimes with a minority of black ones, or
sometimes golden/pale brown 4
3b   Katatergite bristles black or dark brown 6

4a   Scutum with brown medial stripe between definite dorsocentral stripes;
mystax white or golden, often with some light brown bristles dorsally;
abdominal tergites shining brown/black with pale tomentum forming
apicolateral triangles (China: Qinghai Province)
. . . .*L. eichingeri* **Hradský***
4b   Scutum without strong medial stripe, the dorsocentral stripes present or
absent; mystax white or dark with pale setae ventrally; abdominal tergites
broadly grey apically, with or without brown tomentum basally 5

5a   Mystax and occipital setae white; katatergite bristles white; dorsocentral
stripes brown, often faint; bristles of tibiae and tarsi white, often some
light brown; halter knob with brown spot; very small species, the body
length 5-6.5 mm (Russia: Tuvinskaya/Altai region) . .*L. tuvensis* **Richter***
5b   Mystax mostly black with white hairs ventrally; occipital setae black;
katatergite bristles white, often with a few black; scutum grey, dorsocentral stripes obscure or absent; bristles of tibiae and tarsi mostly black;
medium-sized to large species, the body length 9.5-11mm; genitalia as in
figures 173-78 (China: Qinghai Province) . . . . . .*L. qinghaiensis* **sp. nov.**

6a   Basolateral bristles on abdominal tergite 1 all white; halter knob with dark
spot 7
6b   At least some of the bristles on abdominal tergite 1 black; halter knob
with or without dark spot 10

7a   Mystax mostly white, with dark setae dorsally in females; halter spot dark
and definite; genitalia in male bright ferruginous with brown gonocoxite
brush; abdominal tergites grey with basal patches of brown tomentum;

antennal flagellum 1 short and broad, WF1/LF1=0.33-0.42; genitalia as in figures 221-27 (Russia: Primorskiy Kray). . . . .*L. terneicus* **Lehr**

7b   Mystax brown/black, often with a few pale setae ventrally; halter spot definite or often diffuse; colour of genitalia and abdominal tergites variable; shape of antennal flagellum 1 variable. 8

8a   Abdominal tergites with silver-grey tomentum apically, the tomentum extending basally on midline and separating the brown basal tomentum into a pair of patches; lateral setae white, dorsal setulae white on 1 to 5, brown on 6 and 7 (mid-elevation forests of Taiwan). .*L. solox* **Enderlein***

8b   Abdominal tergites largely shining black basally, any brown tomentum thin or absent. *L. hinei* female (*L. leleji* female unknown) with apical grey tomentum covering about half the length of the tergite, the border with the brown basal areas sometimes vague and the grey frequently extending basally on the midline; abdominal setae variable (genitalia as in figures 95-97). 9

9a   Abdominal tergites in male almost completely shining black, the grey apical bands of tomentum covering only the intersegmental membranes; dorsal setulae white on 1 and 2, brown on 3 to 7; genitalia as in figures 91-97; for female see statement 8b, above (European Russia through Siberia to the Pacific Ocean; North America from Alaska east to central Alberta). . . . .*L. hinei* **Cole and Wilcox**

9b   Abdominal tergites in male with apical gold-grey tomentum covering 0.3-0.5 the length; dorsal setulae white, long, hair-like on 1 to 5, brown on 6 and 7; genitalia as in figures 123-26 (female unknown) (Russia: Primorskiy Kray). . . . .*L. leleji* **sp. nov.**

10a   Halter knob dark dorsally; metacoxal pegs usually present; medial stripe often present on scutum; abdominal tergite 1 bristles mostly white with a few black; abdomen largely shining black with pale tomentum bands very narrow apically; male with flattened disc-like lobe on dorsomedial margins of epandrium (cf. figure 245g) (Russia: Lena River region eastward to the Pacific Ocean). . . . .*L. septentrionalis* **Lehr***

10b   Halter knob pale dorsally **11**

11a   Except for some on abdomen and coxae, all setae, including the fine hairs of the gena, proboscis and legs, brown/black; genitalia as in figures 160-63 (female unknown) (China: Qinghai Province).
. . . .*L. phaeothysanotus* **sp. nov.**

11b   Fine hairs of gena, proboscis and much of legs white **12**

12a   Anepimeron without fine hairs; scutellum with a single row of 5 or 6 main setae; mystax black with some white setae ventrally; genitalia ferruginous; female sterite 8 without medial lobes; only female known (China: Qinghai Province). . . . .*L.* **unc-4 sp. nov***

12b Anepimeron with fine hairs: scutellar setae abundant, arranged in two or more irregular rows; female sternite 8 with medial lobes between the hypogynial valves (figure 29) **13**

13a Abdominal tergites shining black dorsally, the brown basal tomentum scarce or absent and the apical bands of pale tomentum very narrow, mostly restricted to the apicolateral margins; female sternite 8 with hypogynial valves slender, knife-like and parallel, the edges oriented dorsoventrally (figure 193); male with gonostylus as in figure 190; genitalia as in figures 189-95 (Japan: Honshu, Kyushu) ....*L. rokuroi* **Hradský**

13b Abdominal tergites with broad apical bands of gold-grey tomentum covering 0.25-0.50 the length of the tergites; tomentum on tergite bases thin gold-brown, the dark cuticle shining through; female and male genitalia variable **14**

14a Male with all abdominal tergites bearing white hairs laterally; gonostylus as in figure 84; subepandrial sclerite longer than broad (figure 86); female sternite 8 with medial lobes much shorter than the hypogynial valves, the valves in ventral view strongly curved medially, their tips almost meeting (figure 87); genitalia as in figures 83-89 ....*L. hasanicus* **Lehr**

14b Male with lateral setae pale on abdominal tergites 1 to 4, dark on 5 to 7; subepandrial sclerite broader than long; female with sternite 8 medial lobes subequal in length to the hypogynial valves, the valves parallel in ventral view, the tips separated (figures 29, 119) **15**

15a Male with phallus dorsal carina broadly disc-like in ventral view (figure 27); gonostylus as in figure 26; female with hypogynial valves including lateral margin of medial lobe (figure 29); genitalia as in figures 25-31 (Japan: Hokkaido. Russia: Kurile Islands) ....*L. akaishii* **Hradský**

15b Male with phallus dorsal carina only narrowly flattened in ventral view (figure 117); gonostylus as in figure 116; female with hypogynial valves including only base of medial lobe (figure 119); genitalia as in figures 115-21 (Russia: Primorskiy Kray) ....*L. lehri* **sp. nov.**

DESCRIPTIONS OF SPECIES IN THE *OPACULUS* SECTION

Lasiopogon akaishii Hradský

Lasiopogon akaishii Hradský, 1981. *Trav. Mus. Hist. nat. Gr. Antipa* 23: 179.

Diagnosis. A medium-sized grey-and-brown species. Mystax and other head bristles dark. Antennae brown, F1 long; F2+3/F1 = 0.52-0.63. Thoracic tomentum dorsally and laterally grey to brown/gold-grey with brown dorsocentral stripes and grey or brown-grey acrostichal stripes. Main thoracic and leg bristles dark; anterior dorsocentral bristles 4 to 5. Abdominal tergites basally with thin brown tomentum; apical bands gold-grey, covering 0.25 to 0.4 times the length of the segments. Bristles on tergite 1 dark; all lateral setae pale on tergites 1 to 4, mostly dark on 5 to 7; dorsal setulae dark. Width of epandrium about 0.35 to 0.4 times the length in lateral view, the dorsal and ventral margins parallel, the apex rounded. In dorsal view strongly concave medially. Female with abdominal grey apical bands about 0.4 to 0.5 times the length of the segments. Terminalia dark brown/black with light brown hairs; sternite 8 with black hypogynial valves parallel, the medial lobes ferruginous with pale brown hairs.

Description. Body length ♂ 8.8-11.2mm; ♀ 10.9-11.7mm.

Head. HW ♂ 1.96-2.20mm; ♀ 2.22-2.52mm. FW ♂ 0.42-0.46mm; ♀ 0.50-0.58mm. VW ♂ 0.96-1.02mm; ♀ 1.07-1.23mm. VW/HW = ♂ 0.46-0.49; ♀ 0.47-0.49. FW/VW = ♂ 0.44-0.47; ♀ 0.47-0.50. VD/VW = ♂ 0.09-0.12; ♀ 0.11. GH/GL = ♂ 0.27-0.33; ♀ 0.30-0.37.

Face with silver or gold tomentum; vertex brown-grey, occiput all grey or grey ventrally, brown/gold-grey dorsally. Beard and labial hairs white, mystax bristles brown/black; all other setae brown/black. Occipital bristles rather strong (to 0.8mm), abundant, those behind the dorsomedial angle of the eye strongly curved anterolaterally; lateral and ventral ones shorter, straighter. Frontal and orbital setae abundant, long (to 0.6mm).

Antennae. Brown; setae brown; F1 setae sometimes present. F1 long, more-or-less parallel-sided or slightly tapered from base WF1/LF1 = ♂ 0.24-0.31; ♀ 0.29-0.35. LF2+3/LF1 = ♂ 0.56-0.61; ♀ 0.52-0.63.

Thorax. Prothorax grey/brown-grey, with gold/brown dorsally, hairs white; postponotal lobes gold-grey, the lateral angle ferruginous, hairs strong, brown. Scutum coloration

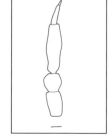

Figure 24. Antenna.
Scale line = 0.1 mm.

variable; tomentum grey with brown highlights, varying to predominantly brown. The grey specimens have dark/medium brown dorsocentral stripes bordered with light brown/gold; they extend to the scutellum. Extensive gold/brown runs ventrolaterally from the postpronotal lobe to the postalar area; the faint brown of intermediate spots is almost invisible. The acrostichal stripes usually with a weak medial stripe of pale gold/brown; stripes only slightly darker grey than the basal tomentum, and are vague, at best, in all specimens. At some angles, posterior notal slope gold-white/gold between the dorsocentral stripes. In other specimens the grey is infused with varying amounts of brown/gold; the acrostichal stripes become grey-brown and the brown-grey intermediate spots are slightly more pronounced between the light brown/gold of the ventrolateral areas and the lateral borders of the dorsocentral stripes. All strong bristles black, finer setae brown. Dorsocentral setae usually abundant; anteriors 4-5, longest to 0.9mm, mixed with only slightly finer setae; 5-7 posteriors. Notal setae abundant, erect, long (to 0.4mm). Postalars 2-3, with shorter setae; supra-alars 2-4 among shorter setae; presuturals 2-3, posthumerals 1. Scutellar tomentum grey with gold highlights or gold-brown dorsally in darker specimens; apical scutellar bristles dark, abundant, in about 2 irregular rows, 3-4 strong ones on each side mixed with many shorter, weaker dark bristles and hairs.

Pleural tomentum silver-grey to gold-grey, gold/brown-grey dorsally in darker specimens. Katatergite setae black, 6-9 among sparse, finer white and dark setae; katepisternal setae white, often long. Anepisternal setae 10-20, dark (to 0.6mm); sometimes a few other pale, weaker ones ventrally; a patch of rather long, erect brown setae on dorsal shelf. Anepimeron with a few white setae.

Legs. Base colour dark brown/black. Tomentum of coxae grey with some gold highlights; tomentum on rest of legs grey. No coxal peg. Main bristles dark brown/black; pale hairs white. Femora dorsally with decumbent hairs in male and female pale basally, dark apically; pale hairs predominate in most, but not all, specimens. Longer erect hairs abundant ventrally and laterally, especially on profemur; pale basally, dark apically; longest pale ventral setae as long as, or longer than, thickness of femur. Femora with dark, abundant, dorsolateral bristles mixed with long hairs; fine, especially on profemur, usually fewer and stronger on meso- and metafemora. Tibiae and tarsi with dark, strong bristles typically arranged, hairs brown/black. Protibia with longest bristles about 4.0 times longer than tibial width.

Wings. Veins medium/dark brown; wing membrane brown in oblique view. DCI = 0.28-0.36; cell M3 open. Halter yellow; knob without dark spot.

Abdomen. Male. Tergite basal colour dark brown/black. Tomentum on tergite bases thin brown/gold-brown, the dark cuticle shining through. Bands of gold-grey tomentum cover 0.25 to 0.40 of each tergite apically; ventrolaterally they cover the length of the tergite. Segment 1 is thinly covered over most of its surface. About 5-7 strong dark bristles on each side of tergite 1. Lateral setae on tergites 1-4 white/pale yellow, erect, some dark dorsally; on 5-7 short, dark, the extreme ventrolateral hairs often pale. Dorsal setulae dark. Sternite tomentum gold-grey, hairs pale.

Female. Grey or gold-grey apical bands about 0.4-0.5 times the tergite length; grey ventrolaterally. Lateral setae on tergites 1-3 white/pale yellow, erect, those on 4 shorter; setae on 5-7 brown. Dorsal setulae dark, more-or-less erect apically. Sternite tomentum gold-grey, hairs pale, mostly dark on 7.

Male genitalia. Epandrium and hypandrium/gonocoxite complex dark chestnut/black and covered with gold-grey tomentum except on hypandrium in some specimens. Setal brush black; other setae brown/dark brown, numerous, prominent. Width of epandrium halves in lateral view about 0.35-0.40 times the length, widest at about midlength; ventral and dorsal margins straight, parallel, the ventral one with a low, obtuse tooth at 0.75 the length; apex rounded. Medial face of epandrium as in figure 25. In dorsal view, medial margins of epandrium strongly concave; basal sclerite absent.

Gonostylus. Medial flange expanded into a large ventrolateral lobe; the dorsal flange has separated from the lateral margin of the gonostylus at the dorsomedial tooth, making the tooth a prominent lobe; lateral and ventrolateral teeth strong. Hypandrium/gonocoxite complex in ventral view about 0.6 times long as wide, transverse slit at 0.6 the length; in lateral view, exposed length of gonocoxal apodeme about 0.5 times basal width of hypandrium; apodeme with small sclerotized web ventrally.

Phallus. Paramere sheath dorsally 0.35 times the length of phallus (excluding ejaculatory apodeme), secondary sclerotization extending it to the tips of the lateral processes. Ventral process long and angled ventrally in lateral view, narrow in ventral view. Apex of paramere sheath broad, the ventrolateral carina short, a ventral flange absent. The dorsal carina bulges dorsally and laterally. Sperm sac width in dorsal view 0.4 times the length of phallus. Ejaculatory apodeme long, curved slightly dorsally in lateral view, moderately spatulate in ventral view; triangular in cross-section, flattened ventrally, the dorsal carina thick and narrow. Subepandrial sclerite as in figure 28. Sclerite broader than long, the triangular unsclerotized area in basal 0.75; spines bluntly acute and densely arranged over whole sclerite.

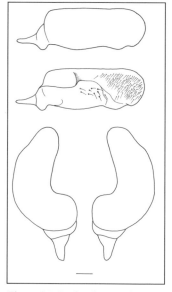

Figure 25. *L. akaishii* epandrium (top to bottom): lateral, medial, dorsal. Scale line = 0.2 mm.

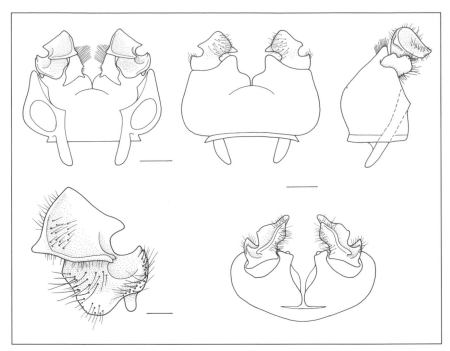

Figure 26. *L. akaishii* hypandrium/gonocoxite complex. Clockwise from top left: dorsal, ventral, lateral and apical (scale lines = 0.3 mm); gonostylus, dorsal (scale line = 0.1 mm).

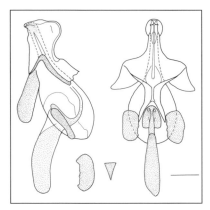

Figure 27. Phallus: lateral (left); ventral (right); lateral ejaculatory process, basal (centre left); ejaculatory apodeme, cross-section (centre right).
Scale line = 0.2 mm.

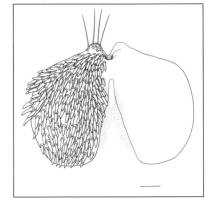

Figure 28. Subepandrial sclerite: ventral. Scale line = 0.1 mm.

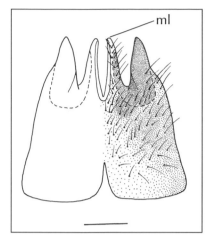

Figure 29. Female sternite 8: ventral.
ml = medial lobe. Scale line = 0.3 mm.

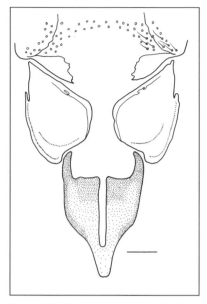

Figure 30. Female sternite 9, tergite 9
and lobes of sternite 10: ventral.
Scale line = 0.1 mm.

Female genitalia. Undissected: Hairs abundant, erect, long and light to medium brown. Tergite 8 dark chestnut/black, often narrowly ferruginous apically. Sternite 8 chestnut basally; hypogynial valves parallel, black, the medial lobes ferruginous with gold/pale brown hair. No strong setae on lateral angles of sternite 8. Cerci hairs white/light brown. Dissected sternite 8 (figure 29) broad, basal width about equal the length; undivided. Sclerite of hypogynial valve includes the lateral margin of the medial lobe; sclerite with a strong carina on inner edge of lateral part. Medial lobes as long as valves. Length of unsclerotized area between medial lobes 0.4 times sternite length.

Tergite 9 sclerites as in figure 30; sternite 9 V-shaped, the medial area largely sclerotized but with a narrow medial slit almost dividing the sclerite. Basal lobes of sternite 10 small. Tergite 10 black, usually with 7-8 black acanthophorite spines on each side. Spermathecae (figure 31) with strongly hooked terminal reservoirs about as thick as reservoir duct; terminal reservoirs with wart-like protuberances on surface. Both striated ducts with fine scales; junction with basal duct scaled, sclerotized and golden; valve not golden; basal duct short and narrow.

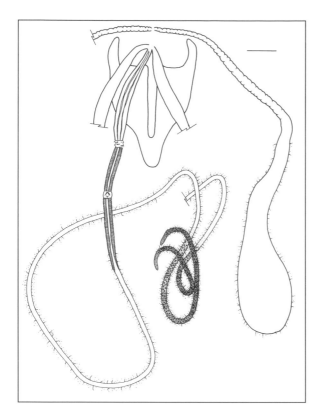

Figure 31. *L. akaishii*
spermatheca: dorsal.
Scale line = 0.1 mm.

Type Material
HOLOTYPE. ♂ (examined but did not copy exact label format) labelled: Japan:
Hokkaido, Mt. Tokachi, Shirogane Hotsprings, 4.viii.1965, Tardaio Akaishi.
MHRC.
ALLOTYPE. ♀ (examined), same data as holotype.
PARATYPES. (2 examined), same data as holotype (2♂, MHRC).

Other Material Examined (35 specimens)
Japan: Hokkaido, Daisetsuzan Nat. Park, Yukomanpetsu, 1100m, 10.vii.1986, D.M.
Wood (1♂, BCPM; 1♀, CNCI); Kamishihoro, Nukabira, 600m, 4.vii.1986, D.M. Wood
(1♀, BCPM; 1♂, 1♀, CNCI); **Russia: Sakhalinskaya Oblast, Kurile Is.**, Kunashir Is.,
on road from Lake Lagunnogo, 4-5.viii.1974, P. Lehr (2♂, 6♀, IBPV); Kunashir Is., 9km
S. of Ozhno-Kurilska, Kislaya R., 25.vii.1989, A. Lelej, (1♂, 1♀, IBPV), 27.vii.1989, A.
Lelej (2♂, 2♀, BCPM; 5♂, 2♀, IBPV); Kunashir Is., Mus Stolbchaty, 28.vii.1989,
Nemkov (1♂, IBPV); Kunashir Is., Golovnina volcano, 13.vi.1973, M. Nesterov (2♂,
IBPV); Kunashir Is., Yuzhno-Kurilsk, 11.vi.1984, V. Makarkin (1♂, IBPV); Shikotan Is.,
Malo-Kurilsk, 9.vi.1984, V. Makarkin (2♂, IBPV; 2♂, 1♀, BCPM).

Type Locality. Japan, Hokkaido, Daisetsuzan Nat. Park, Mt. Tokachi, Shirogane
Hotsprings.

Taxonomic Notes. Aoki (1949) stated that *L. cinctus* has been reported from Hokkaido and Honshu, although he noted that he had not seen any specimens himself. *L. cinctus* does not occur east of European Russia; *L. akaishii* and *L. rokuroi* are the only known Japanese *Lasiopogon* species.

Etymology. Named after Tardaio Akaishi, the collector of the type material.

Distribution. Palaearctic; Japan, Hokkaido; Russia, Kurile Is. See map 3, p. 127.

Phylogenetic Relationships. A member of the *akaishii* species group; sister species of *L. hasanicus*.

Natural History. Riverbanks and stream edges; recorded flight period ranges from 13 June to 4 August.

Lasiopogon apache sp. nov.

Diagnosis. A medium-sized, grey species with grey-gold tomentum laterally. Mystax and most prominent bristles and setae black or dark brown. Antennae brown, base of F1 often ferruginous, ratio F2+3/F1 = 0.49-0.51. Dorsocentral stripes narrow, brown and gold; acrostichal stripes grey, paired, often obscure; anterior dorsocentral bristles 3-5, prominent. Trochanters, tibiae, tarsi, bases and apices of femora ferruginous. Faint brown-gold patches of tomentum basolaterally on dorsum of abdominal segments; bristles on tergite 1 brown to white; lateral setae and dorsal setulae white, the latter mostly brown in female. Epandrium ferruginous, widening from base, in lateral view the widest point about 0.5 times the length; apex broadly rounded dorsally, a right angle ventrally; medial margins moderately concave dorsally. Gonostylus with strong dorsomedial and ventrolateral teeth. Spermathecae with terminal reservoirs straight or slightly curved, small warts at base; sternite 9 Y-shaped; sternite 8 with lateral lobe setae weak, valves with hairs.

Description. Body length ♂ 9.0-9.5mm; ♀ 9.5-10.3mm.

Head. HW ♂ 1.94-2.10mm; ♀ 2.00-2.28mm. FW ♂ 0.40-0.43mm; ♀ 0.40-0.60mm. VW ♂ 0.75mm; ♀ 0.75mm. VW/HW = ♂ 0.19-0.20; ♀ 0.20. FW/VW = ♂ 0.53-0.57; ♀ 0.53-0.56; VD/VW = ♂ 0.19-0.20; ♀ 0.15. GH/GL = ♂ 0.37-0.38; ♀ 0.38-0.44.
   Face and vertex with grey tomentum, mixed with some brown and gold, especially dorsally. Mystax and all major setae black or brown; occipitals strong,

rather sparse (12-16 each side) and shorter than usual
(<0.5mm), usually moderately curved anteriorly, but sometimes rather straight; frontal and orbital setae sparse, rather
strong, longest about as long as scape+pedicel.

Antenna. Brown, some with scape and pedicel and especially base of F1 ferruginous. Setae brown; F1 without
setae. WF1/LF1 = ♂ 0.29-0.30; ♀ 0.29-0.31. LF2+3/LF1 =
♂ 0.49-0.51; ♀ 0.49-0.51.

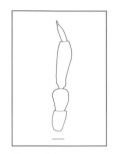

Figure 32.
L. apache sp. nov.
antenna.
Scale line = 0.1 mm.

Thorax. Prothorax grey, hairs white; postpronotal lobes
grey/gold-grey, the lateral angle ferruginous to chestnut,
hairs white/pale brown, fine. Scutum tomentum grey, faintly grey-gold laterally and anteriorly; dorsocentral stripes
narrow, dark brown and gold, disappearing at some angles of view; acrostichal
stripes grey, often obscure. Most bristles and setae brown or black; dorsocentrals
prominent (longest to 0.5mm long), 3-5 anteriors, 2-3 posteriors. Acrostichals
not arranged in rows, brown, fine, very short; scattered notal hairs similar to
acrostichals, longer posteriorly. Postalars 2-4 strong, with several weaker setae;
supra-alars 2; presuturals 1-3, usually 2; posthumerals 0-2, usually 1. Scutellar
tomentum grey, often with some brown-gold highlights; apical scutellar setae
black, 2-4, usually 4 on each side, angled laterally. Few if any discal scutellar
hairs, but fine hairs mixed with setae near margin.
   Pleural tomentum grey, often with brown-gold highlights. Katatergite setae 4-
7, variable in colour from white to dark brown, but mostly brown; katepisternal
setae normal; 2-4 setae (longest 0.2mm) at posterior edge of anepisternum and
patch of dark, fine, short, setulae anterodorsally. Anepimeron without setae.

Legs. Base colour of trochanters, tibiae, tarsi, bases and apices of femora and
parts of coxae ferruginous; most of femora black. Tomentum of coxae grey-gold;
tomentum on rest of legs sparse, grey. No coxal peg. All setae dark brown to
black, typically arranged; finer hairs white to light brown; ventral long setae on
femora sparse, mostly shorter than width of femur, but a few longer, especially
on metafemur. Profemur with 5-10 dorsolateral bristles; mesofemur with 0-2;
metafemur with 3-8. Protibia in male with longest bristle 3.3-3.7 times longer
than tibial width.

Wings. Veins light to medium brown; membrane with very pale yellow-brown
cast when viewed obliquely. DCI = 0.41-0.51. Cell M3 narrowly open. Halter
pale yellow; knob without dark spot.

Abdomen. Ground colour of tergites dark brown, often ferrugious at apex of
segments, especially towards the apex of the abdomen. Most tomentum grey;
each tergite with vague paired dorsal brown/gold patches of tomentum anterolat-

erally. Tergite 1 with 4-8 major bristles brown to white and many smaller white setae. Male with lateral setae white, fine, short and directed posteriorly; dorsal setulae white, short, recumbent. Sternite tomentum grey; hairs white. In the female, most dorsal setulae, lateral setae and sternite hairs brown, especially apically.

Male genitalia. Epandrium and hypandrium/gonocoxite complex ferruginous, without tomentum. Setal brush with very strong brown setae; other setae brown, numerous, prominent. Width of epandrium halves in lateral view about 0.5 times the length, widest at about 0.65 the distance from the base. Posterodorsal angle broadly rounded, posteroventral angle approximately a right angle, ventral edge straight. Medial face of epandrium as in figure 33. In dorsal view, medial margins of epandrium moderately concave; basal sclerite small.

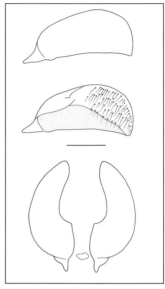

Figure 33. Epandrium (top to bottom): lateral, medial, dorsal. Scale line = 0.2 mm.

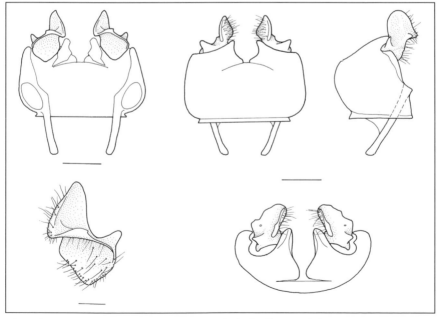

Figure 34. Hypandrium/gonocoxite complex. Clockwise from top left: dorsal, ventral, lateral and apical (scale lines = 0.3 mm); gonostylus, dorsal (scale line = 0.1 mm).

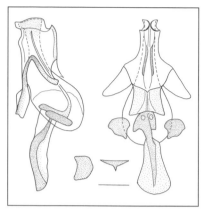

Figure 35. *L. apache* sp. nov. phallus: lateral (left); ventral (right); lateral ejaculatory process, basal (centre left); ejaculatory apodeme, cross-section (centre right). Scale line = 0.2 mm.

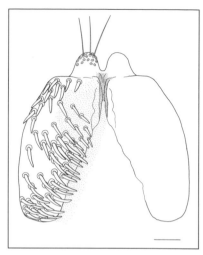

Figure 36. Subepandrial sclerite: ventral. Scale line = 0.1 mm.

Gonostylus. Prominent dorsomedial and ventrolateral teeth present. Hypandrium/gonocoxite complex in ventral view about 0.65 times long as wide, transverse slit at 0.7 the length; In lateral view, exposed length of gonocoxal apodeme about 0.7 times as long as basal width of hypandrium; apodeme with sclerotized web ventrally.

Phallus. Paramere sheath dorsally 0.35 times the length of phallus (excluding ejaculatory apodeme). Ventral process with medial carina. Apex of paramere sheath broad; ventrolateral carina and small ventral flange present; dorsal carina prominent. Sperm sac width in dorsal view 0.4 times the length of phallus. Ejaculatory apodeme almost straight in lateral view, and broadly spatulate in ventral view; in cross-section shallowly triangular with a narrow dorsal carina. Subepandrial sclerite as in figure 36. Broad triangular unsclerotized area in basal 0.75; spines bluntly acute, sparse except basally and medially in basal 0.6.

Female genitalia. Undissected: Tergite 8 ferruginous with brown medial stripe, sometimes vague; setae erect, pale to light brown. Sternite 8 ferruginous; setae scattered, fine, pale; lateral lobe setae weak. Hypogynial valves with fine hairs. Cerci brown/ferruginous with pale setae. Dissected sternite 8 (figure 37) with basal width about 0.65 times the length, undivided but weak along midline. Length of unsclerotized area between hypogynial valves about 0.4 times as long as sternite.

Tergite 9 sclerites with pattern of sclerotization as in figure 38; sternite 9 Y-shaped, medially undivided. Tergite 10 brown or ferruginous, with 7-9 brown acanthophorite spines on each side. Spermathecae (figure 39) with terminal reservoirs about as wide as reservoir duct, straight or gently curved and with

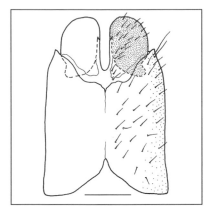

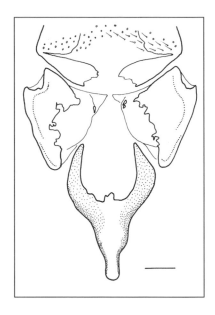

Figure 37 (above). Female sternite 8: ventral. Scale line = 0.3 mm.

Figure 38 (right). Female sternite 9, tergite 9 and lobes of sternite 10: ventral. Scale line = 0.1 mm.

weakly developed warty protuberances basally. Caniculi on duct long. Basal striated duct without fine scales; junction with basal duct without scales; basal duct moderately long, lumen 5-chambered.

Type Material

HOLOTYPE (here designated) ♂ labelled: "[rectangular beige label] White Mtns A.riz.[sic]/VI-29 '49"; "[rectangular beige label] Bee Hive/Spgs."; "[rectangular beige label] J. Wilcox/Coll"; "[rectangular blue label] HOLOTYPE ♂/Lasiopogon/apache/ Wilcox"; "[rectangular green label bordered top and bottom with asterisks] Unpublished/Manuscript Name"; "[rectangular white label bordered top and bottom with asterisks] JOSEPH WILCOX/COLLECTION – 1981/Gift to California/Academy of Sciences"; My Holotype label "HOLOTYPE/Lasiopogon ♂/apache Cannings/des. R.A. Cannings 2002 [red, black-bordered label]" has been attached to this specimen. Dissected genitalia in plastic vial underneath all. There is a female labelled as an allotype (unpublished) by Wilcox with the same collection data as the holotype. I am not designating allotypes and consider this specimen a paratype. CASC.

PARATYPES. (25 designated). **U.S.A.: Arizona.** Apache Co. Greer, Phelps Botanical Area, 23.vi.1957, F. Werner and G. Butler (5♂, 7♀, pair *in copula*, FSCA; 1♂, 1♀, USNM; 1♂, 1♀, BCPM); White Mts., 19.vi.1950, R.H. Beamer (1♀, FSCA), L.D. Beamer (1♀, FSCA), W.J. Arnold (1♀, FSCA); White Mts., Bee Hive Springs, 29.vi.1949, J. Wilcox (1♀, CASC); Greenlee Co. Hannigan Meadows, 9200 ft., 24.vi.1966, J.M. Davidson and M.A. Cazier (2♂, EMFC); San Francisco Mts., 25.vi.1950, R.H. Beamer (1♀, FSCA). Navaho Co. Whiteriver, 19.vi.1950, P.P. Cook (1♂, 1♀, FSCA).

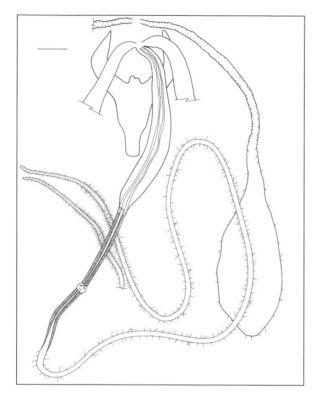

Figure 39. *L. apache* sp. nov. spermatheca: dorsal. Scale line = 0.1 mm.

Type Locality. U.S.A., Arizona, Apache Co., White Mountains, Bee Hive Springs.

Taxonomic Notes. Wilcox gave the species this manuscript name but never published a description.

Etymology. Named for Apache County, Arizona, which contains the type locality; the Apache are aboriginal people in this region.

Distribution. Nearctic; U.S.A., mountains of central and eastern Arizona. See map 2, p. 110.

Phylogenetic Relationships. Member of the *cinereus* species group. Sister to the species pair of *L. cinereus* and *L. shermani*.

Natural History. Habitat: montane forest. Flight recorded between 30 May and 4 July.

Lasiopogon appalachensis sp. nov.

Diagnosis. A medium-sized grey and brown species with dark brown legs and markings; mystax and other head bristles dark. Antennae brown, F2+3 rather long (F2+3/F1 = 0.62-0.79). Thoracic tomentum dorsally grey to brown, laterally gold-grey. Dorsocentral stripes brown; paramedial stripes grey to brown, often obscure, often a thin brown acrostichal stripe. Main thoracic and leg bristles dark, except sometimes katatergite bristles are pale. Anterior dorsocentral bristles 3-4. Abdominal tergites basally with thin dark brown tomentum; apical bands gold-grey, covering 0.1 to 0.25 the length of the segments. Bristles on tergite 1 pale. Epandrium with greatest width 0.45-0.5 times the length in lateral view, apex rounded. In dorsal view only moderately concave medially. Female with black terminalia and dark hairs; strong short setae on hypogynial valves.

Description. Body length ♂ 8.5-10.3mm; ♀ 8.9-11.5mm.

Head. HW ♂ 1.72-2.02mm; ♀ 1.92-2.24mm. FW ♂ 0.30-0.40mm; ♀ 0.38-0.46mm. VW ♂ 0.83-0.92mm; ♀ 0.84-1.05. VW/HW = ♂ 0.43-0.48; ♀ 0.39-0.49. FW/VW = ♂ 0.36-0.46; ♀ 0.44.45; VD/VW = ♂ 0.11-0.17; ♀ 0.15-0.22. GH/GL = ♂ 0.29-0.38; ♀ 0.31-0.38.

Face grey/silver-grey, darker specimens with some gold highlights; vertex brown-grey, occiput grey with brown-grey dorsally. The female has all parts of the head grey with brown highlights. Beard and labial hairs white, mystax bristles brown/black; all other setae brown/black. Occipital bristles rather fine, abundant, those behind the dorsomedial angle of the eye long and strongly curved antero-laterally; lateral and ventral ones shorter, straighter. Frontal and orbital setae abundant, rather fine, some as long as F1+F2+3; ocellars weak, about as long as orbitals.

Antennae. Brown to black, some with base of F1 and F2+3 lighter. Setae brown; F1 sometimes with a seta. F1 short, the dorsal edge straight, the ventral one convex. WF1/LF1 = ♂ 0.31-0.33; ♀ 0.34-0.36. LF2+3/LF1 = ♂ 0.62-0.79; ♀ 0.66-0.71.

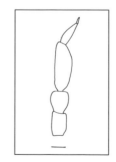

Figure 40. Antenna. Scale line = 0.1 mm.

Thorax. Prothorax grey, often with brown highlights, darker specimens grey-brown dorsally; hairs white. Postpronotal lobes grey to gold/brown-grey, the lateral angle ferruginous, hairs pale to dark brown. Basic scutum tomentum grey-brown; dorsocentral stripes brown, strong, bordered with light brown/gold. Intermediate spots brown; acrostichal stripes grey or grey-brown, sometimes vague or fused, frequently with narrow light or dark brown medial stripe. In browner specimens, the gold-brown is the more dominant basal tomentum, gold is more extensive, and the acrostichal stripes are brown. In some very dark specimens the scutum is all brown with darker brown dorsocentral stripes. The scutum of the lightest specimens is mostly clear grey, with pale gold-brown laterally and on intermediate spots; the strong brown dorsocentral stripes are usually bordered with gold. In these grey forms, acrostichal stripes are all but lacking; a narrow brown medial stripe may or may not be present. All notal setae brown or black. Anterior dorsocentrals 3-4 (longest to 0.8mm), mixed with finer setae; 4-5 posteriors. Notal setae scattered, but concentrated on anterior intermediate spot and acrostichal stripes, as long as shorter dorsocentrals. Postalars 1-2, with shorter hairs; supraalars 1-2; presuturals 2-3, 0-2 weak posthumerals. Scutellar tomentum grey/gold-grey; apical scutellar bristles dark, 3-4 strong ones on each side mixed with shorter, weaker dark bristles and hairs.

Pleural tomentum grey with gold-brown highlights, especially on anepisternum. Katatergite setae 7-8 among finer white hairs, unusually variable in colour; normally black but sometimes with several, or even all, bristles pale. Katepisternal setae sparse, often long. Anepisternal setae 5-12, brown, to 0.4mm, a few white ventrally; a dense patch of erect brown hairs on dorsal shelf. Anepimeron with a few white setae.

Legs. Base colour dark brown/black. Tomentum of coxae grey with some gold highlights; tomentum on rest of legs grey/gold-grey. No coxal peg. Main bristles dark brown/black, finer setae white to black. Femora dorsally with most decumbent hairs dark, some white basally. In females all or most hairs are white. Longer erect pale hairs ventrally and laterally, especially on profemur; many of these can be dark apically; in males longest white ventral setae as long as or longer than thickness of femur. Profemur with 2-6 stronger dorsolateral dark bristles; mesofemur with 2-3; metafemur with 4-11; these bristles mixed with finer bristles and setae. Tibiae and tarsi with dark, strong bristles typically arranged, hairs brown/black. Protibia with longest bristles about 3.0-4.0 times longer than tibial width.

Wings. Veins medium to dark brown; membrane brown in oblique view. DCI = 0.35-0.43; cell M3 open. Halter yellow; knob without dark spot.

Abdomen. Male. Tergite basal colour dark brown/black. Tomentum on tergite bases dark brown, dark cuticle shining through; light brown/gold where brown meets apical bands. Bands of grey/gold-grey tomentum cover 0.1 to 0.25 the length of each tergite apically; tergite 1 is brown dorsally, brown-grey laterally; ventrolateral areas are brown-grey. About 4-5 strong bristles on each side of tergite 1 white/yellow, occasionally 1 or 2 dark. Lateral setae on tergites 1-4 white, erect, those on 5-7 dark. Dorsal setulae dark, decumbent, strong. Sternite tomentum gold-grey; hairs white, dark on 6-7.

Female. As in male except tomentum covers tergites completely. Brown tomentum more dense; apical grey bands wider, about 0.35-0.4 times the length of

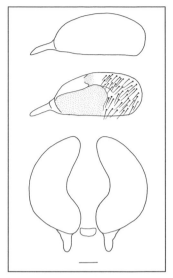

Figure 41. *L. appalachensis* sp. nov. epandrium (top to bottom): lateral, medial, dorsal. Scale lines = 0.2 mm.

each tergite and not extending apically on the midline. Lateral setae on 1-3 white, on 4 white or dark, on 5-7 dark. Sternite tomentum gold-grey; hairs white, dark on 7.

Male genitalia. Epandrium and hypandrium/gonocoxite complex dark brown/black and covered with gold-grey tomentum except on parts of hypandrium in some specimens. Setal brush strong, black; other setae dark brown/black, long. Width of epandrium halves in lateral view about 0.45-0.5 times the length, widest at about 0.5 the distance from the base; ventral margin straight, dorsal margin gently convex; apex rounded. Medial face of epandrium with apical shelf as in figure 41. In dorsal view, medial margins of epandrium only moderately concave; basal sclerite strong.

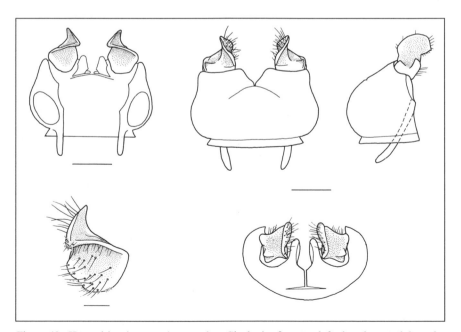

Figure 42. Hypandrium/gonocoxite complex. Clockwise from top left: dorsal, ventral, lateral and apical (scale lines = 0.3 mm); gonostylus: dorsal (scale line = 0.1 mm).

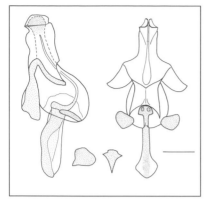

Figure 43. *L. appalachensis* sp. nov. phallus: lateral (left); ventral (right); lateral ejaculatory process, basal (centre left); ejaculatory apodeme, cross-section (centre right). Scale line = 0.2 mm.

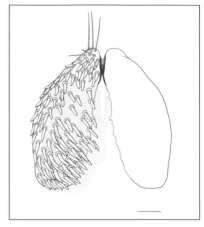

Figure 44. Subepandrial sclerite: ventral. Scale line = 0.1 mm.

Gonostylus. Medial flange low, the height about 0.4 times the length. Lateral and ventrolateral teeth moderately developed; no dorsomedial tooth. Hypandrium/gonocoxite complex in ventral view with length about 0.6 times the width, the transverse slit at 0.75 the length. In lateral view, exposed length of gonocoxal apodeme about 0.5 times the length of basal width of hypandrium; apodeme with sclerotized web ventrally.

Phallus. Paramere sheath dorsally 0.45 times the length of phallus (excluding ejaculatory apodeme). Ventral process with carina. Apex of paramere sheath with ventrolateral carina and ventral flange. Sperm sac width in dorsal view 0.35 times the length of phallus. Ejaculatory apodeme with curved ventral margin in lateral view, moderately spatulate in ventral view; flattened in cross-section with strong dorsal carina. Subepandrial sclerite as in figure 44. Narrowly triangular unsclerotized area in basal 0.80; spines blunt or bluntly acute, dense except for on mediocentral area of each plate.

Female genitalia. Undissected: Hairs brown/black, erect, abundant. Tergite 8 dark brown/black with posterior margin often lighter brown. Sternite 8 dark brown/black, sometimes brown apically; setae scattered, fine; lateral lobe setae strong. Ventral surface of hypogynial valves covered with short, strong bristles. Cerci black with pale setae. Dissected sternite 8 (figure 45) with basal width about 0.75 times the length, strongly divided along midline. Length of unsclerotized area between hypogynial valves 0.3 times the length of sternite.

Tergite 9 sclerites as in figure 46; sternite 9 V-shaped, medially undivided and with a dorsal carina. Tergite 10 brown/black with 5-6 black acanthophorite spines on each side. Spermathecae (figure 47) with terminal reservoirs curled in a com-

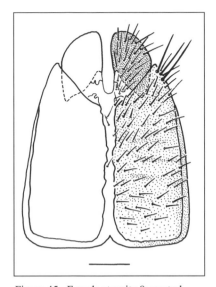 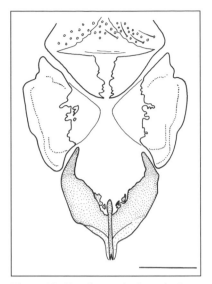

Figure 45. Female sternite 8: ventral.
Scale line = 0.2 mm.

Figure 46. Female sternite 9, tergite 9 and lobes of sternite 10: ventral. Scale line = 0.2 mm.

plete loop and about as wide as reservoir duct; terminal reservoirs without wart-like protuberances on surface. Basal striated duct with fine scales; junction with basal duct sclerotized, golden, scaled; valve sclerotized, golden; striated basal section of reservoir tube with fine scales; basal duct moderately long, thin.

Type Material

HOLOTYPE. (here designated) ♂ labelled: "[rectangular white label] Mingo Co., WV/25 May 1993/Kondratieff and Kirchner/Laurel Fk Hunting Area". My Holotype label "HOLOTYPE/Lasiopogon ♂/appalachensis Cannings/des. R.A. Cannings 2002 [red, black-bordered label]" has been attached to this specimen. Dissected genitalia in plastic vial underneath all. USNM.
PARATYPES (27 designated). **U.S.A.: Kentucky,** Whitley Co., Big Slaughter Creek, jct. Little Dog Creek, 11.v.1988, Baumann, Kirchner, Kondratieff and Nelson (1♀, CRNC). **Tennessee,** Scott Co., Big. South Fork, Hwy.297, 12.v.1988, B. Kondratieff (2♂, 2♀, in cop, CSUC); Big South Fork Nat. River and Rec. Area, 21.v.1990, B. Kondratieff, J.L. Welch and R.F. Kirchner (1♂, 1♀, in cop CASC; 1♂, 1♀, in cop, CNCI; 4♂, 2♀, 2pr of these in cop, CSUC; 2♂, 2♀, in cop, BCPM; 1♂, 1♀, in cop, USNM). **West Virginia,** Mercer Co., Blue Stone R. at I-77 bridge, 16.v.1990, B. Kondratieff, J.L. Welch and R.F. Kirchner (1♂, 1♀, CSUC); Mingo Co., Laurel Fork Hunting Area, 25.v.1993, Kondratieff and Kirchner (1♂, 3♀, CRNC).

Type Locality. U.S.A., West Virginia, Mingo Co., Laurel Fork Hunting Area.

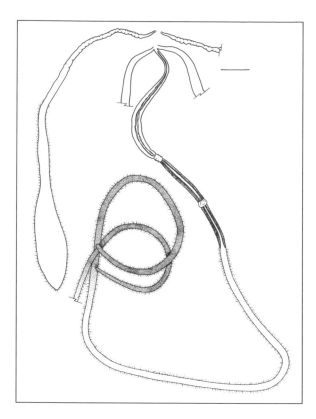

Figure 47. *L. appalachensis* sp. nov. spermatheca: dorsal. Scale line = 0.1 mm.

Etymology. Latin, meaning from the Appalachians.

Distribution. Nearctic; U.S.A., southern/central Appalachian Mountains. See map 6, p. 178.

Phylogenetic Relationships. Member of the *opaculus* species group.

Natural History. Habitat: stream sides in deciduous forest; recorded flight dates 12-25 May.

Lasiopogon chaetosus Cole and Wilcox

Lasiopogon chaetosus Cole and Wilcox, 1938. *Entomologica Americana* 43: 36-38.

Diagnosis. A medium-sized to large grey species with unusually profuse, strong, dark thoracic bristles. Antennae brown, hairs white or mixed brown and white; ratio F2+3/F1 = 0.51-0.65. Postpronotum with short dark bristles. Thoracic tomentum usually grey or grey-gold, sometimes washed with gold-brown on scutum; patches of pale brown or gold tomentum basolaterally on dorsum of abdominal segments. Dorsocentral stripes brown, often bordered with gold; acrostichal stripes grey or brown. Mystax brown mixed with white or all white; occipitals dark. Anterior dorsocentral bristles unusually prominent, especially in males, 7-18. Trochanters and bases of femora and tibiae ferruginous; tarsi mostly pale ferruginous. Epandrium ferruginous to brown, in lateral view greatest width about 0.4 times the length; apex gently down turned, ventral corner a right angle; medial margins strongly concave.

Description. Body length ♂ 9.25-12.0mm; ♀ 9.40-12.5mm.

Head. HW ♂ 1.92-2.40mm; ♀ 2.34-2.40mm. FW ♂ 0.50-0.68mm; ♀ 0.58-0.60mm. VW ♂ 0.80-1.04mm; ♀ 0.86-1.00mm. VW/HW = ♂ 0.42-0.43; ♀ 0.36-0.43. FW/VW = ♂ 0.63-0.65; ♀ 0.58-0.68; VD/VW = ♂ 0.06-0.13; ♀ 0.12-0.18. GH/GL = ♂ 0.45–0.53; ♀ 0.39-0.48.

Face and vertex with grey tomentum, mixed with some gold, especially dorsally; some with vertex golden brown. Mystax variable from all white to brown mixed with white above and below. Occipital bristles brown, longest medially (to 0.8mm), moderately to strongly curved (almost at right angles) anteriorly, 15-30 on each side; dense medially, and more abundant in males than in females. Frontal and orbital setae strong, longest to 0.7mm, all white or mixed brown and white. Ocellar bristles dark mixed with white hairs.

Antennae. Brown, some with basal areas ferruginous. Setae all white or mostly brown on pedicel; F1 sometimes with a few setae. WF1/LF1 = ♂ 0.24-0.28; ♀ 0.24-0.26. LF2+3/LF1 = ♂ 0.51-0.61; ♀ 0.54-0.63.

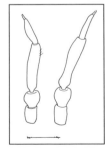

Figure 48. Antenna. Scale line = 0.2 mm.

Thorax. Pronotum grey with white hairs. Postpronotum grey, the lateral angle yellow to dark brown; 2-8 short dark bristles and white hairs. Scutum tomentum grey with gold highlights, sometimes washed strongly with brown-gold; grey-gold laterally. Dorsocentral stripes brown, often edged with gold, faint to strongly coloured. Acrostichal stripes, dark grey or brown, with medial space paler. Intermediate spots faint brown. All strong bristles on scutum dark. Dorsocentral bristles unusually prominent, longest 1.0mm long; 13-18 anteriors in males, 7-10 in females; 3-6 posteriors. Acrostichals and scattered notal hairs fine, very short, white. Postalars 3-4 strong, 2-4 short, mixed with finer hairs; supra-alars 2-4 strong, often with 1-4 weaker; presuturals 3-5, usually 3; posthumerals 2-5, usually 2. Scutellum tomentum grey. Apical scutellar setae dark, 4-10 strong and usually a few smaller; few if any discal scutellar hairs, but fine white hairs mixed with setae near margin.

Pleural tomentum grey-gold. Katatergite setae variably white or brown, 5-9 among finer white hairs; katepisternal setae normal; 1-3 setae (longest 0.6mm) at posterior edge of anepisternum and patch of fine short, white, hairs along dorsal margin. Anepimeron without setae.

Specimens examined from Washington State are more grey and somewhat more setose than those from eastern Oregon, which have more brown and gold tomentum, especially on the scutum. Males are generally more setose than females.

Legs. Base colour of coxae ferruginous to brown; trochanters ferruginous; femora dark brown to black, ferruginous basally and sometimes ventrally; tibiae ferruginous to brown; tarsi yellow or ferruginous to light brown. Tomentum of coxae grey-gold, sparsely grey on rest of legs. No coxal tubercle. Many setae on coxae bristle-like, white. Femora with decumbent white hairs and numerous erect ones; fine ventral bristles sparse, the longest only as long as thickness of femur; apical and dorsolateral bristles strong, numerous; profemur in male with 9-17 dorsolateral bristles, mesofemur with 4-6; metafemur with 16-30. Bristles on tibiae and tarsi strong, rather short; protibia in male with longest bristle 3.3 times longer than tibial width. Hairs on tibiae pale, on tarsi pale to light brown.

Wings. Veins yellow to brown, pale brown microtrichiae lining the veins; membrane with very pale brown cast when viewed obliquely. DCI = 0.30-0.46; cell M3 open. Halter yellow; knob without dark spot.

Abdomen. Tergites brown with grey tomentum; each tergite with rather small, dorsal paired pale brown/gold patches of tomentum anterolaterally, reduced apically. Tergite 1 with 6-10 white to brown major bristles on each side, mixed with many smaller white setae; lateral setae on tergites white, fine, rather short; dorsal setulae white, decumbent. Sternite tomentum all grey; hairs white.

Male genitalia. Epandrium and hypandrium/gonocoxite complex ferruginous to brown; with thin grey-gold tomentum laterally. Setal brush white to brown; other setae pale to light brown, numerous, long, prominent. Width of epandrium halves in lateral view about 0.4 times the length, parallel-sided; apex gently down turned with apicoventral corner a right angle, dorsal corner broadly rounded; setae on external surface rather dense, pale, prominent. Medial face of epandrium as in figure 49. In dorsal view, medial margins of epandrium strongly concave; basal sclerite strong.

Figure 49. *L. chaetosus* epandrium (top to bottom): lateral, medial, dorsal. Scale line = 0.2 mm.

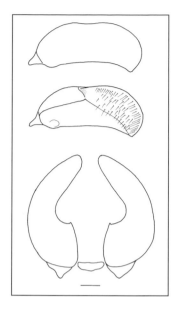

Gonostylus. Dorsal teeth small; ventrolateral tooth moderately developed. Hypandrium/gonocoxite complex in ventral view about 0.6 times long as wide, transverse slit at 0.7 the length. In lateral view, exposed length of gonocoxal apodeme about 0.6 times the basal width of hypandrium; apodeme with small sclerotized web ventrally.

Phallus. Paramere sheath dorsally 0.35 times the length of phallus (excluding ejaculatory apodeme). Apex of paramere sheath with ventrolateral carina; ventral flange absent. Sperm sac width in dorsal view 0.38 times the length of phallus. Ejaculatory apodeme straight in lateral view, narrowly spatulate in ventral view; oval in cross-section with wide, thin dorsal carina.

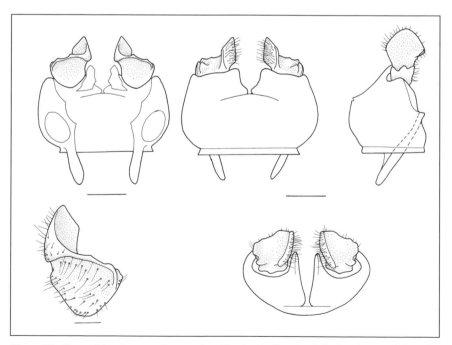

Figure 50. Hypandrium/gonocoxite complex. Clockwise from top left: dorsal, ventral, lateral and apical (scale lines = 0.3 mm); gonostylus, dorsal (scale line = 0.1 mm).

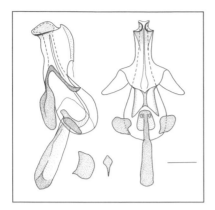

Figure 51. *L. chaetosus* phallus: lateral (left); ventral (right); lateral ejaculatory process, basal (centre left); ejaculatory apodeme, cross-section (centre right). Scale line = 0.2 mm.

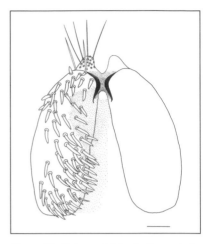

Figure 52. Subepandrial sclerite: ventral. Scale line = 0.1 mm.

Subepandrial sclerite as in figure 52. Triangular unsclerotized area in basal 0.80-0.85; spines bluntly acute, rather sparsely distributed, especially laterally.

Female genitalia. Undissected: All setae pale. Tergite 8 dark brown/black; hairs erect. Sternite 8 colour ferruginous to partly brown/black; hypogynial valves usually darker than the rest, prominent, ventral surface with a few fine hairs. Sternite setae scattered, fine; lateral lobe setae moderately strong. Cerci brown with pale setae. Dissected sternite 8 (figure 53). Basal width about 0.65 times the length; undivided along midline. Length of unsclerotized area between hypogynial valves 0.4 times the length of sternite, lightly haired, without carina.

Tergite 9 sclerites as in figure 54; sternite 9 V-shaped, medially undivided. Tergite 10 brown, usually with 5-7 (usually 6) brown acanthophorite spines on each side. Spermathecae (figure 55) with straight or gently curved terminal reservoirs about as wide as reservoir duct; terminal reservoirs without wart-like protuberances on surface. Surface of basal striated duct rugose in some specimens; junction with basal duct with scales; basal duct moderately long.

Type Material.
HOLOTYPE. ♂ (examined) labelled: "[rectangular white label] Lind, Wash/IV-26-19]"; "[rectangular white label] F.R. Cole/Coll."; "[rectangular red label] HOLOTYPE/Lasiopogon/chaetosus/Cole and Wilcox"; "[rectangular white label] California Academy/of Sciences/Type No. 6432".
ALLOTYPE. ♀ (examined) Adrian, Wash., 29 April 1919 (M.M. Reeher) CASC 6432.

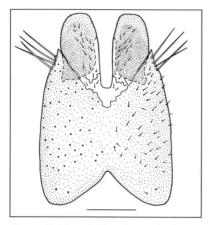

Figure 53 (above). Female sternite 8:
ventral. Scale line = 0.3 mm.

Figure 54 (right). Female sternite 9,
tergite 9 and lobes of sternite 10: ventral.
Scale line = 0.1 mm.

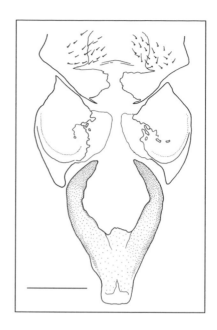

Figure 55.
Spermatheca: dorsal.
Scale line = 0.2 mm.

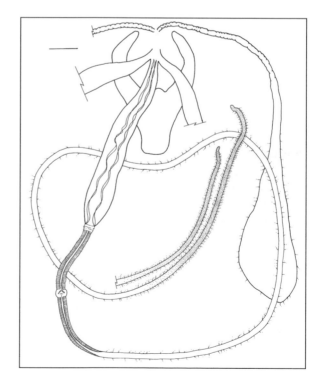

PARATYPES. (52 examined). **U.S.A: Washington**, Moses Coulee, 22.iv.1933, Itol Wilcox (1♂, CASC; 1♂, CNCI; 1♂, FMNH; 1♂, CUIC; 3♂, 1♀, USNM; 1♂, 1♀, EMEC; 1♀, FMNH; 2♀, EMFC), J. Wilcox (1♂, UCRC; 3♂, 6♀, CASC; 3♂, 1♀, EMFC; 2♂, 2♀, OSUC; 3♂, 1♀, MCZC; 1♀, CUIC; 5♂, 2♀, EMEC; 2♂, LGBC; 1♂, 2♀, USNM), 5.v.1935, Itol Wilcox (1♀, EMEC), J. Wilcox (1♀, CNCI; 1♀, UCRC; 1♀, FMNH).

Other Material Examined (61 specimens).
U.S A.: Idaho, Butte Co., Howe, 21 mi NE, 22.vi.1978, W.F. Barr (1♀, ESUW), Howe, 15 mi NE, v.1978, M.W. Hanks (1♀, ESUW); Register Rock State Park, 13.vi.1964, R.J. Lavigne (1♀, ESUW). **Nevada**, Winnemucca, 30.v.1957, T.R. Haig (2♂, FSCA). **Oregon**, Bend, 14.v.1950, C.H. Martin (1♀, AMNH); Lake Co., Abert Lake, N end, sand on E side Hwy 395 (42°44.4'N X 120°06.8'W), 27.v.1995, R.A. Cannings and H. Nadel (25♂, 19♀, BCPM); Lake Co., Alkali Lake, 2.5 km N on Hwy 395 (42°59.5'N X 119°59.6'W), 27.v.1995, R.A. Cannings and H. Nadel (2♂, 1♀, BCPM); Sisters, 19.v.1951, C.H. Martin (1♂, AMNH). **Utah**, Tooel Co., Skull Valley, 9.v.1959, D.E. Johnson (2♀, BYUC). **Washington**, Adrian, 29.iv.1919, Max Reeher (1♂, USNM); Grant Co., O'Sullivan dam, 8.v.1955, M.T. James (1♀, WSUC); Hanford Reserve, Rattlesnake Mtn, Arid Land Ecological Reserve, 2500', 10 mi NW Richland, 3.v.1972, Lee Rogers (1♂, 1♀ *in cop*, USNM); Hanford Reserve, 22.vi.1963, W.W. Cone (1♂, WSUC).

Type Locality. U.S.A., Washington, Adams Co., Lind.

Etymology. From the Greek *chaite* = hair; the Latin *osus* = quality of (abundance); a reference to the unusually strong bristles on the thorax.

Distribution. Nearctic; U.S.A., central Washington State south to Northern Nevada and east to central Idaho. See map 7, p. 192.

Phylogenetic Relationships. Basal member of the *cinereus* species group.

Natural History. Habitat: sandy, dry grasslands and shrub-steppe. At Hanford Reserve in eastern Washington State, habitat includes an *Artemisia, Agropyron* and *Poa* community. Range of dates: 22 April to 22 June. At Abert Lake in southeastern Oregon (4200 ft) on 27 May 1995, the species was common on flat sand and small dunes. Between 1300 and 1400 PDT in sunny but windy conditions (temperature about 25°C) pairs were copulating. They usually rested directly on the sand, but sometimes landed upright, very low on short vegetation. Dominant shrubs were *Atriplex spinosa* (Hook.) Collotzi, *A. confertifolia* (Torr. and Frem.) Wats., *Sarcobatus vermiculatus* (Hook.) Torr., and *Tetradymia spinosa H.andA.* Grasses and forbs were less prominent.

Lasiopogon chrysotus sp. nov.

Diagnosis. Only two males known. A medium-sized gold-grey/brown species with dark brown legs and markings; mystax and other head bristles dark. Antennae brown; F2+3/F1 = 0.57-0.60. Thoracic tomentum dorsally gold-brown, laterally gold-grey. Dorsocentral stripes brown, strong; acrostichal stripes grey/brown to brown. Main thoracic and leg bristles dark. Anterior dorsocentral bristles 2-3. Abdominal tergites basally with dark brown tomentum; apical bands gold-grey, the width 0.5 times the length of the segments and broadly covering ventrolateral areas. Bristles on tergite 1 pale. Epandrium about 0.55 times as wide as long in lateral view, apex rounded; in dorsal view strongly concave medially. Gonostylus with secondary medial flange and very strong dorsomedial tooth; lateral and ventrolateral teeth large.

Description. Female unknown. Body length ♂ 9.6-9.8mm.

Head. HW ♂ 2.00-2.10mm. FW ♂ 0.40-0.46mm. VW ♂ 0.86-0.89mm. VW/HW = ♂ 0.42-0.43. FW/VW = ♂ 0.47-0.52. VD/VW = ♂ 0.19-0.20. GH/GL = ♂ 0.28-0.30.
Face gold-silver/gold-grey, vertex gold-brown, occiput silver-white/grey with gold-brown dorsally. Beard and labial hairs white, mystax bristles brown/black with ventral setae pale-tipped; all other setae brown/black. Occipital bristles rather sparse, moderately strong but short (longest about as long as F1+F2+3) and only moderately curved anterolaterally; lateral and ventral ones shorter, straighter. Frontal and orbital setae sparse; orbitals strong, longest about 0.5mm and reaching tips of ocellars.

Antennae. Brown to black; setae brown, F1 without a seta. F1 long, widest at midlength. WF1/LF1 = ♂ 0.29-0.33; LF2+3/LF1 = ♂ 0.57-0.60.

Thorax. Prothorax gold-grey, brown dorsally on antepronotum; hairs white. Postpronotal lobes gold-grey, lateral angle dark brown, hairs sparse, pale to dark brown. Basic scutum tomentum gold-brown; dorsocentral stripes, intermediate spots and acrostichal stripes brown. All notal setae brown or black. Anterior dorsocentral bristles 2-3 (longest to 0.8mm), mixed with finer setae; 3 posterior dorsocentrals. Notal setae rather short, scarce, mostly in posthumeral area. Postalars 1-2, supra-alars 1-2, presuturals 2, 0-1 weak posthumerals. Scutellar tomentum gold-grey; apical scutellar bristles dark, 2 strong ones on each side mixed with a few shorter, weaker dark setae.

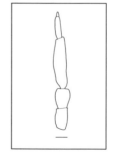

Figure 56. Antenna. Scale line = 0.1 mm.

    Pleural tomentum gold-grey. Katatergite setae 5-6 among finer white hairs; mostly black, 1-2 smaller pale ones.

Katepisternal setae sparse. Anepisternal setae 1-2, reduced (to 0.2mm), brown; a small patch of brown hairs on dorsal shelf. Anepimeron without setae.

Legs. Base colour dark brown/black. Tomentum of coxae and rest of legs gold-grey. No coxal peg. Main bristles dark brown/black, finer setae white to black. Femora dorsally with decumbent hairs dark apically, white basally. Longer erect pale hairs sparse, mostly ventral; longest ventral setae as long as or longer than thickness of femur. Profemur with 5-6 stronger dorsolateral dark bristles; mesofemur with 1-3; metafemur with 5-9. Tibiae and tarsi with dark, strong bristles typically arranged, hairs brown/black. Protibia with longest bristles about 4.0 times longer than tibial width.

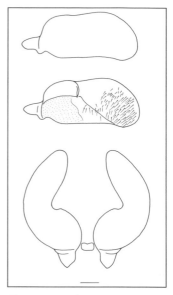

Figure 57. *L. chrysotus* sp. nov. epandrium (top to bottom): lateral, medial, dorsal. Scale lines = 0.2 mm.

Wings. Veins medium brown; membrane light brown in oblique view. DCI = 0.35; cell M3 open. Halter white/yellow; knob without dark spot.

Abdomen. Male. Tergite basal colour dark brown/black. Tomentum on tergite bases dark brown. Bands of gold-grey tomentum broad, covering at least half the length of each tergite apically and weakly extending anteriorly along the midline; ventrolateral areas gold-grey. Tergite 1 is brown only along apical margin, otherwise gold-grey. Four bristles on each side of tergite 1 white/yellow. Lateral setae all white, long on 1-3, short on 4-7. Dorsal setulae mostly pale; dark on brown areas of tergites. Sternite tomentum gold-grey, hairs white.

Male genitalia. Epandrium and hypandrium/gonocoxite complex dark brown/black and covered with gold-grey tomentum except on parts of hypandrium. Setal brush black; other setae dark brown/black, long. Width of epandrium halves in lateral view about 0.55 times the length, widest about midlength; ventral margin straight, dorsal margin gently convex, apex rounded. Medial face of epandrium with apical shelf as in figure 57. In dorsal view, medial margins of epandrium strongly concave; basal sclerite strong.

Gonostylus. Prominent secondary medial flange linked by a ridge to very strong dorsomedial tooth; lateral and ventrolateral teeth strong; a small tooth present on dorsal flange near junction of medial flange. Hypandrium/gonocoxite complex in ventral view with length about 0.6 times the width, transverse slit at

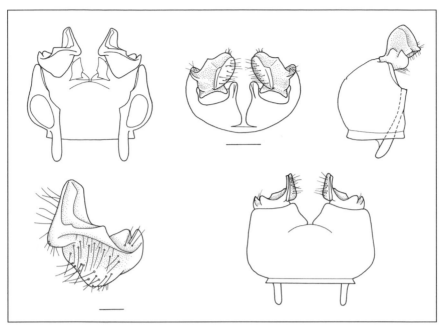

Figure 58. Hypandrium/gonocoxite complex. Clockwise from top left: dorsal, ventral, lateral and apical (scale lines = 0.3 mm); gonostylus, dorsal (scale line = 0.1 mm).

0.75 the length. In lateral view, exposed length of gonocoxal apodeme about 0.3 times the basal width of hypandrium; apodeme with sclerotized web ventrally.

Phallus. Paramere sheath dorsally 0.4 times the length of phallus (excluding ejaculatory apodeme). Apex of paramere sheath narrow, ventrolateral carina present, ventral flange absent. Sperm sac width in dorsal view 0.35 times the length of phallus. Ejaculatory apodeme straight in lateral view, spatulate in ventral view; flattened in cross-section with strong, thick dorsal carina. Subepandrial sclerite as in figure 60, with a central unsclerotized area triangular in basal 0.3, narrow and parallel-sided in central 0.5; spines blunt, dense except subbasally.

Female unknown.

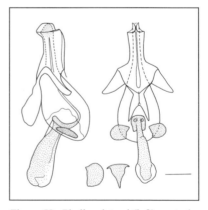

Figure 59. Phallus: lateral (left); ventral (right); lateral ejaculatory process, basal (centre left); ejaculatory apodeme, cross-section (centre right). Scale line = 0.2 mm.

Type Material.
HOLOTYPE. (here designated) ♂
labelled: "[rectangular white label]
TENNESSEE Scott Co.,/Big S Fork
Cumberland/River, Hwy
297,/Leatherwood Crossing"."
[rectangular white label] 12 May 1988,
#5186/Baumann, Kirchner/Kondratieff,
Nelson". My Holotype label "HOLO-
TYPE/Lasiopogon ♂/chrysotus
Cannings/des. R.A. Cannings 2002 [red,
black-bordered label]" has been attached
to this specimen. Dissected genitalia in
plastic vial underneath all. USNM.
PARATYPES (1 designated). **U.S.A.:**
Tennessee, Scott Co., Big S Fork
Cumberland River, Hwy 297,
Leatherwood Crossing, 12 May 1988,
Baumann, Kirchner, Kondratieff, Nelson
(1 ♂, CSUC).

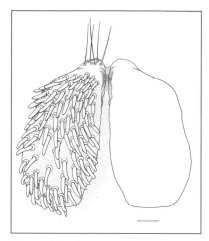

Figure 60. *L. chrysotus* sp. nov.
subepandrial sclerite: ventral.
Scale line = 0.1 mm.

Type Locality. U.S.A., Tennessee, Scott Co., Big S Fork Cumberland River, Hwy 297, Leatherwood Crossing.

Etymology. Greek: chrysotos = gilt. The tomentum of the face, pleura and wide abdominal bands is a rich golden-grey; the basic tomentum of the scutum is rich golden-brown.

Distribution. Nearctic; U.S.A., Tennessee; probably more widespread in the central Appalachians. See map 12, p. 266.

Phylogenetic Relationships. Member of the *tetragrammus* group; in an unresolved trichotomy with *L. flammeus* and *L. woodorum*.

Natural History. Habitat: stream sides in mountain forests; recorded date, 12 May.

Lasiopogon cinereus Cole

Lasiopogon cinereus Cole, 1919. *Proc. Calif. Acad. Sci.*, Ser. 4, 9: 229.
Lasiopogon atripennis Cole and Wilcox, 1938. *Entomologica Americana* 43: 27-29 (CASC). **New Synonymy.**

Diagnosis. A medium-sized to large, typically pale grey species with main head, thoracic and leg bristles dark. Specimens from the Coast Mountains from southern Washington to northern California are usually browner dorsally and otherwise different as detailed below. Antennae brown, F2+3/F1 = 0.44-0.60. Acrostichal stripes usually absent but, if present, grey; brown medial stripe usually present. Anterior dorsocentral bristles 3-5. Abdominal tergites basally with brown tomentum; apical bands gold-grey, narrow, the width 0.1 to 0.25 times the length of the segments and broadening laterally. Bristles on tergite 1 dark. Epandrium narrow and distinctively shaped, the width about 0.30-0.40 the length in lateral view, sometimes slightly broader at the base; low broad tooth ventrally before the rounded apex. In dorsal view only weakly concave medially. Gonostylus compressed apically and with large lateral flange. Female with apical grey/gold-grey bands, the width usually about 0.25-0.5 times the length of each tergite. Tergite 8 hairs pale; sternite 8 dark, often ferruginous/chestnut basally; hypogynial valves of sternite 8 haired and without carinae. Sternite 9 divided; central spermathecal tube forked.

Description. Body length ♂ 7.7-11.1mm; ♀ 9.6-13.0mm.

Head. HW ♂ 1.60-2.40mm; ♀ 1.82-2.52mm. FW ♂ 0.34-0.50mm; ♀ 0.42-0.60mm. VW ♂ 0.80-1.04mm; ♀ 0.92-1.20mm. VW/HW = ♂ 0.40-0.50; ♀ 0.48-0.51. FW/VW = ♂ 0.40-0.53; ♀ 0.45-0.50; VD/VW = ♂ 0.11-0.16; ♀ 0.14-0.18. GH/GL = ♂ 0.30-0.38; ♀ 0.33-0.40.

Face grey/silver-grey, vertex grey to gold/brown-grey; occiput grey with variable amounts of gold/brown dorsally. Beard and labial hairs white, mystax bristles brown/black; all other setae brown/black. Occipital bristles fine to moderately strong, those behind the dorsomedial angle of the eye to 0.6mm and moderately curved anterolaterally; lateral and ventral ones shorter, straighter. Frontal and orbital setae to 0.4mm, usually about as long as, or a little shorter than, F1.

Antennae. Brown to black, some with base of F1 and parts of F2+3 chestnut. Setae brown; F1 sometimes with a seta. F1 variable, sometimes rather long; margins almost parallel or, more commonly, dorsal margin more-or-less straight, the ventral one gently convex, widest around midlength. WF1/LF1 = ♂ 0.23-0.33; ♀ 0.26-0.33. LF2+3/LF1 = ♂ 0.44-0.58; ♀ 0.47-0.60.

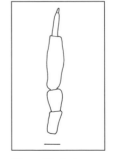

Figure 61. Antenna.
Scale line = 0.1 mm.

Thorax. Prothorax grey, some gold/brown dorsally, hairs white; postpronotal lobes grey, the lateral angle ferruginous to chestnut, hairs usually brown, sometimes some white. Scutum tomentum variable, the majority of specimens from northern British Columbia to Colorado and Utah rather evenly grey with the dorsocentral stripes medium/dark brown bordered by gold, somewhat expanded anteriorly. Acrostichal stripes absent or, if present, grey; narrow brown/gold medial stripe usually present. Ventrolateral areas with variable amounts of gold-brown tomentum. Other specimens, especially many of those from western Washington, Oregon and northern California, have varying amounts of brown infused in the grey. In the darkest specimens the notal pattern is grey-brown; the brown dorsocentral stripes and the brown-grey acrostichal stripes and intermediate spots are all faint. All notal setae brown or black. Anterior dorsocentral bristles 3-5 (to 0.9mm) mixed with finer setae; 4-5 posteriors. Notal setae scattered, to about 0.2mm; the acrostichals irregularly arranged, about the same length and density. Some specimens with a weak row or two of hairs along the median stripe. Postalars 1-3, with very few shorter hairs; supra-alars 2-3, with very few shorter hairs; presuturals 2-3 (usually 2), posthumerals 1-3. Scutellar tomentum grey/gold-grey; apical scutellar bristles dark, very variable in number, 1-7 strong ones on each side mixed with shorter, weaker dark bristles and hairs. Pleural tomentum grey to gold/brown-grey. Katatergite setae brown/black, sometimes one pale and a few weak, 7-10 among a few finer white hairs; katepisternal setae sparse, long, white. Anepisternal setae 6-12, usually strong, to 0.7mm; a patch of short erect brown hairs on dorsal margin of sclerite. Anepimeron with sparse white setae.

Legs. Base colour dark brown/black; tomentum of coxae and rest of legs grey/gold-grey. No coxal peg. Main bristles dark brown/black, finer setae white to black. Most specimens with shorter, decumbent hairs on dorsum of femora white; some specimens with almost all hairs dark, often some white ones at base; Longer more erect hairs mostly ventral, mixed white and dark, in most specimens the pale ones predominating except apically, but in coastal Oregon and California sometimes all are dark; in males longest ventral setae as long as, or longer than, thickness of femur. Dark dorsolateral bristles fine on profemur, 10-25 in males, fewer in females; stronger on mesofemur (1-4) and metafemur (6-12). Tibiae with dark, strong bristles typically arranged; hairs pale, dark or mixed. Hairs on tarsi dark. Protibia with longest bristles about 4.0-4.5 times longer than tibial width.

Wings. Veins medium/dark brown; membrane transparent and very pale brown to smoky brown when viewed obliquely. DCI = 0.35-0.46; cell M3 open. Halter yellow; knob with vague dark smudge or sometimes clear.

Abdomen. Male. Tergite basal colour dark brown/black. Tomentum on tergite bases dark brown, thin, the dark cuticle showing through. Apical bands of grey/

Figure 62. *L. cinereus* epandrium (clockwise from top left): lateral, four variations; dorsal; medial. Scale lines = 0.3 mm.

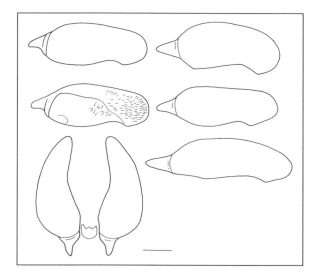

grey-gold tomentum variable, the width usually about 0.25 times the length of each tergite (but about only 0.1 or less in specimens from coastal Oregon and California), expanding laterally and covering the length of the tergite ventrolaterally, narrowing basally. 4-8 strong dark bristles on each side of tergite 1. Lateral setae long, erect, sparse, often white on all segments with darker hairs dorsolaterally; in many other specimens hairs are white on 1-4 only, shorter and dark on 5-7. Dorsal setulae all dark. Sternite tomentum grey to grey-brown, hairs all white or white on 1-5, dark on 6-7.

Female. Apical grey/gold-grey bands with width usually about 0.25-0.5 times the length of each tergite, but in coastal Oregon and California many specimens have bands as narrow as 0.1 times the length; basal tomentum light to dark brown. Lateral setae pale on tergites 1 to 3 or 4, the dorsolateral ones dark, sometimes bristle-like; lateral hairs short and dark on 4 or 5-7. Dorsal setulae dark, short, more-or-less erect and bristle-like. Sternites gold-grey to grey-brown, hairs dark on 5 or 6 to 7.

Male genitalia. Epandrium and hypandrium/gonocoxite complex chestnut/black with gold-grey tomentum except on the hypandrium in most specimens. Setal brush and other setae brown/black, abundant. Epandrium halves in lateral view long and narrow, the width about 0.30-0.4 times the length; the ventral and dorsal margins mostly parallel, usually gently arched dorsally. Apex rounded, a broad, shallow tooth ventrally in the apical 0.25. Medial face of epandrium as in figure 62. Variation in shape of epandrium, mainly in the ratio of length to width, is greater than in most other species. In dorsal view, medial margins very weakly concave.

Gonostylus. Medial flange reduced, somewhat flattened apically; dorsomedial tooth reduced, ventrolateral tooth well developed. The most prominent feature is

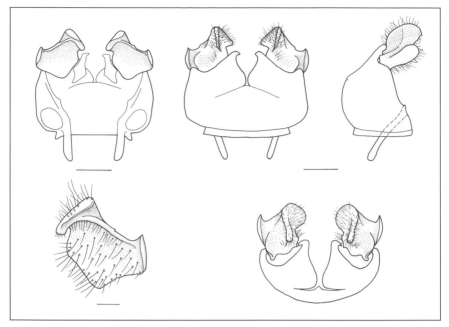

Figure 63. *L. cinereus* Hypandrium/gonocoxite complex. Clockwise from top left: dorsal, ventral, lateral and apical (scale lines = 0.3 mm); gonostylus, dorsal (scale line = 0.1 mm).

the unusually developed lateral tooth, which forms a broad, projecting flange; this flange often overlaps the ventral edge of the epandrium and is visible even in undissected specimens. Hypandrium/ gonocoxite complex in ventral view with length 0.7 times the width, transverse slit at 0.6 the length. In lateral view, the exposed length of the gonocoxal apodeme is about 0.5 times the basal width of the hypandrium; the apodeme without sclerotized web ventrally.

Phallus. Paramere sheath dorsally 0.4 times the length of phallus (excluding ejaculatory apodeme). Apex of paramere sheath broad; ventrolateral carina present, with two secondary carinae basal to the primary one. Ventral flange absent. Sperm sac width in dorsal view 0.4 times the length of phallus. Ejaculatory apodeme broad and ventrally angled in lateral view, broadly spatulate in ventral

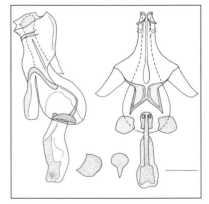

Figure 64. Phallus: lateral (left); ventral (right); lateral ejaculatory process, basal (centre left); ejaculatory apodeme, cross-section (centre right).
Scale line = 0.2 mm.

Figure 65. Subepandrial sclerite: ventral (usual form, left; variation, right). Scale line = 0.1 mm.

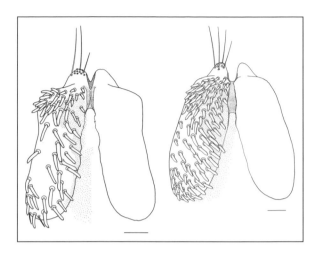

view; round in cross-section with a strong dorsal carina. Subepandrial sclerite usually as in figure 65 (left), with a variant shown (right). Narrow triangular un-sclerotized area in basal 0.7. Spines blunt, often narrowly spatulate, concentrated basally and especially apicolaterally (this pattern is accentuated in specimens from coastal California and Oregon where the central medial area is almost bare).

Female genitalia. Undissected: Tergite 8 dark brown/black often with pale apical band; hairs pale, erect. Sternite 8 dark brown/black, often ferruginous/chestnut basally. Hypogynial valves usually a prominent keel, black, the basal area some-times ferruginous; lateral lobe setae dark, moderately strong. Cerci black with gold setae. Dissected sternite 8 (figure 66) with basal width about 0.65 times the length; undivided but often weak along midline. Length of unsclerotized area between hypogynial valves 0.4 times the length of sternite; valves lightly haired; carina absent.

Tergite 9 sclerites as in figure 67; sternite Y- or V-shaped, medially divided. Tergite 10 dark brown/black with 6-9 (usually 7-8) black acanthophorite spines on each side. Spermathecae (figure 68) with curved terminal reservoirs about as wide as reservoir duct and medial one forked; terminal reservoirs with wart-like protuberances on surface. Basal striated duct without fine scales; junction with basal duct vague and without scales; basal duct short.

Variation. Over most of its large range, the coloration of *L. cinereus* varies in only minor ways; in the Coast Range of Oregon and northern California, how-ever, specimens tend to be darker. The thorax is darker grey/brown and the brown of the abdominal tergites is darker and more extensive than that of most other populations. In some of these specimens the leg hairs are almost completely dark and the wing membrane is darker brown than usual. In these coastal specimens antennal segment F1 is often shorter and broader than that of other populations

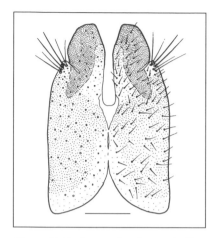

Figure 66. *L. cinereus* female sternite 8:
ventral. Scale line = 0.3 mm.

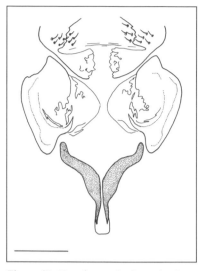

Figure 67. Female sternite 9, tergite 9
and lobes of sternite 10: ventral.
Scale line = 0.2 mm.

(WF1/LF1 = 0.26-0.34; mean of 0.31, n = 10 compared to 0.23-0.33; mean of
0.29, n = 10) but the overlap is considerable. The subapical broad tooth on the
ventral margin of the epandrium is often slightly larger in the coastal specimens
and there is some greater than usual variation in the structure of the gonostyli and
subepandrial sclerite.

Type Material.
HOLOTYPE. ♂ (examined) labelled: "[rectangular white label] Hood
River/Ore. IX-24-17]"; "[rectangular white label] F R Cole/Coll."; "[rectangular
white label] ♂"; "[rectangular red label] Holotype/cinereus"; "[rectangular
white, black-bordered label] Lasiopogon/cinereus/Type Cole"; "[rectangular
white label] California Academy/of Sciences/Type No.476". The holotype of *L.
atripennis* Cole and Wilcox, here synonymized with *L. cinereus,* is a male:
U.S.A., Oregon, Douglas Co., Smith River, 14.ix.1932, D.K. Frewing. CASC.
ALLOTYPE. ♀ (examined) labelled: "[rectangular white label] Hood
River/Ore. VII-28-17]"; "[rectangular white label] F R Cole/Coll."; "[rectangu-
lar white label] ♂"; "[rectangular red label] Allotype/cinereus"; "[rectangular
white, black-bordered label] Lasiopogon/cinereus/Type Cole";"[rectangular
white label] California Academy/of Sciences/Type No.476".
PARATYPES (2 examined). **U.S.A.: Oregon,** Joseph (1 ♀, EMEC); Parkdale,
18.vi.1917, F.R. Cole (1 ♀, EMEC).

Figure 68. Spermatheca: dorsal. Scale line = 0.2 mm.

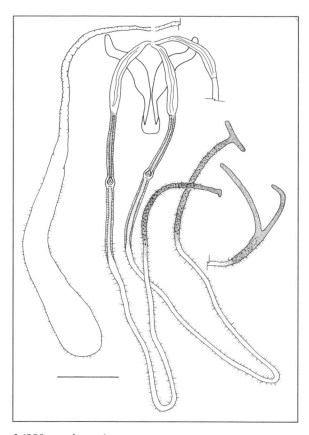

Other Material Examined (880 specimens).

CANADA: Alberta, Banff, N.B. Sanson (1 ♀, CNCI); Bow River Forest, Wilkinson Cr., 31.vii.1963, S. Adisoemarto and J. and C. Sharplin (2 ♂, 3 ♀, UASM); Cochrane, 26mi NE, Waiparous Creek Rec. Area, 30.vii.1991, A.W. Hook (5 ♂, 4 ♀, CRNC); Crowsnest Forest, Dutch Cr., 31.vii.1963, S. Adisoemarto and J. and C. Sharplin (4 ♂, 2 ♀, UASM); Frank, 15.viii.1926, E.H. Strickland (1 ♂, BCPM); Lundbreck Falls, Hwy.#3, 68km W Ft. McLeod, 18.vii.1985, ROME field party (1 ♀, ROME); Luscar, McLeod R., 10.viii.1963, S. Adisoemarto and J. and C. Sharplin (6 ♂, 3 ♀, 2pr in cop; UASM); Nordegg, 25.vii.1936, E.H. Strickland (1 ♂, UASM; 1 ♀, EMEC); Red Deer, Red Deer R., 8.viii.1964, S. Adisoemarto (6 ♂, 4 ♀, UASM); Waterton, 17.vii.1923, H.E. Grey (1 ♀, UASM); Waterton Lakes Nat. Pk, Blakiston Brook, 2.viii.1963, S. Adisoemarto and J. and C. Sharplin (4 ♂, UASM); Waterton Lakes Nat. Pk, Cameron L. Rd., 16.viii.1991, R.A. Cannings (1 ♂, BCPM). **British Columbia,** Ashnola R., Ewart Cr., 30.viii.1983, S.G. Cannings (2 ♂, 1 ♀, SMDV); Dean R., 52°75'N x 126°05'N, 20-24.viii.1983, W.W. Middlekauff (2 ♀, EMEC); Dean R., 52°75'N x 126°05'W, 25-30.viii.1986, W.W. Middlekauff (1 ♀, 'CASC); Dean R., 52°90'N x 126°40'W, 22-27.viii.1987, W.W. Middlekauff (5 ♀, CASC); Dean R., 52°49'N x 126°46'W, 20-25.viii.1988, W.W. Middlekauff (4 ♀, CASC); Fernie, 23.vii.1934, H.B. Leech (1 ♀, CASC); Flathead, 20km by air NW Storm Cr., 5500', 27.vii.1989, C.S. Guppy (1 ♂, 1 ♀, BCPM); Hedley, West Forest Rec. Site, 28.vii.1988, C.S. Guppy (2 ♂, 1 ♀, BCPM), G.E. Hutchings (2 ♂, 1 ♀,

BCPM); Hope, Othello, Coquihalla R., 25.viii.1994, R.A. Cannings and H. Nadel (28♂, 10♀, BCPM); Hope, River Park, Coquihalla R., 8.viii.1989, P.H. Arnaud, Jr (1♀, CASC); Kitimat, 19.vii.1960, B. Heming (1♀, CNCI); Kitimat R., 24mi S Terrace, 19.vii.1960, C.H. Mann (2♂, 1♀, CNCI; 1♂, 1♀, BCPM); Lilloett R., Meager Cr. Hotsprings, 19.vii.1988, C.S. Guppy (1♂, BCPM); Merritt, Juliet Cr., Hwy.#5, 24.viii.1989, C.S. Guppy (1♂, 2♀, BCPM); Peace River region, Kinuseo Cr. at Murray R., 10U 612775E 6080465N, 10.viii.1997, H. Nadel (1♀, BCPM); Peace River region, Redwillow Cr. at crossing of Redwillow Forest Rd., 10U 676878E 6092470N, 12.viii.1997, H. Nadel (1♂, 1♀, BCPM); Peace River region, Sakunka R., 10U 58384E 614246N, 8.viii.1997, H. Nadel (6♂, 8♀, BCPM); Sakunka R, Windfall Cr. campground, 10U 5576E 61143N, 9.viii.1997, H. Nadel (2♂, 2♀, BCPM); Similkameen R., Copper Cr., 30.viii.1983, S.G. Cannings (1♂, BCPM); Terrace, 10mi E at Copper Slough, 14.viii.1960, C.H. Mann, BCPM) (2♀, CNCI).

   U.S.A.: California, El Dorado Co., Echo L., 2256m, 23-30.vi.1985, W. Middlekauff (1♀, CASC); Del Norte Co., Jed. Smith State Park, 11.vii.1989, E.M. Fisher (1♀, EMFC); Fresno Co., Heart L., 10500', 1.ix.1952, E.I. Schlinger (1♀, CNCI); Humboldt Co., Humboldt Co., Grizzly Creek State Park, 11.viii.1953, P.H. Arnaud, Jr (3♂, 2♀, CASC); Crannell, 3.vii.1969, C. Slobodichikoff (2♂, EMEC); Humboldt Co., Humboldt Redwoods St. Pk, S. Fork Eel R., 30-31.vii.1970, S. Frommer and L. La Pre (4♂, 1♀, UCRC); Humboldt Co., Eel R., 5mi S. Shirley, 3.vii.1985, L. Bezark (1♀, LGBC); Humboldt Co., Weott, 1mi N, 23.viii.1969, J. Powell (1♂, EMEC); Lake Mary, 8.viii.1957, J. Wilcox (2♂, 1♀, CASC); Lane Co., Frog Meadows, 4300', 18.vii.1932, D. K. Frewing (1♀, EMEC); Madera Co., Garnet L., 9.viii.1952 (1♀, EMEC); Madera Co., 1000 Island L., 5.viii.1952 (1♀, EMEC; 1♀, EMFC), 9.viii.1952 (1♀, EMEC); Mendocino Co., Branscomb, 3mi N, 1400', 21-23.v.1982, M. Hochberg (1♀, EMEC); Mendocino Co., Eel R., Smithe Redwood Grove, 18.vi.1979, L.G. Bezark (2♂, 1♀, 1pr in cop, LGBC); Mendocino Co., Laytonville, 10mi N, 1400', 5.viii.1982, E.M. Fisher (8♂, 5♀, EMFC); Mendocino Co., Navarro R., 3mi SE P.M. Dimick St. Pk, 25-30.vii.1971, P. Rude (17♂, 13♀, EMEC; 2♂, EMFC); Mono Co., Devil's Post Pile Nat. Mon., 23.viii.1958, J. Wilcox (1♂, CASC); Mono Co., Sonora Pass, 14.vii.1985, L.G. Bezark (22♂, 5♀, LGBC); Mono Co., Tioga L., 3.viii.1956, J.L. Herring (1♂, EMEC), C.D. MacNeill (1♀, EMEC); Sierra Co., Independence Lake., 25.vi.1974, R.O. Moon (1♂, EMEC); Siskiyou Co., South Fork Salmon R., Big Flat Camp, 1510m, 15.viii.1980, P.H. Arnaud, Jr (1♀, CASC), 19.viii.1980, P.H. Arnaud Jr (1♀, CASC), 20.viii.1980, P.H. Arnaud, Jr (1♂, 2♀, CASC), 21.viii.1980, P.H. Arnaud, Jr (2♂, 3♀, CASC); Trinity Co., Boulder Cr., Goldfield Campgr., 926m, 15.viii.1980, P.A. Arnaud, Jr (5♂, 1♀, CASC), 16.viii.1980, P.A. Arnaud. Jr (4♂, CASC); 17.viii.1980, P.A. Arnaud. Jr (1♂, CASC); 18.viii.1980, P.A. Arnaud. Jr (1♂, CASC); Trinity Co., Grey Falls Campground, Trinity R., 16.viii.1980, P.A. Arnaud, Jr (1♂, 2♀, CASC); Tuolumne Co., Dana Meadows, 17.vii.1949, L.L. Jensen (1♂, EMEC), Tuolumne Co., Mt. Lyall, 18.viii.1947, L.L. Jensen (1♀, EMEC); Yolo Co., Elkhorn Ferry, 17.v.1951, G. Mitchell (1♂, FSCA); Yosemite Nat. Pk, Fletcher L., 3880-4000', 27.vii.1940, E.G. Linsley (1♂, 1♀, CASC); Yosemite Nat. Pk, Saddlebags Lake, 2.viii.1936, W.B. Herms (1♀, AMNH; 1♀, BPBM; 1♂, 1♀, CSU; 6♀, EMEC); Yosemite Nat. Pk, Tuolumne Meadows, 4.vii.1927, J.M. Aldrich (1♂, 1♀, CASC; 2♀, EMEC). **Colorado,** Gunnison Co., Gunnison R., NPS picnic area, 31.viii.1991, G. DeJong and B. Kondratieff (2♂, 2♀, CSUC); La Plata Co., Hesperus, 8000', 6.viii.1935, C.R. Rotger (1♀, AMNH; 2♂, CSUC); Larimer Co., Fort Collins, 30.ix.1990, S. Gannaway and L. Martinez Park (1♀, CSUC); Larimer Co.,

Lyons Peak, 12.ix.1991, B. Kondratieff (1♂, CSUC); Pitkin Co., Redstone, Crystal R. off
Rte. 133, 7.viii.1992, B. Kondratieff (1♂, CRNC); Rockwood, 10.vii.1937, R.H. Beamer
(1♂, SEMC; 1♂, 2♀, USNM); Routt Co., Yampa R., Rte. 131, 17.vii.1987, B.
Kondratieff (1♂, 1♀ in cop; CSUC). **Idaho,** Elmore Co., Willow Cr. near Snake Cr.
G.S., 1.vii.1975, W.F. Barr (2♂, ESUW), 1.viii.1975, W.F. Barr (6♂, 5♀, ESUW), J.K.
Wangberg (1♂, ESUW); Moscow Mt., 26.v.1918, A.L. Melander (2♀, USNM),
17.vi.1918, A.L. Melander (2♀, USNM). **Montana,** Flathead Nat. For., Middle Fork
Flathead R., 30.vii.1977, R.J. Lavigne (3♂, 2♀, ESUW); Gallatin Nat. For., Slough Cr.,
4mi N Yellowstone Nat. Pk boundary, 17.vii.1971, S.L. Jenkins (1♂, EMFC); Glacier
Peak, 8.viii.1923, A.L. Melander (1♀, SEMC); Glacier Nat. Pk, Apgar, 15mi NE,
18.viii.1969, E.M. and J.L. Fisher (2♀, EMFC); Three Forks, 1.viii.1918, A.L. Melander
(1♀, USNM); Wheatland Co., Winnecook, edge of Musselshell R., 15.viii.1924, S.S.
Berry (3♂, EMEC), 24.viii.1925, S.S. Berry (1♂, LGBC). **Oregon**, Brookings, 8mi E,
30.vi.1965, K. Goeden (1♂, EMFC); Deschutes Co., Bend, Meadow campgr., 3640',
30.vii.1970, P.A. Arnaud, Jr (1♂, CASC), Bend, Shevlin Park, 4.ix.1975, Westcott and
Penrose (2♂, EMFC; 3♂, 3♀, ODAC); Deschutes Co., Mt. Batchelor, 6300',
24.vii.1966, P. Rude (1♀, LGBC); Dodge Park, 21.vi.1933, R.E. Dimick (1♂, OSUO);
Douglas Co., Smith River, 14.ix.1932, D.K. Frewing (1♂, CASC [holotype of *L. atripennis*]; 1♂, USNM; 3♂, 1♀, CASC [paratypes of *L. atripennis*]), 15.ix.1932, D.K. Frewing
(3♂, 2♀, CASC; 1♀, USNM; 1♀ OSUC [paratypes of *L. atripennis*]); Douglas Co.,
Scottsburg, 19.ix.1932, D.K. Frewing (1♂, OSUC; 1♂, 2♀, CASC [paratypes of *L.
atripennis*]; 1♂, BEZA; 1♂, EMFC); Hood R., 24.ix.1917, Childs (1♂, 1♀, CASC);
Hood R., E. Fork, 19.vii.1947 (2♂, FSCA); Hood R., E. Fork, 3mi S Parkdale,
4.viii.1968, E.M. Fisher (4♂, 4♀, EMFC); Josephine Co., Graves Creek near Galice,
6.viii.1970, K.J. Goeden (1♂, CASC; 3♂, 3♀, EMFC); Lebanon, 27.v.1931, J. Wilcox
(2♂, CASC); Linn Co., Santiam Pass, 12.viii.1946, C.H. and D. Martin (1♂, 1♀,
FSCA); Little Dearborn Isl., (9 mi E Blue River, 24.viii.1984, W.W. Middlekauf (1♂,
CASC); Marion Co., Mehama, 19.vi.1932, J. Wilcox (1♂, 1♀, AMNH; 2♂, 2♀, MCZC;
1♀, CUCC; 1♂, 1♀, CSUC; 1♂, 2♀, CASC; 1♂, EMFC; 1♂, 1♀, TAMU; 1♂, 1♀,
WSUC; 1♂, CUCC; 1♀, MTEC; 1♂, 1♀, EDNC; 2♀, EMEC; 1♀, USNM); Mehama,
6mi ESE, 21.vii.1968, K. Goeden (1♂, 1♀, ODAC); Mount Hood (1♀, OSUC),
26.vi.1932, J. Wilcox (2♂, 1♀, CASC; 1♂, EMEC), H.K. Morrison (1♂, EMEC);
Mount Hood, Hood Rapids, 29.vii.1921, A.L. Melander (1♂, USNM); Parkdale, vi.1917,
F.R. Cole (1♀, EMEC); 30.vi.1938, K. Grey and J. Schuh (2♂, 2♀, FSCA; 4♂, 2♀,
WSUC; 3♂, 5♀, USNM), 1.vii.1938, K. Grey and J. Schuh (1♀, WSUC), 14.vii.1940,
K. Grey and J. Schuh (1♂, 5♀, OSUO); Unity, 4.vii.1933, F.H. Schirk (1♀, CASC);
Wallowa L., 9.ix.1932, J. Wilcox (2♀, CASC; 1♀, USNM); Wasco Co., Dufur City Park,
18.vii.1978, R.L. Westcott (1♀, ODAC); Willamette Nat. For., Lost Prairie, 9.viii.1949,
S.E. Crumb, Jr (3♂, 1♀, FSCA), C.R. Amen (1♀, FSCA). **Utah,** Beaver, 6mi E.,
7.ix.1963, J. Wilcox (5♂, 7♀, 1pr cop, CASC; 1♂, EMFC); Daggett Co., Sheepcreek,
12.viii.1981 (1♂, 1♀, EMUS); Duchesne, vii.1926, V.M. Tanner (1♀, BYUC); Duchesne
Co., Red Creek, 3mi E Fruitland, 21.vi.1990, Whiting and Clark (3♂, 2♀, BYUC);
Duchesne Co., Sheep Creek, C. Cottam (1♀, BYUC); Duchesne Co., Uintah Canyon,
11.vii.1972, G.F. Knowlton and W.J. Hanson (1♂, EMUS), 11.viii.1981, Hanson and
Keller (2♂, 3♀, EMUS; 1♂, BCPM); Manila, 6.ix.1939, G.F. Knowlton and F.C.
Harmston (1♀, EMUS); Morgan Co., Lost Cr. Reservoir, 19.vii.1991, R.W. Baumann
(2♂, 4♀, BYUC), R. Findlay (1♀, BYUC); Provo, D.E. Johnson (4♂, 3♀, BYUC), M.
Hammond (1♂, 1♀, BYUC); Roosevelt, 24.vi.1941, G.F. Knowlton and F.C. Harmston

(1 ♀, EMUS); Uintah Co., Merkeley Park, 16.viii.1972,W.J. Hanson and G.F. Knowlton
(1 ♀, BCPM); Uintah Co., Whiterocks Canyon, 11.viii.1981, Hanson and Keller (1 ♂,
BYUC; 2 ♂, 3 ♀, EMUS), 22.vii.1939, G.F. Knowlton and F.C. Harmston (2 ♂, EMUS);
Uintah Mts., vi.1926, T. Swallow (1 ♂, CASC); Utah Co., Hobble Cr. Canyon, Kelly's
Brove, 26.vii.1994 A.J. Evans (1 ♀, BYUC); Weber Canyon, 4.vii.1932, J. Nottingham
(1 ♂, SEMC). **Washington,** American R., Indian Flat, 10.vii.1932, C.H. and D. Martin
(1 ♂, EMFC; 1 ♂, FSCA); Blewett Pass, 12.vi.1932, J. Wilcox (1 ♀, CNCI; 1 ♂, CASC;
1 ♂, SEMC), C.H. and D. Martin (1 ♀, EMFC; 4 ♂, 1 ♀, FSCA; 1 ♀, SEMC); Buckley,
14.vi.1932, J. Wilcox (1 ♂, 1 ♀, CASC); Cle Elum, 4.vii.1932, Wm. W. Baker (1 ♂,
CASC); 5.vii.1932, C.H. and D. Martin (1 ♂, EMEC); 19.viii.1935, S.E. Crumb, Jr (6 ♂,
5 ♀, CASC; 1 ♂, LGBC; 3 ♀, USNM), J. Wilcox (3 ♂, 1 ♀, CASC; 1 ♂, EMFC; 1 ♂, 1 ♀,
EMEC); Ellensberg, 2.viii.1928, M.D. Leonard (1 ♂, CUIC); Gaynor, 3.ix.1933, C.H.
Martin (2 ♂, 1 ♀, FSCA); Glacier, 19.vi.1965 (1 ♀, LGBC); Goldendale, 23.vi.1935, S.E.
Crumb, Jr (4 ♂, CASC), J. Wilcox (1 ♀, UCRC; 6 ♂, 5 ♀, CASC; 1 ♀, LGBC); Kalama,
21.vii,1931, J, Nottingham (1 ♂, 1 ♀ in cop SEMC); Klickitat Co., Little Klickitat R., 3mi
NE Goldendale, 21-22.vi.1969, E.M. Fisher (15 ♂, 12 ♀, EMFC); Lake Chelan, Stehekin,
30.vii.1919, A.L. Melander (1 ♂, USNM); Mason Co., Lake Cushman, 6.viii.1919, F.M.
Gaige (1 ♂, OSUC), 8.viii.1919, F.M. Gaige (1 ♂, OSUC); Mt. Rainier, 15.vii.1936, S.E.
Crumb, Jr (1 ♂, CASC); Mt. Rainier, Carbon Glacier, 26.viii.1935, J. Wilcox (1 ♀,
CASC); Mt. Rainier, Ipsut Cr. Camp, 14.viii.1932, Wm. W. Baker (1 ♂, 3 ♀, CASC; 1 ♂,
EMFC), 23.vii.1935, J. Wilcox (1 ♂, CASC), Wm. W. Baker (1 ♂, FSCA); Mt. Rainier,
Lodgepole Forest Camp, 16.viii.1932, Wm. W. Baker (2 ♂, CASC), S.E. Crumb, Jr (1 ♂,
EMEC), 10.ix.1935, J. Wilcox (1 ♂, ESUW); Mt. Rainier, Sunrise, 6318', 24.vii.1932, J.
Wilcox (1 ♀, EMEC), Sunrise, 6380', 27.vii.1932, Wm.W. Baker (1 ♂, EMEC),
31.vii.1932, J. Wilcox (1 ♀, EMEC), Sunrise, 6400', 30.viii.1933, C.H. Martin (1 ♀,
FSCA); Mt. Rainier, White R. entrance, 14.viii.1931, Wm. W. Baker (1 ♂, CASC), J.
Wilcox (1 ♀, EMFC; 1 ♀, EMEC), 31.vii.1932, J. Wilcox (2 ♀, CASC), 3.ix.1932, J.
Wilcox (1 ♀, CASC; 1 ♂, 2 ♀ EMFC), 4.ix.1932, J. Wilcox (1 ♀, CASC); 27.viii.1933,
C.H. Martin (1 ♀, FSCA); Natches, 10.vii.1932, J. Wilcox (1 ♂, 3 ♀, CASC); Rainier Nat.
For., Indian Flat Camp, 10.vii.1932, J. Wilcox (1 ♀, CASC), 8.viii.1938, S.E. Crumb, Jr
(2 ♂, FSCA); Rainier Nat. For., Pleasant Valley, 9.viii.1938, S.E. Crumb, Jr (1 ♂, FSCA);
Rainier Nat. For., Current Flat, 27.vii.1940, S.E. Crumb, Jr (1 ♂, FSCA); North Yakima,
18.viii.1903, E. Jenne (1 ♂, USNM); Satus Cr., 23.vi.1935, J. Wilcox (1 ♂, UCRC), S.E.
Crumb, Jr (1 ♀, CASC), 6.viii.1969, F.C. Harmston (3 ♂, 2 ♀, EMFC); Tipsoo L.,
28.vii.1932, J. Wilcox (1 ♀, EMEC); Virden, 5.ix.1931, J. Wilcox (1 ♂, USNM);
4.vii.1932, J. Wilcox (1 ♂, 3 ♀, CASC; 1 ♀, ESUW), 5.vii.1932, C.H. and D. Martin (4 ♂,
4 ♀, FSCA); 8.vii.1932, C.H. and D. Martin (1 ♂, ESUW), 4.ix.1932, C.H. Martin (1 ♂,
CUIC), D. Martin (2 ♂, FSCA); 5.ix.1932, J. Wilcox (1 ♂, CNCI; 1 ♂, 1 ♀, CASC; 2 ♂,
2 ♀, EMEC; 1 ♂, USNM); 4.ix.1933, C.H. Martin (1 ♀, FSCA), D. Martin (3 ♀, D.
Martin; 1 ♀, ESUW); Walla Walla, 3.vii.1922, A. Spuler (1 ♂, WSUC), 18.viii.1923, A.L.
Melander (2 ♂, 1 ♀, SEMC; 1 ♂, 1 ♀, BCPM; 1 ♀, CNCI; 1 ♀, CASC; 3 ♂, 5 ♀, USNM);
Walla Walla, Mill Cr. (3 ♂, 3 ♀, USNM), 16.viii.1923, V. Argo (2 ♂, DEI; 2 ♂, 6 ♀,
USNM); 2.ix.1923, V. Argo (1 ♂, 5 ♀, DEIC; 1 ♂, 1 ♀, USNM), 2-6.vii.1922, A.L.
Melander (2 ♂, 10 ♀, USNM), 18.viii.1923, A.L. Melander (4 ♂, USNM); Yakima Co.,
Mt. Adams, Bench L., 5000', 6.ix.1975, R.L. Westcott (1 ♀, EMFC), Yakima Co.,
Yakima, Eschbach Pk, 7.viii.1941, Reeves and Brookman (1 ♂, 1 ♀ in cop; CASC);
Yakima Co., Rimrock, 29.viii.1941, B. Brookman (1 ♂, CASC). **Wyoming,** Big Horn
Co., 4.1km NE Shell Canyon End Resort, 30.vii.1980 (2 ♂, ESUW); Grand Teton Nat.

Pk, Pilgrim Cr., 17.viii.1963, R.J. Lavigne (1♂, ESUW), 14.viii.1964, F. Holland and
R.J. Lavigne (8♂, 3♀, ESUW); Greybull, Greybull R., 14.viii.1975, R.J. Lavigne (2♂,
1♀, ESUW); Hudson, Little Popo Agie, 21.viii.1964, R.J. Lavigne and F. Holland (2♂,
6♀, 1pr in cop, ESUW); Jackson, 14.viii.1964, R.J. Lavigne (6♂, 1♀, ESUW); Lander,
5-8000', R. Moodie (1♂, 1♀, OSUC); Moran, Jackson Hole Biol. Stn., 25.vii.1964, H.E.
Evans (1♂, FSCA), Moran, 14.viii.1964, F.R. Holland (1♂, 1♀ in cop); Shoshone Nat.
For., Pahaska Tepee, 8.viii.1964, F.R. Holland (3♂, ESUW), 11.viii.1964, R.J. Lavigne
and F.R. Holland (4♂, 9♀, ESUW), 13.viii.1964, R.J. Lavigne and F.R. Holland (1♂,
EMFC; 4♂, 5♀, 2pr in cop, ESUW; 1♂, 1♀, LGBC; 1♂, 1♀, USNM), Pahaska Tepee,
5mi N, Middle Cr., 8.viii.1964, F.R. Holland (1♂, 1♀; ESUW), 11.viii.1964, F.R.
Holland (1♂, 1♀ in cop, ESUW), 12.viii.1964, R.J. Lavigne and F.R. Holland (1♀,
EMFC; 5♂, 3♀; ESUW), 13.viii.1964, R.J. Lavigne (2♂, 2♀ in cop, ESUW; 1♀,
IBPV), Pahaska, Pahaska campgr., 21.viii.1963, R.J. Lavigne (1♂, 3♀, 1pr in cop); Swan
Valley, 11mi N, 12.viii.1967, R.J. Lavigne (2♂, ESUW); Sweetwater Co., Green R. x
Rte. 187, 18.vii.1976, R. Lavigne (1♀, ESUW); Teton Nat. For., Teton Overlook,
14.viii.1964, F. Holland and F.R. Lavigne (3♂, 2♀, 2pr in cop, ESUW); Thumb. Station,
V.M. Tanner (1♂, BYUC); Yellowstone Nat. Pk, Continental Divide, 8200', 8.viii.1918,
A.L. Melander (1♀, USNM); Yellowstone Nat. Pk, Lamar Picnic Area, 2.viii.1990, R.J.
Lavigne (2♂, 1♀, ESUW).

Type Locality. U.S.A., Oregon, Hood River Co., Hood River.

Taxonomic Notes. The darkest specimens, which come from the central Coast
Range of Oregon (Douglas Co.), were described as *L. atripennis* by Cole and
Wilcox (1938). Wilcox subsequently determined material from several localities
in southwestern Oregon (Josephine Co.) and northwestern California (Humboldt
Co.) as this species. In the original description he noted that *L. atripennis* appeared at first glance to be a melanic form of *L. cinereus*, but he felt that the
shape of the epandrium and antennal segments warranted separate status for
these populations. However, I can find no consistently distinct differences in either the coloration or structure. The antennal measurements that Wilcox believed
are significant lie well within the variation of the species as a whole, as noted
above in the description. Even the dark coloration is widely variable in these
coastal populations, and it apparently simply represents one extreme of a range of
patterns. In my experience, in this region coastal populations of many disparate
taxa are frequently darker than their inland conspecifics; some examples are Fox
Sparrows (*Passerella iliaca)*) and Hairy Woodpeckers (*Picoides villosus*) in birds
and the Western River Cruiser (*Macromia magnifica)* in dragonflies. *L. atripennis* is here synonymized with *L. cinereus*.

Etymology. From the Latin meaning ash-coloured, grey; refers to the colour of
the extensive tomentum in most specimens.

Distribution. Nearctic; northern British Columbia and the mountains of Alberta
south to California, Utah and Colorado. Map 2, next page.

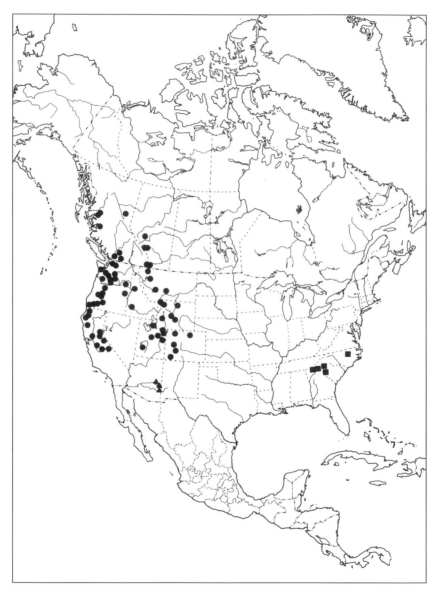

Map 2. Distribution records for
- *Lasiopogon cinereus*
- *L. shermani*
- *L. apache*

Phylogenetic Relationships. Member of the *cinereus* species group; sister species of *L. shermani*.

Natural History. *L. cinereus* is a common and locally abundant species within its range. It typically flies along the edges of mountain streams at mid or high elevations, perching on stones and boulders in, or at the edge of, the water. Adults feed mostly on small aquatic insects; specimens examined with prey include *Rhyacophila* (Trichoptera: Rhyacophilidae); Ephemeroptera: Ametropodidae; *Ravinia duplicata* (Hall) (Diptera: Sarcophagidae); *Pegomya cauduca* (Huckett) (Diptera: Anthomyiidae); *Dicranota* (Diptera: Tipulidae). Flight dates range from 17 May (Yolo Co., California) to 24 September. Dates from the northern and southern parts of the range are, for the most part, similar; the vast majority of the records are from July and August.

Lasiopogon coconino sp. nov.

Diagnosis. A large, grey species with bristles and hairs minimal, bristles black in both sexes. Antennae brown, ratio F2+3/F1 = 0.54-0.62. Thoracic tomentum laterally faintly grey-gold and faint patches of brown tomentum basolaterally on dorsum of abdominal segments. Dorsocentral stripes faint, brown; acrostichal stripes grey. Anterior dorsocentral bristles 3-5, insignificant; posterodosocentrals 1-3, usually prominent. Trochanters and bases of femora and tibiae mostly without tomentum, ferruginous to brown; tarsi pale ferruginous to brown. Epandrium brown with grey tomentum, halves in dorsal view strongly emarginate medially; in lateral view greatest width at midlength, about 0.6 times the length; dorsal margin and apicodorsal angle broadly rounded, ventral margin straight. Gonostylus with very large medial flange and short, strong secondary medial flange. Hypogynial valves weakly haired basally and with ventral carina.

Description. Body length ♂ 10.4-11.0mm; ♀ 12.5-13.5mm.

Head. HW ♂ 2.10-2.46mm; ♀ 2.40-2.60mm. FW ♂ 0.44-0.54mm; ♀ 0.56-0.64mm. VW ♂ 0.88-0.96mm; ♀ 1.00-1.20mm. VW/HW = ♂ 0.39-0.42; ♀ 0.42-0.46. FW/VW = ♂ 0.50-0.56; ♀ 0.53-0.56; VD/VW = ♂ 0.21-0.28; ♀ 0.18-0.28. GH/GL = ♂ 0.28-0.36; ♀ 0.30-0.36.

    Face silver, vertex grey; occiput grey, some gold-grey dorsally. Beard and labial hairs white, all other setae black. Occipital bristles short but strong (to 0.4mm), straight to moderately curved anterolaterally, dense, 13-18 on each side, with additional finer bristles interspersed. Frontal and orbital setae moderately strong, sparse, the longest longer than scape+pedicel.

Antennae. Brown, setae brown/black; F1 sometimes without setae. WF1/LF1 = ♂ 0.26-0.30; ♀ 0.26-0.28. LF2+3/ LF1 = ♂ 0.56-0.62; ♀ 0.54-0.60.

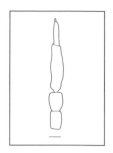

Thorax. Prothorax grey, hairs white; postpronotal lobes grey, the lateral angle ferruginous to chestnut; short, fine light brown to black bristles mixed with fine white hairs. Scutum tomentum pale grey, faint grey-gold highlights laterally; dorsocentral stripes gold/brown, rather faint; area lateral to dorsocentral stripes dark grey. Acrostichal stripes grey, with medial space pale grey. All bristles black; anterior dorsocentrals 3-5, short and weak, longest 0.7mm long; 3 posteriors, prominent, strong. No acrostichals distinct from notal setulae, which are dense, short and appressed.

Figure 69.
L. coconino sp. nov.
antenna.
Scale line = 0.1 mm.

Postalars 1-2, strong, with several weaker bristles; supra-alars 2-3; presuturals 2-3; posthumerals 0-1. Scutellar tomentum grey; apical scutellar bristles 2-3 on each side, mixed with finer bristles and scattered hairs; none to a few scattered white or brown discal hairs.

Pleural tomentum grey-gold; katatergite bristles black, 5-7 among finer white hairs; katepisternal setae long, sparse; 3-4 black setae (longest 0.4mm) at posterior edge of anepisternum and patch of very short, black setulae long dorsal margin. Anepimeron without setae.

Legs. Base colour of femora dark brown or black; tibiae the same or lighter brown; trochanters and tarsi brown or ferruginous in younger specimens. Tomentum of coxae grey or grey-gold, tomentum on rest of legs sparse, grey, absent on much of trochanters. No coxal peg. All bristles black, typically arranged. Femora with decumbent white hairs (in some specimens dorsal decumbent hairs are mostly brown) and erect ones, most numerous on profemur; fine ventral bristles sparse, the longest as long as thickness of femur, except a few longer on metafemur. Profemur in male with 4-8 dorsolateral bristles; mesofemur with 0-5; metafemur with 3-8. Bristles on tibiae and tarsi strong; protibia in male with longest bristle 3 times longer than tibial width. Hairs on tibiae and tarsi mixed white and brown.

Wings. Veins light brown; membrane light brown in oblique view. DCI = 0.40-0.53; cell M3 open. Halter pale yellow; knob without dark spot.

Abdomen. Tergites with grey tomentum, sometimes with golden highlights; each with obscure dorsal paired brown patches of tomentum anterolaterally. Tergite 1 basolaterally with 5-10 black bristles and finer white hairs; lateral setae on tergites white, short and directed posteriorly; in some females lateral setae are black and hardly distinguished from dorsal setulae. Dorsal setulae mostly black or brown, but mixed with pale ones, especially laterally. Sternite tomentum grey or grey/gold; hairs sparse, mostly white, in some specimens increasingly brown towards the posterior.

Male genitalia. Epandrium and hypandrium/ gonocoxite complex ferruginous to dark brown with grey-gold tomentum, absent on hypandrium and bases of gonocoxites and epandrium. Setal brush black; other setae brown/black. Width of epandrium halves in lateral view about 0.6 times the length, widest at midlength; dorsal margin and apicodorsal angle broadly rounded, ventral margin straight. Medial face of epandrium as in figure 70. In dorsal view, medial margins of epandrium abruptly indented; basal sclerite prominent.

Gonostylus. Medial flange very large relative to dorsal flange and basal area; short but prominent secondary medial flange present. Hypandrium/gonocoxite complex in ventral

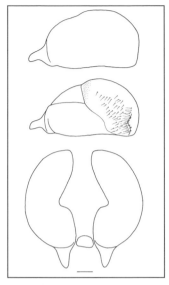

Figure 70. Epandrium (top to bottom): lateral, medial, dorsal. Scale lines = 0.2 mm.

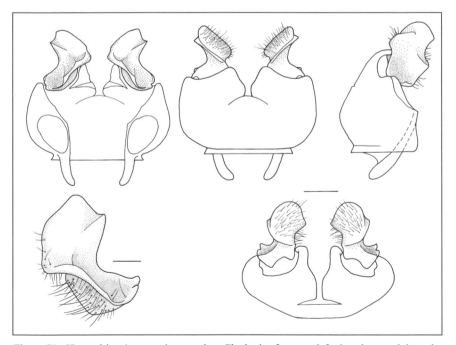

Figure 71. Hypandrium/gonocoxite complex. Clockwise from top left: dorsal, ventral, lateral and apical (scale lines = 0.3 mm); gonosylus, dorsal (scale line = 0.1 mm).

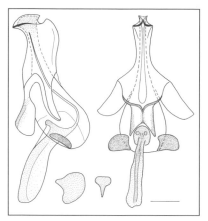

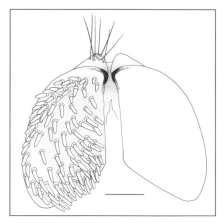

Figure 72. *L. coconino* sp. nov. Phallus: lateral (left); ventral (right); lateral ejaculatory process, basal (centre left); ejaculatory apodeme, cross-section (centre right). Scale line = 0.2 mm.

Figure 73. Subepandrial sclerite: ventral. Scale line = 0.2 mm.

view the width about 0.65 times the length, transverse slit 0.65 the distance from base to apex. In lateral view, gonocoxal apodeme stout and curved ventrally, webbed ventrally; exposed length of apodeme about 0.55 times the length of basal width of hypandrium.

Phallus. Paramere sheath dorsally 0.4 times the length of phallus (excluding ejaculatory apodeme). Apex of paramere sheath with ventrolateral carina, the ventral lip strongly projecting; no ventral flange. Sperm sac width in dorsal view 0.3 the length of phallus. Ejaculatory apodeme gently curved ventrally in lateral view, weakly spatulate in ventral view; oval in cross-section with thin dorsal carina. Subepandrial sclerite as in figure 73. Broad triangular unsclerotized area in basal 0.25, narrowing to parallel-sided gap in central portion; spines more-or-less parallel-sided, blunt, densely arranged apically and basally.

Female genitalia: Undissected: All setae pale. Tergite 8 dark brown/black with posterior margin ferruginous; hairs rather abundant, erect. Sternite 8 dark brown/black to mostly ferruginous in some specimens; lateral lobe setae weak to moderately strong. Hypogynial valves with hairs basally. Cerci yellow-brown with pale setae. Dissected sternite 8 (figure 74) broad, with basal width about equal to the length; undivided but weak along midline. Length of unsclerotized area between hypogynial valves 0.4 times the length of sternite; lateral lobe setae weak to moderately strong.

Tergite 9 sclerites as in figure 75; sternite 9 V-shaped, medially undivided and often with short lateral projections. Tergite 10 brown/black with 8-10 black acanthophorite spines on each side. Spermathecae (figure 76) with straight or gently

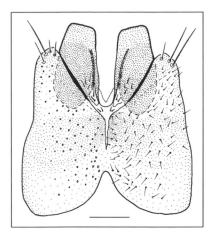

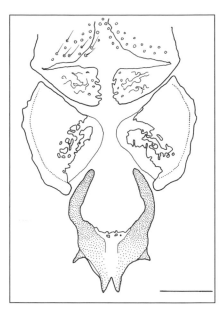

Figure 74 (above). Female sternite 8:
ventral. Scale line = 0.2 mm.

Figure 75 (right). Female sternite 9, tergite 9
and lobes of sternite 10: ventral.
Scale line = 0.2 mm.

curved terminal reservoirs about half as wide as reservoir duct; terminal reservoirs without wart-like protuberances on surface. Basal striated duct without fine scales; junction with basal duct with scales; basal duct moderately long.

Type Material.
HOLOTYPE. (here designated) ♂ labelled: "[rectangular beige label] Oak Cr. Cyn. Ariz./V-29 1963"; "[rectangular beige label] J. Wilcox/Coll"; "[rectangular blue label] HOLOTYPE ♂/Lasiopogon/coconino/Wilcox"; "[rectangular green label bordered top and bottom with asterisks] Unpublished/Manuscript Name"; "[rectangular white label bordered top and bottom with asterisks] JOSEPH WILCOX/COLLECTION – 1981/Gift to California/Academy of Sciences"; My Holotype label "HOLOTYPE/Lasiopogon ♂/coconino Cannings/des. R.A. Cannings 2002 [red, black-bordered label]" has been attached to this specimen. Dissected genitalia in plastic vial underneath all. CASC.
PARATYPES (76 designated). **U.S.A.: Arizona**, Apache Co., Rock Creek, 6700', 30.vi.1995, H. Greeney III (1♂, UAIC); Coconino Co., Oak Creek Canyon, 29.v.1963, J. Wilcox (2♂, 2♀, BCPM; 15♂, 7♀, CASC; 2♂, 2♀, CNCI); 12.vi.1948, J. Wilcox (3♂, 3♀, CASC); 19.vi.1949, Chas. H. Martin (2♂,1♀, FSCA); 19.vi.1949, Dorothy Martin (3♂, 2♀, pair *in copula*, FSCA); 5.vi.1963, C.H. Spitzer (1♂, EMFC; 1♀, USNM); 5200 ft, 11.vi.1958, J.M. and S.N. Burns (1♀, EMEC); 11.vi.1953, A. and H. Dietrich (1♀, COR); Cave Springs campground, 5400 ft, 22.v.1971, P.H. and M. Arnaud (2♂, EMFC); 5000 ft., 17.vi.1978, M.E. Buegler (1♂, 1♀, EMEC); Coconino Co., 7 mi N of

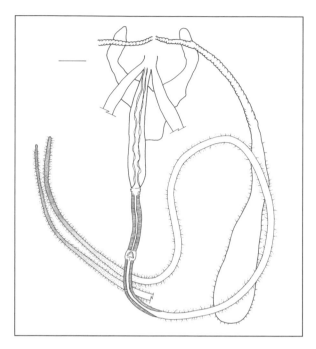

Figure 76. *L. coconino* sp. nov. spermatheca: dorsal. Scale line = 0.1 mm.

Sedona, 16.vi.1978, R.C. Miller (1♂, UCDC); White Mtns, 19.vi.1950, R.H. Beamer (6♂, 1♀, pair *in copula*, FSCA; 1♂, EMFC); 19.vi.1950, W.J. Arnold (1♂, 1♀, pair *in copula*, EMFC; 1♀, FSCA); White Mtns, South Fork Camp, 19.vi.1947, G.H. and J.L. Sperry (1♂, USNM); 20.vi.1947, G.H. and J.L. Sperry (2♂, 2♀, USNM). **New Mexico**, Catron Co., San Franciso River near Arizona, 20.vi.1987, Baumann, Kondratieff, Sargent and Wells (3♂, 1♀, BYUC; 1♂, 1♀, BCPM).

Type Locality. U.S.A., Arizona, Coconino Co., Oak Creek Canyon.

Taxonomic Notes. Wilcox gave this species a manuscript name and labelled type material (CASC) but never published a description.

Etymology. Named for Coconino County, Arizona, the type locality.

Distribution. Nearctic; U.S.A.; mountains of central and eastern Arizona. See map 11, p. 258.

Phylogenetic Relationships. Member of the *tetragrammus* species group; sister to the species pair of *L. quadrivittatus* and *L. lavignei*.

Natural History. Habitat: stream sides; specimen dates range from 22 May to 20 June.

Lasiopogon flammeus sp. nov.

Diagnosis. Only three males known. A medium-sized grey species with mystax and abdominal segment 1 bristles white but all other major bristles brown or black. Antennae brown, ratio F2+3/F1 = 0.47-0.51. Golden brown tomentum on scutum; broad gold-grey apical bands contrasting with dark brown-gold basal areas on the dorsum of abdominal segments. Dorsocentral stripes dark brown, acrostichal stripes brown. Several anterior dorsocentral bristles, but only most posterior one prominent. Trochanters, bases of femora, tibiae and tarsi ferruginous. Leg setation reduced. Epandrium ferruginous, in lateral view two times as long as wide at widest point and dorsal margin convex; in dorsal view strongly concave basomedially.

Description. Body length ♂ 8.5-9.0mm; ♀ unknown.

Head. HW 2.00-2.2mm. FW 0.40-0.48mm. VW 0.84-0.90mm. VW/HW = 0.42; FW/VW = 0.48-0.52; VD/VW = 0.20-0.22; GH/GL = 0.38-0.40.
Face silver, vertex gold-brown; occiput gold-grey, more golden dorsally. Beard and labium white-haired, mystax white to yellow-white, all other setae brown. Ocellar setae short, about as long as length of F1+F2+3; frontal and orbital setae sparse, the longest about as long as length of F1. Occipital setae sparse, 10-12 on each side in main series, moderately strong but short, the longest about 4.5mm; most rather straight, but those along dorsal margin of eye strongly curved anteriorly.

Antennae. Brown, with base of F1 paler. All setae brown; F1 without setae. F1 moderately long, widest at midlength. WF1/LF1 = 0.28-0.30; LF2+3/LF1 = 0.47-0.51.

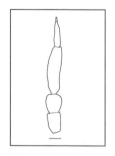

Figure 77. Antenna.
Scale line = 0.1 mm.

Thorax. Prothorax silver-grey with gold highlights, especially dorsally on antepronotum; all hairs white. Postpronotal lobes silver-grey, lateral angle ferruginous; setae short, brown. Scutum tomentum grey-gold, dark gold anteriorly. Dorsocentral stripes and intermediate spots medium to dark brown; acrostichal stripes brown, bordering narrow grey-gold medial area. Most mesothoracic setae brown or black. Anterior dorsocentrals weak except for posterior one or two; most only as long as scape+pedicel. Posterior dorsocentrals stronger; three main setae present. Presutural area with numerous short notal setae; acrostichals not differentiated. Postalars 1-2, supra-alars 2, presuturals 2, posthumerals 1-2, weak or strong. Scutellar tomentum silver-grey with gold highlights, dorsum especially golden. Apical scutellar bristles black, 2-3 on each side, a few dark hairs intermixed.

Pleural tomentum silver-grey with gold highlights. Katatergite with scattered white hairs and 5-8 bristles, mostly brown, some white; 2-3 dark setae on posterior margin of anepisternum, extensive patch of short dark setulae on dorsal shelf. Katepisternal setae sparse. Anepimeron without setae.

Legs. Base colour of trochanters, tibiae, tarsi, bases and apices of femora and parts of coxae ferruginous; most of femora dark brown or black. Tomentum of coxae silver with some gold highlights; tomentum on rest of legs sparse, gold-grey. No coxal tubercle. All bristles brown or black, typically arranged, finer hairs white to brown. Femora dorsally with relatively long decumbent hairs mostly brown, white basally; fine long white ventral setae sparse, the longest shorter than thickness of femur. Profemur in male with 5-8 dorsolateral bristles; mesofemur with 2-4; metafemur with 5-7. Bristles on tibiae and tarsi rather short, protibia in male with longest bristle about 2.5 times longer than tibial width.

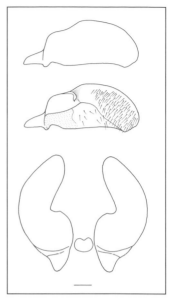

Figure 78. *L. flammeus* sp. nov. epandrium (top to bottom): lateral, medial, dorsal. Scale lines = 0.2 mm.

Wings. Veins medium brown, light brown to ferruginous near anterior margin; membrane pale red-brown in oblique view. DCI = 0.32-0.41; cell M3 open. Halter yellow; knob without dark spot.

Abdomen. Tergite basal colour dark brown, ferruginous apically on apical segments. Tomentum on tergite bases brown with much gold in lateral and anterior views. Bands of gold-grey tomentum cover about half of each tergite apically (segment 1 is almost all covered) and all the ventrolateral areas; bands almost disappear in posterior view. Tergite 1 with 6-8 major white bristles on each side and many smaller white setae; lateral setae on tergites 1-3 white, erect, sparse but moderately long; those on posterior segments much shorter and directed posteriorly. Dorsal setulae mixed white and brown, short and hair-like, decumbent. Sternite tomentum grey; hairs white.

Male genitalia. Epandrium and hypandrium/ gonocoxite complex ferruginous with thin golden tomentum on epandrium and gonocoxites, more dense apically on epandrium. Setal brush dark brown; other setae dark brown, numerous, prominent. Width of epandrium halves in lateral view about 0.5 the length, widest about midlength; ventral margin rather straight, dorsal margin moderately convex; apex rounded, slightly excavated at ventral corner. Medial face of epandrium as in figure 78. In dorsal view, medial margins of epandrium strongly concave; basal sclerite prominent.

Gonostylus. Prominent secondary medial flange linked by a ridge to very strong dorsomedial tooth; lateral and ventrolateral teeth strong; a small tooth present on dorsal flange near junction of medial flange. Hypandrium/gonocoxite complex in ventral view about 0.6 times long as wide, transverse slit at 0.75 the length. In lateral view, exposed length of gonocoxal apodeme about 0.65 as long as basal width of hypandrium; apodeme with sclerotized web ventrally.

Phallus. Paramere sheath dorsally 0.45 the length of phallus (excluding ejaculatory apodeme). Apex of paramere sheath with ventrolateral carina, ventral flange absent. Sperm sac width in dorsal view 0.40 the length of phallus. Ejaculatory apodeme in lateral view, broad, ventral margin slightly curved; strongly spatulate in ventral view; flattened in cross-section with prominent dorsal carina. Subepandrial sclerite as in figure 81. Central unsclerotized area triangular in basal 0.3, narrow and parallel-sided in central 0.5; spines blunt, dense except subbasally.

Female unknown.

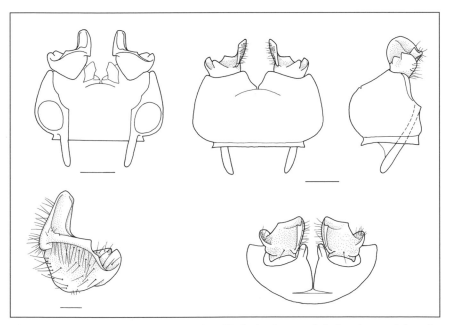

Figure 79. Hypandrium/gonocoxite complex. Clockwise from top left: dorsal, ventral, lateral and apical (scale lines = 0.3 mm); gonostylus, dorsal (scale line = 0.1 mm).

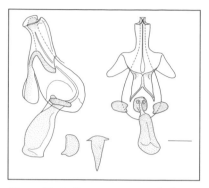

Figure 80. *L. flammeus* sp. nov. phallus: lateral (left); ventral (right); lateral ejaculatory process, basal (centre left); ejaculatory apodeme, cross-section (centre right). Scale line = 0.2 mm.

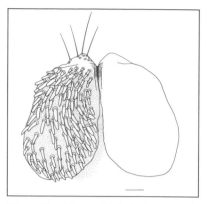

Figure 81. Subepandrial sclerite: ventral. Scale line = 0.1 mm.

Type Material

HOLOTYPE. (here designated) ♂ labelled: "[rectangular white label] N.C."; "[rectangular blue and green banded label] A.E Pritchard/Collection/1962"; "[rectangular white label] Lasiopogon/shermani/Det. 1940 C&W/A.E. Pritchard". My holotype label "HOLOTYPE/Lasiopogon ♂/flammeus Cannings/des. R.A. Cannings 2002 [red, black-bordered label]" has been attached to this specimen. Dissected genitalia in plastic vial underneath all. USNM.

PARATYPES (2 designated). **U.S.A.: North Carolina,** Blantyre, early May 1908, F. Sherman (1 ♂, CASC); no other data (1 ♂ USNM).

Type Locality. U.S.A., North Carolina.

Taxonomic Notes. Cole and Wilcox (1938) did not distinguish this taxon from *L. shermani*, which shares the grey and red coloration and also occurs in the Carolinas.

Etymology. Latin meaning "flame"; the legs, wing veins and genitalia are reddish and the scutum is golden.

Distribution. Nearctic; U.S.A., North Carolina. See map 11, p. 258.

Phylogenetic Relationships. Member of the *tetragrammus* species group; in an unresolved trichotomy with *L. chrysotus* and *L. woodorum*.

Natural History. Habitat: unknown, but most likely stream sides in mid-elevation forests. No recorded flight dates, but probably March to May.

Lasiopogon hasanicus Lehr

Lasiopogon hasanicus Lehr, 1984. *Zoologicheskii Zhurnal* 63:703

Diagnosis. A medium-sized to large, setose grey and dark brown species. Mystax and other head bristles dark. Antennae brown, F1 long and narrow; F2+3/F1 = 0.49-0.64. Thoracic tomentum dorsally and laterally grey/gold-grey with brown dorsocentral stripes and grey or brown-grey acrostichal stripes. Main thoracic and leg bristles dark; anterior dorsocentral bristles 5-8, moderately strong. Abdominal tergites basally with only thin brown tomentum, the dark cuticle shining through. Apical bands gold-grey, their width 0.3 to 0.4 times the length of the segments. Bristles on tergite 1 dark; all lateral setae pale, dorsal setulae dark. Epandrium dark brown/black, about 0.45-0.6 as wide as long in lateral view, the greatest width at about half the length; apex rounded. In dorsal view strongly concave medially. Female with abdominal pattern similar to that of male; tergites 4-7 with brown lateral setae. Terminalia dark brown/black with light brown hairs.

Description. Body length ♂ 9.7-11.2mm; ♀ 10.3-12.8mm.

Head. HW ♂ 2.04-2.40mm; ♀ 2.32-2.58mm. FW ♂ 0.50-0.58mm; ♀ 0.54-0.64mm. VW ♂ 0.95-1.14mm; ♀ 1.04-1.20mm. VW/HW = ♂ 0.46-0.48; ♀ 0.45-0.47. FW/VW = ♂ 0.51-0.55; ♀ 0.51-0.53. VD/VW = ♂ 0.07-0.12; ♀ 0.11-0.16. GH/GL = ♂ 0.31-0.37; ♀ 0.36-0.44.

Face with thin gold/brown-grey tomentum; vertex brown-grey, occiput grey ventrally, brown-grey dorsally. Beard and labial hairs white, mystax bristles brown/black; all other setae brown/black. Occipital bristles rather strong (to 1.0mm), abundant, those behind the dorsomedial angle of the eye strongly curved anterolaterally; lateral and ventral ones shorter, straighter. Frontal and orbital setae abundant, long (to. 0.6mm).

Antennae. Brown; setae brown; F1 without setae. F1 long, narrow, parallel-sided or slightly tapered from base. WF1/LF1= ♂ 0.20-0.21; ♀ 0.21-0.25. LF2+3/ LF1= ♂ 0.49-0.64; ♀ 0.49-0.59.

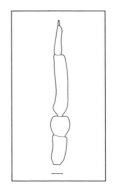

Figure 82. Antenna. Scale line = 0.1 mm.

Thorax. Prothorax gold-grey, browner dorsally, hairs white; postponotal lobes grey or gold-grey, the lateral angle ferruginous, hairs strong, brown. Scutum tomentum usually light brown-grey, the dorsocentral stripes dark/medium brown bordered with light brown/gold. Brown dark around presutural and supra-alar bristles. Acrostical stripes brown-grey. Colours darkening at some angles of view, especially from the posterior. Some specimens with scutum more grey with gold highlights, the acrostichal stripes dark grey. A setose species; all strong bristles black, finer setae brown.

Dorsocentral setae abundant; anteriors 5-8, rather strong (longest to 1.0mm), mixed with only slightly finer setae; 3-6 posteriors. Notal setae abundant, erect, long (to 0.5mm), as long as shorter dorsocentrals. Postalars 3-5, with shorter setae; supra-alars 2-3 among shorter setae; presuturals 2-4, 1-2 posthumerals. Scutellar tomentum grey with gold highlights; apical scutellar bristles dark, abundant, in 2-3 irregular rows, 5-8 strong ones on each side mixed with many shorter, weaker dark bristles and hairs.

Pleural tomentum grey to gold/brown-grey. Katatergite setae black, 9-12 among finer white hairs; katepisternal setae white, long, abundant. Anepisternal setae 15-25, dark (to 0.7mm); sometimes a few other pale, weaker ones ventrally; a dense patch of rather long, fine, erect brown setae on dorsal shelf. Anepimeron with a few white setae. Female thorax similar to that of male.

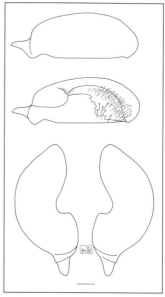

Figure 83. Epandrium (top to bottom): lateral, medial, dorsal. Scale lines = 0.2 mm.

Legs. Base colour dark brown/black. Tomentum of coxae grey with some gold highlights; tomentum on rest of legs gold-grey. No coxal peg. Main bristles dark brown/black; pale hairs white to yellow. Femora dorsally with decumbent hairs in male pale basally, dark apically; usually all pale in female. Longer erect hairs abundant ventrally and laterally, especially on profemur; pale basally, dark apically; longest pale ventral setae as long as, or longer than, thickness of femur. Femora with abundant, fine, dorsolateral bristles mixed with long hairs. Tibiae and tarsi with dark, strong bristles typically arranged, hairs on tibiae mostly black/brown; tarsal hairs dark. Protibia with longest bristles about 4.0 times longer than tibial width.

Wings. Veins medium/dark brown; membrane pale medium brown in oblique view. DCI = 0.30-0.40; cell M3 open. Halter yellow; knob without dark spot.

Abdomen. Male. Tergite basal colour dark brown/black. Tomentum on tergite bases thin brown/gold-brown, the dark cuticle shining through. Bands of gold-grey tomentum cover 0.3 to 0.4 the length of each tergite apically; ventrolaterally they cover the length of the tergite. Segment 1 is thinly covered on about the apical 0.6. About 9-11 strong dark bristles on each side of tergite 1. Lateral setae on tergites white/yellow, abundant, erect, rather long on 1-3, becoming shorter posteriorly. Dorsal setulae more-or-less erect, dark. Sternite tomentum gold-grey, hairs pale.

Female. Width of grey or gold-grey apical bands about 0.35 times the tergite length, wider on 6-7; grey ventrolaterally. Lateral setae on tergites 1-3 white/yellow, abundant, erect, rather long, those on 4 shorter and mixed with brown ones; setae on 5-7 brown. Dorsal setulae more-or-less erect, dark.

Male genitalia. Epandrium and hypandrium/ gonocoxite complex dark brown/black and covered with gold-grey tomentum except on hypandrium in some specimens. Setal brush black; other setae brown/dark brown, numerous, prominent. Width of epandrium halves in lateral view about 0.45-0.6 times the length, widest at about midlength; ventral margin straight, dorsal margin often gently convex, apex rounded. Medial face of epandrium as in figure 83. In dorsal view, medial margins of epandrium strongly concave; basal sclerite vague.

Gonostylus. Medial flange expanded into a large ventrolateral lobe; the dorsal flange angled apically at its medial end, separated from the lateral margin of the gonostylus; lateral and ventrolateral teeth produced into long curved points. Hypandrium/gonocoxite complex in ventral view with length about 0.8 times the width, transverse slit at 0.6 the length; distal to the slit the gonocoxites are concave. In lateral view, exposed length of gonocoxal apodeme about 0.55 times the basal width of hypandrium; apodeme with small sclerotized web ventrally.

Figure 84. *L. hasanicus* hypandrium/gonocoxite complex. Clockwise from top left: dorsal, ventral, lateral and apical (scale lines = 0.3 mm); gonostylus, dorsal (scale line = 0.1 mm).

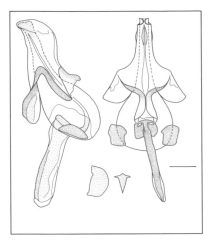

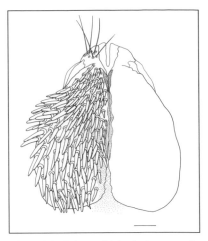

Figure 85. Phallus: lateral (left); ventral (right); lateral ejaculatory process, basal (centre left); ejaculatory apodeme, cross-section (centre right). Scale line = 0.2 mm.

Figure 86. Subepandrial sclerite: ventral. Scale line = 0.1 mm.

Phallus. Paramere sheath dorsally 0.45 times the length of phallus (excluding ejaculatory apodeme), secondary sclerotization extending it to the tips of the lateral processes. Ventral process angled ventrally in lateral view, compressed in ventral view. Apex of paramere sheath broad, the ventrolateral carina short, a very small ventral flange present. The dorsal carina bulges dorsally and is slightly inflated laterally. Sperm sac width in dorsal view 0.45 times the length of phallus. Ejaculatory apodeme long and straight in lateral view, slightly spatulate in ventral view; flattened ventrally in cross-section with a prominent dorsal carina. Subepandrial sclerite as in figure 86. Very narrow triangular unsclerotized area in basal 0.4; spines bluntly acute, dense over all sclerite.

Female genitalia. Undissected: Hairs light to dark brown, erect, abundant. Tergite 8 flared ventrolaterally; dark brown/black, often narrowly brown apically. Sternite dark brown/black, midline and medial lobes of hypogynial valves yellow/medium brown with pale hairs. Lateral lobe setae weak. Hypogynial valves black, vertically oriented and approximated apically. Cerci black with light brown setae. Dissected sternite 8 (figure 87) with basal width about 0.75 times the length, undivided along midline. Length of unsclerotized area between medial lobes 0.3 times the sternite length. Medial lobe not included in hypogynial valve sclerite and shorter than valve. Valve with strong carina, medial in apical half, on midline in basal half.

Tergite 9 sclerites as in figure 88; sternite 9 V-shaped, medially undivided. Tergite 10 black, usually with 8 black acanthophorite spines on each side. Basal

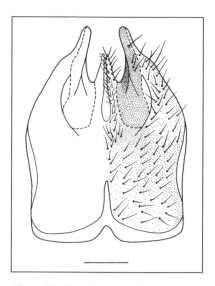

Figure 87. Female sternite 8: ventral.
Scale line = 0.3 mm.

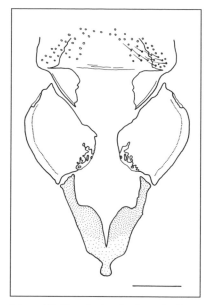

Figure 88. Female sternite 9, tergite 9
and lobes of sternite 10: ventral. Scale
line = 0.2 mm.

lobes of sternite 10 small. Spermathecae (figure 89) with strongly hooked terminal reservoirs about as thick as reservoir duct; terminal reservoirs with wart-like protuberances on surface. Both striated ducts with fine scales; junction with basal duct scaled, sclerotized and golden; valve not golden; basal duct short.

Type Material

HOLOTYPE. ♂ (examined) labelled [in Russian, in Lehr's handwriting]: "[rectangular white label with black border] Primorsk.[iy] Kr.[ay],/12 km N Khasan,/"Golubinoi Cliff,"/28 V 1974/Lehr";"[rectangular red label] Holotypus 1981/Lasiopogon/hasanicus". IBPV.

PARATYPES. (47 examined). **Russia: Primorskiy Kray**, 12 km N. of Khasan Station, Golubinoi Cliff, 28.v.1974, P. Lehr (8♂, 8♀, IBPV; 2♂, MHRC; 3♂, 1♀, BCPM), 11 km N. of Khasan Station, Golubinoi Cliff, 28.v.1974, P. Lehr (1♂, 1♀, IBPV; 1♂, 1♀, MHRC; 1♀, BCPM), 29.v.1974, P. Lehr (5♂, 9♀, IBPV; 2♀, MHRC; 2♀, BCPM), 30.v.1974, P. Lehr (1♀, IBPV). Khasan region, 7.vi.1975, N. Kurzenko (1♀, IBPV).

Other Material Examined (42 specimens).

Russia: Primorskiy Kray, Kievka, 8km E, Lazovskiy Reserve, 25.v.1979, S. Storozhenko (1♂, IBPV); Livadiya, 22.vi.1981, A. Lelej (1♂, 1♀, IBPV), 24.vi.1981, A. Lelej (1♂, IBPV); Oblachnaya Mt., 45km NE Lazo, 17.viii.1984, P. Lehr (1♀, IBPV);

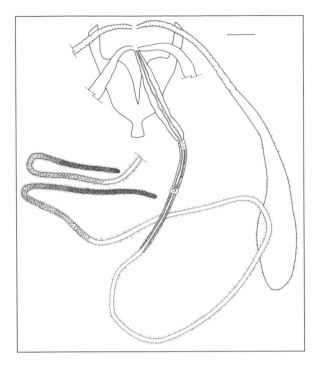

Figure 89. *L. hasanicus*
spermatheca: dorsal.
Scale line = 0.1 mm.

Ol'ga, 13.vi.1986, A. Lelej (13♂, 7♀, IBPV; 3♂, 3♀, BCPM), 23.vi.1986, A. Lelej (1♂, IBPV; 1♂, BCPM); Vladivostok District, 28.v.1978, Kurzenko (1♂, IBPV); Vladivostok District, Manchuzhur Bay, 30.v.1987, P. Lehr (1♂, 1♀, IBPV); Vysokogorsk, 25km ENE Kavalerovo, 13.vi.1986, P. Lehr (4♂, 2♀, IBPV).

Type Locality. Russia, Primorskiy Kray, Khasanskiy District, Golubinoi Cliff, 12 km N. of Khasan Station.

Etymology. Refers to the type locality in the Khasanskiy District, Primorskiy Kray.

Distribution. Palaearctic; Russia, Primorskiy Kray (Ussuri R./Vladivostok region). Map 3.

Phylogenetic Relationships. Member of the *akaishii* species group; sister species to *L. akaishii*.

Natural History. Lives in dry, open areas from stony ridges and cliff summits to coastal sand dunes; sometimes found on tree trunks; can be abundant at some sites (Lehr 1984a). Specimen dates range from 25 May to 17 August; the majority are from late May and the first half of June. Prey identified includes muscid flies and a specimen of *Bibio* (Diptera: Bibionidae).

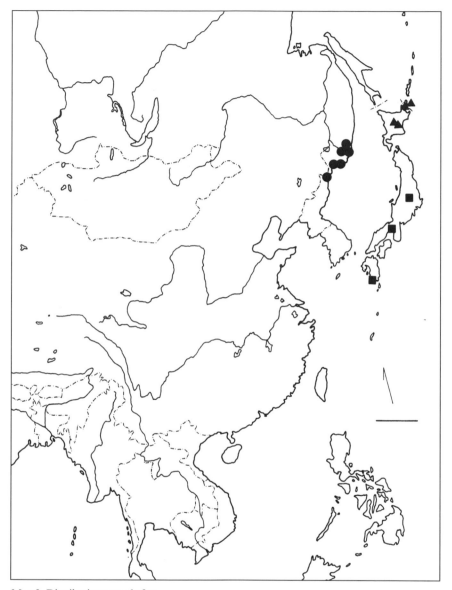

Map 3. Distribution records for
- *L. hasanicus*
- ■ *L. rokuroi*
- ▲ *L. akaishii*

Lasiopogon hinei Cole and Wilcox

Lasiopogon hinei Cole and Wilcox, 1938. *Entomologica Americana* 43: 51-53.
Lasiopogon sibiricus Lehr, 1984. *Zoologicheskii Zhurnal* 63: 696-706 (p. 704-5).

Diagnosis. A small to medium-sized grey and black species with shining black abdomen in the male. Mystax dark with a few pale bristles ventrally; other head bristles dark. Antennae brown, F2+3/F1 = 0.45-74. Thoracic tomentum dorsally and laterally grey/gold-grey with brown dorsocentral stripes and paired grey/brown acrostichal stripes. Main thoracic and leg bristles dark. Halter with brown patch on knob. Dorsocentral setae abundant, strong anterior bristles 3-5, moderately strong. Abdominal tergites mostly shining black with abundant white hair laterally on basal segments, dark hair apically. Apical bands grey, very narrow, covering little more than the intersegmental membrane. Bristles on tergite 1 pale, rarely a few black. Epandrium chestnut/dark brown, the width about 0.4-0.45 times the length in lateral view, the greatest width in basal third; apex ventrally angular and turned ventrally. In dorsal view strongly concave medially. Female with broad grey apical bands covering 0.4-0.7 the length of the tergites, the grey sometimes running vaguely forward mid-dorsally dividing the brown basal tomentum. Tergite 8 black, hairs abundant, brown; sternite 8 mainly ferruginous.

Description. Body length ♂ 8.4-11.8mm; ♀ 9.0-12.4mm. (A complete drawing of a male *L. hinei* is featured on the cover of this book and in figure 1, p. vi.)

Head. HW ♂ 1.68-2.16mm; ♀ 1.74-2.56mm. FW ♂ 0.32-0.48mm; ♀ 0.38-0.58mm. VW ♂ 0.75-0.98mm; ♀ 0.69-1.05mm. VW/HW = ♂ 0.43-0.45; ♀ 0.40-0.44. FW/VW = ♂ 0.43-0.50; ♀ 0.53-0.55. VD/VW = ♂ 0.17; ♀ 0.16-0.17. GH/GL= ♂ 0.30-0.39; ♀ 0.31-0.40.

Face grey with brown highlights; vertex brown-grey, occiput grey, often with gold or brown highlights dorsally. Beard and labial hairs white (pale brown in a few Kurile Island specimens), mystax bristles brown/black, usually with a few pale setae ventrally; all other setae brown/black. Occipital bristles moderately strong (to 0.7mm), abundant, those behind the dorsomedial angle of the eye strongly curved anterolaterally; lateral and ventral ones shorter, straighter. Frontal and orbital setae abundant, long (to 0.6mm); ocellar bristles fine, hardly distinguished from orbitals.

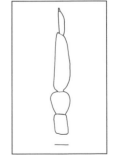

Antennae. Brown, sometimes with various parts (for example, apex of pedicel, base of F1, or parts of F2+3) dark ferruginous. Setae brown; some specimens with setae on F1. WF1/LF1 = ♂ 0.24-0.34; ♀ 0.23-0.31. LF2+3/LF1 = ♂ 0.45-0.74; ♀ 0.56-0.62; most 0.60-0.70.

Figure 90. Antenna. Scale line = 0.1 mm.

Thorax. Prothorax grey/brown-grey, often darker on antepronotum, with white/pale brown hairs; postponotal lobes gold-grey, the lateral angle ferruginous, hairs white to brown. Scutum tomentum variable. Some specimens with rather clear grey tomentum and dorsocentral stripes medium brown, the acrostichal stripes grey and obscure. Most specimens brown-grey, the dorsocentral stripes dark brown, weakly bordered with gold/brown; stripes change in intensity with angle of view, darkest in anteriorlateral view. Acrostichal stripes brown-grey, usually vague. Intermediate spots faint brown, ventrolateral areas with varying amounts of brown/gold. At the other extreme, brown is more prevalent, with the darker intermediate spots and the brown acrostichal stripes spreading to merge with the dorsocentral stripes. All strong bristles black, finer setae brown. *L. hinei* is a rather setose species. Dorsocentral setae abundant, strong anteriors 3-5 (to 0.9mm) mixed with many finer setae; 4-6 posteriors. Notal setae rather abundant on intermediate spots and posthumeral areas, rather long (to 0.3mm), about as long as shorter dorsocentrals. Postalars 2-3 among shorter setae; supraalars 2-3 with shorter setae; presuturals 2-3, 1-3 posthumerals, usually weak. Scutellar tomentum grey; apical scutellar bristles dark, 3-5 strong ones on each side mixed with shorter, weaker dark bristles and hairs.

Pleural tomentum grey to gold/brown-grey. Katatergite setae brown/black, 7-8 among finer hairs; katepisternal setae sparse, often long (usually white as normal, but brown in some Kurile Island specimens). Anepisternal setae 6-12, dark (to 0.6mm), sometimes a few pale ones ventrally; a patch of rather long (to 0.2mm), erect brown setae on dorsal shelf. Anepimeron without setae. Female thorax similar to that of male.

Legs. Base colour dark brown/black; tibiae sometimes distally dark chestnut, tarsi ferruginous to dark chestnut. Tomentum of coxae grey with some gold highlights; tomentum on rest of legs grey. No coxal peg. Main bristles dark brown/black. Femora dorsally with short hairs white basally, brown/black apically, the dark hairs predominanting in males (except those from Buryatskaya ACCP) and the white much more extensive in females. In some Kurile Island specimens, especially males, all pale hairs (even many on coxae) are light/medium brown. Longer erect white/gold and dark hairs abundant ventrally and laterally, especially on profemur; the colour pattern is similar to that of the shorter hairs. Brown/black dorsolateral bristles on femora are numerous, long, and fine and are almost indistinguishable from the surrounding hairs, at least on the profemur. In males the longest pale ventral setae are longer than the width of the femur. Tibiae and tarsi have dark, strong bristles typically arranged; hairs pale brown to black, especially abundant on tibiae. Protibia with longest bristles about 5 times longer than tibial width.

Wings. Veins medium/dark brown; membrane faint light brown in oblique view, darker in some Kurile Island specimens. DCI = 0.32-0.41; cell M3 open. Halter yellow; knob with a light brown, often diffuse patch.

Abdomen. Male. Tergite basal colour shining dark brown/black, without tomentum over most of tergites. Very narrow bands of thin grey tomentum hardly cover more than the intersegmental membrane of each tergite apically; segment 1 is thinly covered except mid-dorsally. Ventrolateral areas are narrowly covered with thin grey tomentum. About 5-7 pale bristles on each side of tergite 1; rarely a few of these are black. Lateral setae prominent on all segments, those on tergites 1-4 white, on 5-7 gold/brown; in some specimens the brown hairs extend basally to tergite 3. In specimens from Buryatskaya ACCP white hairs are more extensive. Dorsal setulae long, hair-like, dense and more-or-less erect; white on 1-2, pale brown on 3-7. Sternite tomentum grey, hairs white on 1-4, gold/brown on 5-7. Some specimens from the Kurile Islands have all abdominal hairs brown.

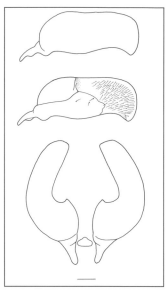

Figure 91. *L. hinei* epandrium (top to bottom): lateral, medial, dorsal. Scale lines = 0.2 mm.

Female. Tomentum covers tergites completely; gold-brown area generally in basal half or restricted to basolateral patches, the grey or gold-grey apical bands covering 0.4-0.7 the tergite length and sometimes extending mid-dorsally to the tergite base. The boundary between the grey and brown is sometimes obscure, the grey predominating. Extensive grey ventrolaterally. Erect white hairs on 1-3, white ventrally on 4 but dark dorsolaterally, all dark on 5-7. Dorsal setulae erect, dark on 2-7, white on 1 and on base of 2. Sternites grey, hair white, brown or white on 7.

Male genitalia. Epandrium and hypandrium/gonocoxite complex ferruginous to dark brown (lightest on hypandrium) and covered with grey tomentum except on hypandrium. Setal brush only 1-2 rows, sparse, very long setae laterally, dark brown/black; other setae gold/brown, numerous, prominent, especially basally on epandrium. Width of epandrium halves in lateral view about 0.4-0.45 times the length, widest in the basal third where dorsal margin is slightly convex. Apex rounded dorsally; ventrally angular and down turned. Medial face of epandrium as in figure 91. In dorsal view, medial margins of epandrium strongly concave; basal sclerite strong.

Gonostylus. Medial flange a rounded lobe; dorsal flange short, linked to prominent lateral tooth by apical ridge; ventrolateral tooth absent. Hypandrium/gonocoxite complex in ventral view with length about 0.75 times the width, the transverse slit at 0.4 the length. In lateral view, exposed length of gonocoxal apodeme about 0.4 times the basal width of hypandrium; apodeme with small sclerotized web ventrally.

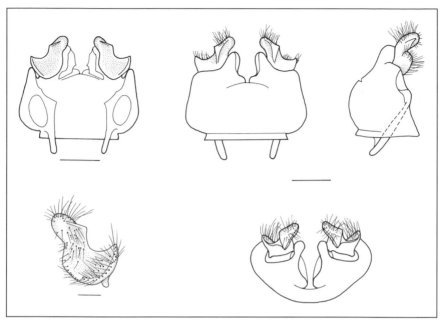

Figure 92. Hypandrium/gonocoxite complex. Clockwise from top left: dorsal, ventral, lateral and apical (scale lines = 0.3 mm); gonostylus, dorsal (scale line = 0.1 mm).

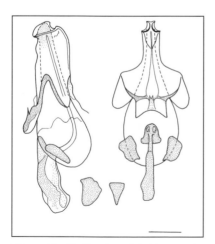

Figure 93. Phallus: lateral (left); ventral (right); lateral ejaculatory process, basal (centre left); ejaculatory apodeme, cross-section (centre right). Scale line = 0.2 mm.

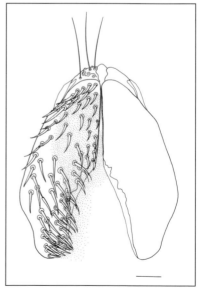

Figure 94. Subepandrial sclerite: ventral. Scale line = 0.1 mm.

Phallus. Paramere sheath dorsally 0.4 times the length of phallus (excluding ejaculatory apodeme). Ventral process flat, short and projecting ventrally; small projections basally. Apex of paramere sheath with ventrolateral carina and small ventral flange. Lateral process very short. Sperm sac width in dorsal view 0.35 times the length of phallus. Ejaculatory apodeme curved dorsally in lateral view, weakly spatulate in ventral view; triangular in cross-section, flattened ventrally with thick dorsal carina. Subepandrial sclerite as in figure 94. Broad triangular unsclerotized area in basal 0.55, the margins of the plates narrowly and irregularly unsclerotized; a narrow unsclerotized portion in central 0.3; spines slender and attenuate, most dense apically and basally.

Female genitalia. Undissected: Hairs abundant, erect, long and light to medium brown (white in Buryatskaya Oblast specimens). Tergite 8 dark brown/black, often narrowly brown apically. Sternite 8 normally ferruginous/chestnut; variable amounts of darker brown basally and laterally; hypogynial valves moderately haired. Lateral lobe setae strong. Cerci brown with golden setae. Dissected sternite 8 (figure 95) with basal width about 0.7 times the length; undivided but weak along midline. Length of unsclerotized area between hypogynial valves 0.4 times the length of sternite. Lateral lobe setae strong.

Tergite 9 sclerites as in figure 96; sternite 9 Y-shaped, medially undivided. Tergite 10 brown/black usually with 7-9 black acanthophorite spines on each side. Spermathecae (figure 97) with straight or slightly curved terminal reservoirs slightly narrower than reservoir duct; terminal reservoirs without wart-like protuberances on surface. Basal striated duct with fine scales; junction with basal duct strongly sclerotized, golden with scales; basal duct short.

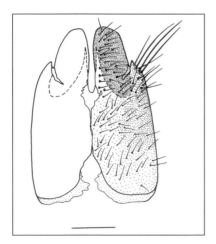

Figure 95. *L. hinei* female sternite 8: ventral. Scale line = 0.3 mm.

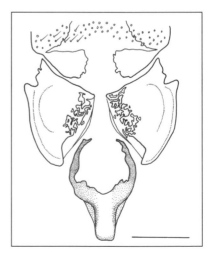

Figure 96. Female sternite 9, tergite 9 and lobes of sternite 10: ventral. Scale line = 0.2 mm.

Figure 97.
Spermatheca: dorsal.
Scale line = 0.1 mm.

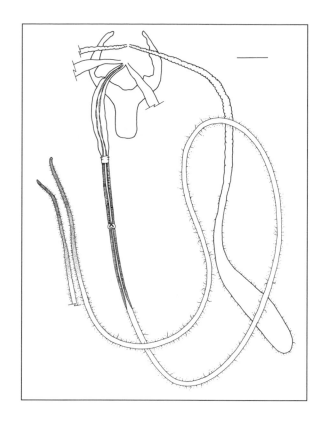

Variation. There is significant variation in setae colour in *L. hinei* across its extensive range. Specimens from Buryatskaya ACCP south of Lake Baikal in central Asia are paler than normal; hairs on the apical abdominal segments and on the apical parts of the femora are largely white. Conversely, many flies from Kunashir in the Kurile Islands are unusually dark. Hairs all over the body, but most noticeably those on the legs and abdomen, that are white in most other specimens are brown in *L. hinei*. This variation is a good example of the tendency for species to be pale on sunny, warm grasslands and dark in cooler, marine-influenced habitats.

Type Material.
HOLOTYPE. ♂ (examined) labelled: "[rectangular beige label] Katmai/Alaska/July '17";"[rectangular white label] Jas S Hine/Collector";"rectangular red label] HOLOTYPE/Lasiopogon/hinei/Cole and Wilcox" OSUC.
ALLOTYPE. ♀ (examined). Same data as holotype (OSUC).
PARATYPES. (8 examined). **U.S.A.: Alaska**, Katmai, vii.1917, J.S. Hine (2♂, 4♀, OSUC; 1♂, 1♀, CASC).

Other Material Examined (313 specimens).
CANADA: Alberta, Edmonton, Rainbow Valley, 18.vi.1963, S. Adisoemarto (3♂, 4♀,
UASM); Edmonton, Whitemud Park, 17.vi.1963, S. Adisoemarto (1♂, 1♀, UASM);
Rocky Mountain House, 29.vi.1963 (1♂, UASM). **British Columbia**, Alaska Hwy, km
281.6, Buckinghorse Provincial Campground, 27.vi.1978, P.H. Arnaud, Jr(1♀, CASC).
Yukon, Alaska Hwy. km 1787, Sakiw Creek, 61°29'N x 139°16'W, 9.vi.1979
(ROME#791008b), ROME field party (1♀, ROME); Carcross, sand dunes, 16-
18.vii.1982, G. and M. Wood (2♀, CNCI); Dempster Hwy., mi 40, 13-18.vii.1973, G.
and D.M. Wood (2♀, CNCI); Dempster Hwy., km 140.5 (mi 87), 1-4.vii.1973, G. and
D.M. Wood (3♂, 1♀, CNCI), 1-4.vii.1973, G. and D.M. Wood (1♂, BCPM), 5-
7.vii.1973, G. and D.M. Wood (1♂, CNCI), 5-7.vii.1973, G. and D.M. Wood (1♀,
BCPM), 8-12.vii.1973, G. and D.M. Wood (3♂, CNCI), 8-12.vii.1973, G. and D.M.
Wood (1♂, BCPM), 8-12.vii.1973, G. and D.M. Wood (1♂, IBPV), 16-17.vii.1973, G.
and D.M. Wood (3♂, 1♀, CNCI), 16-17.vii.1973, G. and D.M. Wood (1♂, BCPM), 21-
23.vii.1980, G. and M. Wood (1♂, 1♀, CNCI), 27-29.vii.1980, Lafontaine and Wood
(1♀, CNCI), 11-13.vii.1981, D. Lafontaine and G. and M. Wood (5♂, CNCI), 17-
18.vii.1981, D. Lafontaine and G. and M. Wood (5♂, 1♀, CNCI), 17-18.vii.1981,
D.Lafontaine and G. and M. Wood (1♀, BCPM), 17-18.vii.1981, D. Lafontaine and G.
and M. Wood (1♀, IBPV); Dempster Hwy. km 141, 24-28.vi.1982, G.and M. Wood (1♂,
CNCI); Dempster Hwy. km 379, Eagle River, 66°27'N x 136°45'W, 6.vi.1979
(ROME#791101), ROME field party (1♀, ROME); Dempster Hwy. km 200, Ogilvie
River, 65°22'N x 138°19'W, 29-30.vi.1979 malaise (ROME#791077), ROME field party
(1♀, BCPM); Dempster Hwy., Eagle River, 9.vii.1985, S.G. Cannings (1♀, SMDV);
Klondike Hwy. km 466, Pelly Crossing campground, 62°49'N x 136°45'W, 14-17.vi.1979
malaise (ROME#791038a), ROME field party (1♂, 5♀, ROME); Klondike Hwy. km
382, Tachun Creek campground, 62°16'N x 136°17"W, 14.vi.1982, ROME field party
(1♀, ROME).

   RUSSIA: Altaiskiy Kray, Kosh-Agach, 1750m, 16.vi.1964, Grunin (1♀, IBPV).
Amurskaya Oblast, Baikal-Amur Railway, NW Amurskaya Obl., 12.vii.1975, Petrova
and Soboleva (1♂, 1♀, IBPV); Beytonovo, near Dzhalinda, 2.vi.1915, Popov (1♀,
IBPV); Khorogochi [75km W Tynda], 3.vii.1975, B. Petrova (1♂, 1♀, IBPV);
Klimoutsy, 40km W Svobodnyy, 16.v.1959, Zinov'ev (1♂, IBPV), 19.vi.1959, Kerzhner
(1♀, IBPV); 27.v.1959, Borisova (1♀, IBPV); Malaya Pera R. at Semenovka, 3.vii.1975,
P. Lehr (1♀, IBPV); Natal'ino [70km N of Blagoveshtshensk], 6.vi.1975, P. Lehr (1♂,
IBPV); Nyukzha R., 16.vii.1975, Bogrovayu (1♀, IBPV). **Arkangelskaya Oblast**,
Malozemel'skaya Tundra, 2.vii.1934, Ahrens (1♀, IBPV). **Buryatskaya ASSR**,
Khoronkhoi [10km N Naushki], 19.v.1980, P. Lehr (1♂, BCPM), 20.v.1980, P. Lehr (1♀,
IBPV), 23.v.1980, P. Lehr (6♂, 5♀, IBPV; 1♂, 2♀, BCPM), 24.v.1980, P. Lehr (2♂,
1♀), 25.v.1980, P. Lehr (2♀, IBPV); Ust-Zaza, Vitim R., Rasnichin (1♂, 2♀, IBPV).
Chitinskaya Oblast, Zabaykal'sk, 10.vi.1961, Redikortsev (1♀, IBPV). **Kamchatskaya
Oblast**, Kosrevsk, 8km S, 16.vii.1915 (1♂, IBPV). **Khabarovskiy Kray**, Gorin R.
mouth at Amur R., 16.vi.1988, V. Mutin (2♀, IBPV), 17.vi.1988, V. Mutin (1♂, IBPV),
18.vi.1988, V. Mutin (5♂, 2♀, IBPV); Bureinskiy Mtns, Suluk R. mouth, 1.vii.1988, V.
Mutin (1♀, IBPV), 7.vii.1988, V. Mutin (3♀, IBPV); Bureinskiy Mtns, Suluk R., middle
part, 6.vii.1988 (1♂, IBPV); Amur R., left shore near Savinskoye, 8.viii.1991, P. Lehr
(1♀, IBPV). **Krasnoyarskiy Kray**, Minusinska, 11.v.1912, Sushkin and Redikortsev
(1♂, IBPV); Nizhnyaya Tunguska R., 26.vi.1873, Chekanovskiy (1♂, IBPV).
Primorskiy Kray, Anisimovka, 3.vi.1977, P. Lehr (1♀, IBPV); Anisimovka, 2km W,

27.v.1974, A. Lelej (1♀, IBPV), 28.v.1974, A. Lelej (8♂, 8♀, IBPV [includes holotype of *L. sibiricus*]; 1♂, 2♀, BCPM); Anuchino, 10km W, 7.vi.1986, A. Lelej (1♀, IBPV), 8.vi.1986, A. Lelej (1♂, 2♀, IBPV); Barabash-Levada, 50km W Khanka L., 6.vi.1980, A. Lelej (1♀, IBPV); Brovnichi, Serebryanka R., 15.vi.1975, P. Lehr (2♀, IBPV; 1♀, BCPM); Dzhigitovka R., 23.vi.1979, P. Lehr (1♂, 1♀, IBPV; 1♂, BCPM), 4.vi.1985, P. Lehr (3♀, IBPV), 5.vi.1985, P. Lehr (1♀, IBPV); Khasaniskiy District, Zanadvorovka, 22.vi.1987, P. Lehr (2♀, IBPV), P. Lelej (1♂, IBPV), 23.vi.1987, P. Lehr (3♀, IBPV); Lazovskiy Reserve [near Kievka], 1.vi.1980, T. Oliger (1♂, IBPV), 10.vi.1980, A. Egorov (1♀, IBPV); Margaritovka R., 15km SSW Margaritovka, 14.vi.1986, A. Lelej (1♂, 3♀, IBPV; 1♀, BCPM); Novomikhailovka, Ussuri R., 3.vi.1985, P. Lehr (1♂, 1♀, IBPV), 9.vi.1986, P. Lehr (1♀, IBPV); Milogradovo, 14.vi.1986, P. Lehr (1♀, IPBV); Novovarvarovka, 10 km W Anuchino, 7.vi.1986, P. Lehr (3♂, IBPV); Sikhote-Alinskiy Reserve, 21.vi.1979, P. Lehr (1♀, IBPV); Tigrovogo, 17.vi.1975, A. Lelej (1♀, IBPV); Tigrovogo, 15km NE, 15.vi.1975, A. Lelej (1♂, IBPV); Tigrovogo, 3km NE, Serebryanka R. 15.vi.1975, A. Lelej (2♀, IBPV); Tigrovogo, 5km N, 16.vi.1975, Lehr (1♀, IBPV); Ussuriyiskiy Reserve, 3.vii.1985, P. Lehr (1♀, IBPV), 3.vi.1989, Lelej (1♀, IBPV), 23.vi.1990, P. Lehr (1♂, IBPV), 4.vi.1991, P. Lehr (4♂, 4♀, IBPV), 14.vi.1991, P. Lehr, 1♂, 3♀, IBPV), A. Lelej (3♂, 1♀, IBPV; 3♂, 2♀, BCPM); 19.vi.1991, P. Lehr (1♂, IBPV), 29.vi.1991, A. Lelej (1♀, IBPV), 1.vii.1991, A. Lelej (2♀, IBPV), 4.vii.1993, R.A. Cannings (2♀, BCPM); Varvarovka, 7.vi.1985, P. Lehr (2♂, IBPV); Yakolevka, 21.vi.1926 (1♂, 1♀, IBPV). **Sakhalinskaya Oblast, Kurile Is.**, Kunashir Is., Lagunnoe, 4.viii.1974, P. Lehr (2♂, 1♀, IBPV; 1♂, 1♀, BCPM), 5.viii.1974, P. Lehr (2♂, 3♀, IBPV; 1♂, BCPM); Neva, 23.vii.1984, Basarukin (1♂, IBPV), 24.vii.1984, Basarukin (3♂, 1♀, IBPV; 1♂, BCPM); Yuzhno-Kuril'ska, Lesnaya R., 25.vii.1989, Lelej (1♀, IBPV; 1♂, 1♀, BCPM); Stolochatyy, 28.vii.1989, A. Lelej (1♂, 2♀, IBPV); Tretyakovo, 10.viii.1974, P. Lehr (1♀, IBPV), 11.viii.1974, P. Lehr (1♀, IBPV). **Sakhalin Is.**, Yuzhno-Sakhalinsk, Vintis R., 5.vi.1981, B. Kuznetsov (2♀, IBPV). **Yakutskaya ASSR**, Lena R. right side at Byosyuke R., 70°5'N, 17.vii.1984, P. Lehr (1♀, IBPV); Verkhoyansk, vi.1903, Rozhnovskiy (1♀, IBPV); Yakutsk, 9.vi.1928 (1♂, IBPV); Zhigansk (Lena R.), 28.vi.1875, Chekanovskiy (1♀, IBPV). **Magadanskaya Oblast**, Bulun River near Karkadon, 29.vi.1982, P. Lehr (3♂, 5♀, IBPV; 1♂, BCPM), 30.vi.1982, P. Lehr (1♂, IBPV), 3.vii.1982, P. Lehr (1♂, IBPV), 4.vii.1982, P. Lehr (1♂, 4♀, IBPV; 2♂, 1♀, BCPM), 5.vii.1982, P. Lehr (2♂, 7♀, IBPV; 2♂, 4♀, BCPM) (paratypes of *L. sibiricus*); Debin, 19.vi.1963, Zhelokhovtsev (1♀, IBPV), 22.vii.1963, Zhelokhovtsev (1♀, IBPV); Kolymskaya Rd. at Arga-Yurakh R., 18.vii.1993, R.A. Cannings (1♀, BCPM); Palatka, 13.vii.1972, K. Elberg (1♀, IBPV). **U.S.A.: Alaska**, Alaska Hwy., Moon Lake, 8.vii.1978, P.H. Arnaud, Jr (2♂, CASC); Anchorage, 15.vi.1921, J.M. Aldrich (1♀, EMEC); Dalton Hwy., mp272, small lake on tundra, 68°26'Nx149°25'W, 6.vii.1982, R. Jaagumagi (1♀, ROME); Elliott Hwy., Fox, 17.1km N, Oines Pond, 1.vii.1996, P.H. Arnaud, Jr and M.M. Arnaud (1♀, CASC); Fairbanks, 30.vi.1921, J.M. Aldrich (1♂, EMFC; 1♀, EMEC), 4.vii.1921, J.M. Aldrich (1♀, EMEC); Glenn Hwy., Chugach Mtns, Glacier Park Resort, Matanuska Glacier, 19.vii.1996 (1♀, CASC); Heal, 26.vi.1921, J.M. Aldrich (1♀, EMEC), 27.vi.1921, J.M. Aldrich (1♀, EMEC); King Salmon, Naknek R., 19.vii.1952, W.R. Mason (1♀, CNCI); Mount McKinley Nat. Park, Horseshoe Lake, 470m, 14.vii.1978, P.H. Arnaud, Jr (3♀, CASC); Nogahabara Dunes, 65 mi N Galena, 23-30.vi.1989, M. Polak and D.M. Wood (1♂, 5♀, CNCI); Richardson Hwy., Donnelly Cr., 15.vii.1985, S.G. Cannings (1♂, SMDV); Steese Hwy., Arctic Circle Hotsprings at Circle Hotsprings, 13.4km from

Central, 900m, 6-7.vii.1989, P.H. Arnaud, Jr (1 ♀, CASC); Taylor Hwy., Walker Fork Campground, 7.vii.1978, P.H. Arnaud, Jr (1 ♂, CASC), 17.vi.1996, P.H. Arnaud, Jr and M.M. Arnaud (2 ♂, CASC; 1 ♂, BCPM), 18.vi.1996, P.H. Arnaud, Jr and M.M. Arnaud (1 ♂, 2 ♀, CASC; 1 ♀, BCPM).

Type Locality. U.S.A.; Alaska, Katmai.

Taxonomic Notes. *L. sibiricus* Lehr, the name originally used for the Eurasian populations, was synonymized with *L. hinei* by Cannings (1997).

Etymology. Named after J.S. Hine (1866-1930), prominent dipterologist and professor at Ohio State University, who collected the type series.

Distribution. Holarctic (Palaearctic-East Beringian). In the Palaearctic the species ranges in the Russian taiga and tundra east from Arkangelskaya Oblast (known as far north as about 68°N on the Malozemel'skaya Tundra and 70°5'N on the Lena River) to Kamchatka and Chukotka on the Pacific Ocean, south in Siberia to the Selenga River on the Mongolian Border, east to southern Primorskiy Kray (43°N), Sakhalin Island and Kunashir in the Kurile Islands. Its presence on southern Sakhalin and Kunashir suggests that it might occur on the island of Hokkaido in Japan. In the Nearctic it ranges from eastern and north-central Alaska (68°26'N on the Dalton Highway) east to the Yukon Territory, south to northeastern British Columbia and central Alberta (52°N). Map 4.

Phylogenetic Relationships. Member of the *hinei* species group; sister species to *L. kjachtensis.*

Natural History. Habitat: often near streams in northern forests; in grasslands and sand dunes, or on paths, stones and logs in forests. *L. hinei* extends into the southern tundra along the northern edge of its range and into temperate forest at its southern limits.

Most Alaskan and Yukon habitats range from open, dry Lodgepole Pine woods on sandy soil to riverbanks in moist riparian woods of willow, spruce and poplar; its most common congener in these places is *L. canus.* Most of the Yukon specimens of *L. hinei* come from along the Dempster Highway in the Ogilvie Mountains and on the Eagle Plain, but it also ranges to the south (Yukon/Tintina region, Shakwak Trench, Southern Lakes and into northeastern British Columbia and Alberta. Around Edmonton, Alberta, near the southern limit of its range in the New World, *L. hinei* lives along riverbanks in mixed grassland and aspen, cottonwood and spruce woods; *L. quadrivittatus* and *L. boreas* share these sites. In the Ussuriyiskiy Reserve north of Vladivostok, near the southern limit of its known distribution in Asia, I collected *L. hinei* along streams in woodlands dominated by Japanese Elm (*Ulmus davidiana* Planch.) and Ostrich Fern (*Matteuccia struthiopteris* (L.) Todaro). It flies there with *L. lehri* sp.nov. Farther north in the

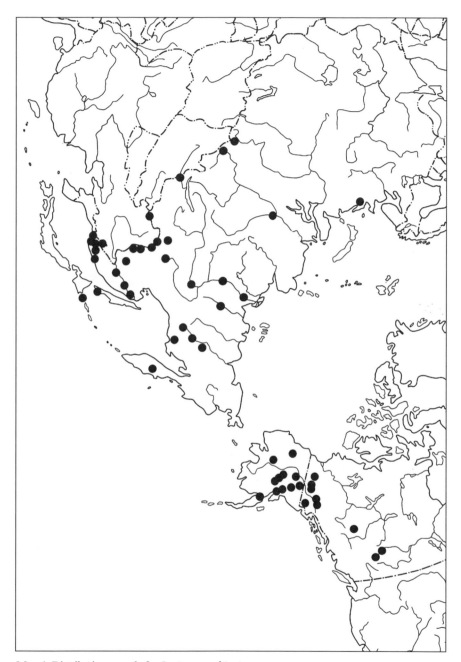

Map 4. Distribution records for *Lasiopogon hinei*.

Kolyma District of the Magadanskaya Oblast, along rivers flowing through a region of mixed larch taiga and tundra, *L. hinei* flew with *L. septentrionalis*.

Recorded flight period in Yukon is 6 June to 29 July; 18 of 23 records are from July. In Siberia dates in the literature are 11 May-5 August. *L. hinei* is the most widespread *Lasiopogon* species in Siberia and Far-eastern Russia. Based on its tolerance of a wide variety of habitats, Lehr (1984a) called *L. hinei* an ecological analogue to *L. cinctus*, which replaces it in northern Europe.

Prey associated with specimens examined include Diptera (Tipulidae and Muscidae) and Ephemeroptera.

Lasiopogon kjachtensis Lehr

Lasiopogon kjachtensis Lehr, 1984. *Zoologicheskii Zhurnal* 63: 696-706 (p. 703) (IBPV).

Diagnosis. A medium-sized, grey/gold-brown species with white setae and ferruginous genitalia, tarsi and leg joints. Mystax mixed white and light brown. Antennae brown, partly ferruginous; F1 long, narrow and tapered from base (LF2+3/LF1 = 0.68-0.82). Scutum grey with gold-brown highlights or more heavily infused with gold-brown; thorax laterally gold/brown-grey. Dorsocentral stripes dark brown, acrostichal stripes grey-brown, sometimes obscure. Thorax and legs setose; anterior dorsocentral bristles 5-7. Abdomen with thin grey tomentum over most of tergites, the black cuticle shining through, especially basally, the tomentum concentrated apicolaterally. Lateral areas broadly grey. All hairs white, abundant, long; bristles on tergite 1 pale. Epandrium ferruginous, the width about 0.4 times the length in lateral view, apex broadly rounded dorsally. In dorsal view strongly concave medially. Female with gold-grey apical tomentum merging into gold-brown basal tomentum. Tergite 8 black, apex ferruginous; sternite 8 brown basally, ferruginous apically. Hairs white, acanthophorites and spines ferruginous.

Description. Body length ♂ 7.7-9.6mm; ♀ 8.6-10.1mm.

Head. HW ♂ 1.80-2.04mm; ♀ 1.90-2.02mm. FW ♂ 0.40-0.46mm; ♀ 0.40-0.46mm. VW ♂ 0.72-0.84mm; ♀ 0.71-0.83mm. VW/HW = ♂ 0.40-0.42; ♀ 0.35-0.41. FW/VW = ♂ 0.50-0.57; ♀ 0.53-0.62. VD/VW = ♂ 0.17-0.19; ♀ 0.17-0.20. GH/GL = ♂ 0.30-0.31; ♀ 0.28-0.33.

Face and vertex gold/brown-grey, occiput grey, gold/brown-grey dorsally. Beard and labial hairs white; mystax bristles white mixed with brown. All other setae white, except strong ocellars often brown/black. Occipital bristles abundant, moderately strong; longest occipitals to 0.6mm (shorter than F1+F2+3), those behind the dorsomedial angle of the eye strongly curved anterolaterally; lateral and ventral ones shorter, straighter. Frontal and orbital setae rather short, about 0.3-0.4mm long.

Antennae. Brown; pedicel, base of F1 and most of F2+3 usually chestnut/ferruginous. Setae white; no setae on F1. F1 long and tapered, widest at base; WF1/LF1 = ♂ 0.25-0.33; ♀ 0.29-0.30. LF2+3/LF1 = ♂ 0.70-0.79; ♀ 0.65-0.82.

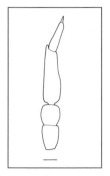

Figure 98. Antenna.
Scale line = 0.1 mm.

Thorax. Prothorax gold/brown-grey, hairs white. Postpronotal lobes gold/brown-grey, the lateral angle ferruginous/chestnut, hairs white. Scutum tomentum gold/brown grey. Dorsocentral stripes dark brown, changing to medium brown/gold at some angles of view; most gold tomentum concentrated near dorsocentral area. Acrostichal stripes brown-grey, distinct to obscure. Intermediate spots medium brown, faint. All strong bristles white/gold with a few pale brown; finer setae white. A strongly hirsute species. Dorsocentrals abundant, anterior dorsocentrals 6-7 (longest to 0.8mm), mixed with many slightly shorter setae; 4-5 posteriors. Notal setae abundant, long, to 0.2mm. Postalars 3-4, with shorter setae; supra-alars 2-3 with shorter setae; presuturals 2-3, posthumerals 1. Scutellar tomentum grey with gold highlights; apical scutellar bristles pale, abundant, 4-7 strong ones on each side mixed with shorter, weaker bristles and hairs.

Pleural tomentum gold/brown-grey; all setae white. Katatergite setae 7-11 among a few finer white hairs; katepisternal setae long, sparse. Anepisternal setae 4-7 (to 0.5mm); a patch of erect setae on dorsal shelf. Anepimeron without hairs.

Legs. Base colour dark brown/black; trochanters brown/ferruginous, shining. Joints of femora and tibiae ferruginous, otherwise these segments brown. Tarsal segments ferruginous basally, usually light/medium brown apically. Tomentum of coxae and rest of legs gold-grey. No coxal peg. Main bristles white/gold, all hairs white. Femora dorsally with decumbent hairs; longer erect pale hairs abundant ventrally and laterally; in males longest white ventral setae as long as, or longer than, width of femur. Profemur with 10-15 fine dorsolateral bristles; mesofemur with 5-6; metafemur with 10-15, all mixed with long hairs. Tibiae and tarsi with bristles typically arranged; protibia with longest bristles about 3.5 times longer than tibial width.

Wings. Veins yellow to medium brown; when present, yellow most prominent basally; membrane very pale yellow brown in oblique view. DCI = 0.37-0.50; cell M3 open. Halter yellow; knob without dark spot.

Abdomen. Male. Tergite basal colour dark brown/black. Thin grey tomentum covers most of tergites, concentrated apicolaterally, the black cuticle shining through, especially basally. Lateral areas broadly grey. All hairs white, abundant, long; about 6-7 pale bristles on each side of tergite 1. Dorsal setulae long, appressed. Sternite tomentum gold-grey, hairs white.

Female. Cuticle at tergite apices ferruginous. Gold-grey tomentum on apical part of tergites grades into gold-brown basal tomentum; ventrolateral areas grey/brown-grey. Lateral hairs white; dorsal setulae white, erect, very short.

Male genitalia. Epandrium and hypandrium/ gonocoxite complex brown basally, ferruginous apically, with sparse gold-grey tomentum. Setal brush only 1-2 rows, sparse, very long laterally, golden; other setae abundant, white, long. Width of epandrium halves in lateral view about 0.4 times the length, the ventral and dorsal margins parallel, apex broadly rounded dorsally, more angular ventrally. Medial face of epandrium as in figure 99. In dorsal view, medial margins of epandrium strongly concave; basal sclerite absent.

Gonostylus. Setose; medial flange a low, flattened ridge; dorsal flange short, linked to prominent lateral tooth by apical ridge; ventrolateral tooth present. Hypandrium/gonocoxite complex in ventral view with length about 0.7 times the width, the transverse slit at 0.75 the length. Gonocoxal apodemes short, in lateral view, exposed length about 0.3 times the basal width of hypandrium; apodeme without sclerotized web ventrally.

Phallus. Paramere sheath dorsally 0.35 times the length of phallus (excluding ejaculatory apodeme). Ventral process with small projections basally. Apex of paramere sheath with ventrolateral carina and small ventral flange. Sperm sac width in dorsal view 0.4 times the length of phallus. Ejaculatory apodeme curved dorsally in lateral view, weakly spatulate in ventral view; triangular in cross-section, flattened ventrally with thick dorsal carina. Subepandrial sclerite as in figure 102. Broad triangular unsclerotized area in basal 0.50, the margins of the plates irregularly unsclerotized; a very narrow unsclerotized portion in central 0.3; spines slender and attenuate, most dense apically and basally.

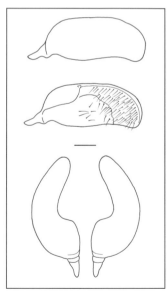

Figure 99. *L. kjachtensis* epandrium (top to bottom): lateral, medial, dorsal. Scale lines = 0.2 mm.

Female genitalia. Undissected: Hairs white, erect. Tergite 8 dark brown/black, narrowly ferruginous apically. Sternite 8 ferruginous apically, brown basally, the hypogynial valves with hairs ventrally. Lateral lobe setae moderately

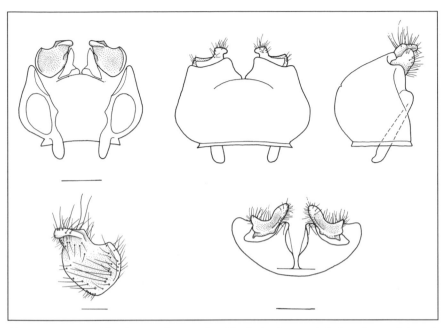

Figure 100. Hypandrium/gonocoxite complex. Clockwise from top left: dorsal, ventral, lateral and apical (scale lines = 0.3 mm); gonostylus, dorsal (scale line = 0.1 mm).

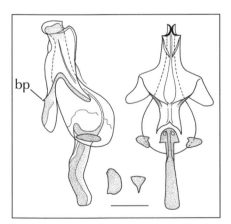

Figure 101. Phallus: lateral (left); ventral (right); lateral ejaculatory process, basal (centre left); ejaculatory apodeme, cross-section (centre right). bp = basal process. Scale line = 0.2 mm.

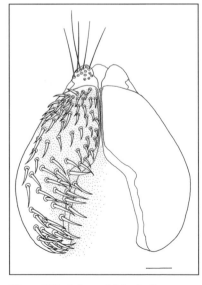

Figure 102. Subepandrial sclerite: ventral. Scale line = 0.1 mm.

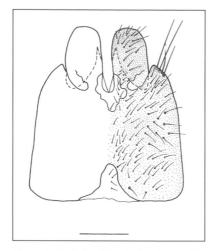

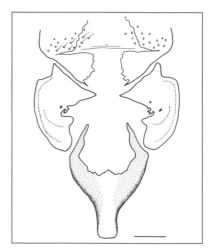

Figure 103. *L. kjachtensis* female sternite 8: ventral. Scale line = 0.3 mm.

Figure 104. Female sternite 9, tergite 9 and lobes of sternite 10: ventral. Scale line = 0.1 mm.

strong. Cerci ferruginous/brown with pale setae. Dissected sternite (figure 103) very flat, the basal width about 0.9 times the length, undivided along midline. Length of unsclerotized area between hypogynial valves 0.3 times the length of sternite.

Tergite 9 sclerites as in figure 104; sternite 9 Y-shaped, medially undivided. Tergite 10 ferruginous, usually with 7-9 ferruginous acanthophorite spines on each side. Spermathecae (figure 105) very long (about 5.75mm) with hooked terminal reservoirs about as wide as reservoir duct; terminal reservoirs without wart-like protuberances on surface. Basal striated duct without fine scales; junction with basal duct sclerotized, golden, and with scales; basal duct short.

Type Material.
HOLOTYPE. ♂ (examined) labelled [in Russian]: "[rectangular white label] Buryatskaya ASSR/Khoronkhoi station, 20.v.1980/Lehr";"[rectangular red label] Holotypus 1981/Lasiopogon/kjachtensis Lehr". IBPV.
PARATYPES. (172 examined). **Russia: Buryatskaya ASSR**, Khoronkhoi Station, 18.v.1980, P. Lehr (10♂, 8♀, IBPV; 3♂, 1♀, BCPM), 19.v.1980, P. Lehr (12♂, 12♀, IBPV; 1♀, BCPM), 20.v.1980, P. Lehr (14♂, 15♀, IBPV; 1♂, 2♀, BCPM), 21.v.1980, P. Lehr (13♂, 12♀, IBPV; 3♂, BCPM), 22.v.1980, P. Lehr (3♂, 4♀, IBPV; 1♀, BCPM), 23.v.1980, P. Lehr (2♂, 12♀, IBPV; 1♀, BCPM), 24.v.1980, P. Lehr (8♂, 13♀, IBPV; 1♂, 2♀, BCPM), 25.v.1980, P. Lehr (2♂, 15♀, IBPV); Kjachta (formerly Troitskosavsk), 20.v.1928, Th. Lukjanovich (2♂, IBPV).

Figure 105.
Spermatheca: dorsal.
Scale line = 0.1 mm.

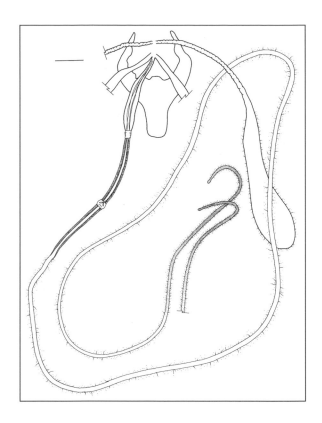

Type Locality. Russia; Buryatskaya ASSR, Khoronkhoi station. This locality is near the Selenga River north of the Russian/Mongolian border.

Etymology. From Kjachta, a town on the Russian/Mongolian border, near where the first collection of the species was made (not type locality).

Distribution. Palaearctic; Russia, Buryatskaya ASSR (Selenga R. region, south of Lake Baikal). See map 5, p. 155.

Phylogenetic Relationships. Member of the *hinei* species group; sister species to *L. hinei*.

Natural History. Habitat: bare, sandy areas, from riversides to crests of hills. Lehr (1984a) noted that the species hunts on bare, usually sandy, soil. At night and in bad weather the flies shelter under dead, dry plants; on warm days they begin to hunt at 08:00-09:00 when the temperatures are around 10-15°C (night-time lows about 8°C). At these sites *L. kjachtensis* is preyed on by larger robber flies of the genus *Tolmerus*. Dates of collections range from 18 to 24 May.

Lasiopogon lavignei sp. nov.

Diagnosis. A large grey species with only faint grey-gold tomentum laterally and light brown patches of tomentum basally on dorsum of abdominal segments. Antennae brown, F2 paler than rest, some white hairs mixed with brown on scape, ratio F2+3/F1=0.47-0.64. Dorsocentral stripes narrow, brown and gold, rather weak; acrostichal stripes brown-grey, often obscure. Mystax and most prominent bristles and setae black or dark brown, except for katatergites, which are predominantly pale; anterior dorsocentral setae 1-5, often weak. Tarsi and sometimes tibiae brown, contrasting with dark femora. Epandrium halves in dorsal view only slightly concave medially; in lateral view greatest width about 0.65 times the length; broadly rounded apically.

Description. Body length ♂ 10.4-10.8mm; ♀ 11.7-11.8mm.

Head. HW ♂ 2.18-2.32mm; ♀ 2.16-2.46mm. FW ♂ 0.46-0.50mm; ♀ 0.50-0.56mm. VW ♂ 1.04-1.16mm; ♀ 1.12-1.20mm. VW/HW = ♂ 0.48-0.50; ♀ 0.49-0.52. FW/VW = ♂ 0.43-0.44; ♀ 0.45-0.47; VD/VW = ♂ 0.18-0.20; ♀ 0.20-0.22. Gibbosity height/length = ♂ 0.30-0.37; ♀ 0.38-0.40.

Face silver/white; vertex and occiput grey. Beard and labial hairs white; mystax and all major setae black or brown, with a few white setae below on mystax; occipitals strong, rather sparse (14-20 on each side) and shorter than usual (to 0.4mm), usually moderately curved anterolaterally, but sometimes almost straight; frontal and orbital setae sparse, rather strong, longest frontal about as long as scape, longest orbital about as long as scape+pedicel.

Antennae. Brown, lighter on F2. Setae brown, some white on scape; none on F1. F1 long, usually rather straight on dorsal margin, gently curved ventrally, widest in basal half or at midlength. WF1/LF1 = ♂ 0.28-0.30; ♀ 0.25-0.34. LF2+3/LF1 = ♂ 0.47-0.64; ♀ 0.48-0.62.

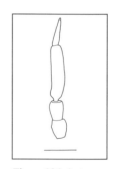

Figure 106. Antenna. Scale line = 0.2 mm.

Thorax. Prothorax grey, hairs white. Postpronotal lobes grey, the lateral angle yellow to ferruginous; setae mixed short, weak, brown bristles and fine white hairs. Scutum tomentum grey, faintly grey-gold laterally; dorsocentral stripes narrow, not strong, brown and gold, almost reaching scutellum; acrostichal stripes brown-grey, sometimes obscure. Most bristles and setae brown or black; dorsocentrals variable, rather weak and sometimes hardly represented anteriorly (longest to 0.8mm), prominent posteriorly (longest to 1.0mm), 1-5 anteriors (some only 2-4 times as long as surrounding notal setae), 1-5 posteriors. Acrostichals indistinguishable from surrounding notal setae, brown, very short, strongly directed posteriorly; scattered notal hairs similar to acrostichals.

Postalars 2, strong, with several short, weak setae; supra-alars 2-3; presuturals 2, sometimes a weak posthumeral. Scutellar tomentum grey, with some faint brown-gold highlights. Apical scutellar setae black, 2-4 on each side, angled laterally. No discal scutellar hairs, but some brown hairs and weak setae mixed with primary setae on margin.

Pleural and scutellar tomentum grey, with some faint brown-gold highlights. Katatergite setae 6-7, long (to 1.2mm), variable in colour from white to medium brown, but mostly white, with numerous white hairs on sclerite; katepisternal setae white, rather long (to 0.4mm); 4-9 setae (longest 0.5mm) at posterior edge of anepisternum and patch of dark, fine, short, setulae along dorsal margin usually dense, but sometimes as few as 6. Anepimeron without setae.

Legs. Base colour of trochanters, femora and tibiae dark brown, tarsi and sometimes tibiae light brown. Tomentum of coxae grey; tomentum on rest of legs sparse, grey. No coxal peg. All bristles black, typically arranged. Femora with short white decumbent hairs, the fine white ventral bristles sparse, the longest about as long as thickness of femur. Apical and dorsolateral bristles strong, scattered, short; profemur in male with 5-12 dorsolateral bristles, mesofemur with 3-5, metafemur with 8-9. Bristles on tibiae and tarsi rather short; protibia in male with longest bristle 2-3 times longer than tibial width. Hairs on tibiae and tarsi brown to white.

Wings. Veins light to medium brown; membrane pale brown when viewed obliquely. DCI = 0.0.30-0.35; cell M3 narrowly open. Halter pale yellow; knob without dark spot.

Abdomen. Tergites with apical band of grey tomentum, its width about 0.3-0.5 the length of the tergite, often extending weakly basally along midline. Basal tomentum brown. Tergite 1 with small brown patch basally. Lateral margins of tergites with grey tomentum. Tergite 1 basolaterally with 3-5 major bristles brown to white, mostly brown. In both sexes dorsal setulae white, long; lateral hairs white, rather short and sparse, longest on segments 1-2. Sternite tomentum grey; hairs white.

Male genitalia. Epandrium and hypandrium/gonocoxite complex dark chestnut with grey tomentum except on hypandrium, which is shining. Setal brush brown/black; other setae dark brown, prominent. Width of epandrium halves

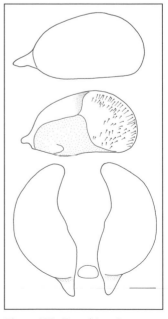

Figure 107. Epandrium (top to bottom): lateral, medial, dorsal. Scale lines = 0.2 mm.

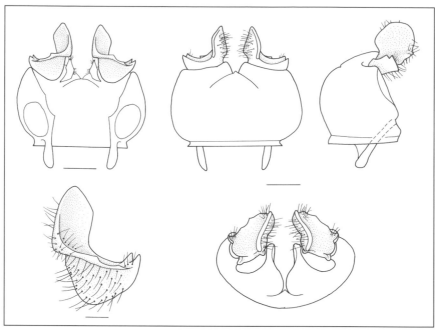

Figure 108. *L. lavignei* sp. nov. hypandrium/gonocoxite complex. Clockwise from top left: dorsal, ventral, lateral and apical (scale lines = 0.3 mm); gonostylus, dorsal (scale line = 0.1 mm).

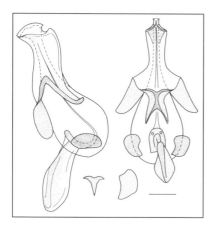

Figure 109. Phallus: lateral (left); ventral (right); lateral ejaculatory process, basal (centre left); ejaculatory apodeme, cross-section (centre right).
Scale line = 0.2 mm.

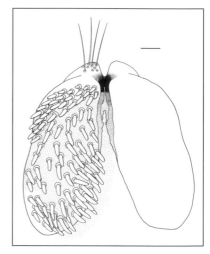

Figure 110. Subepandrial sclerite: ventral. Scale line = 0.1 mm.

in lateral view about 0.65 times the length, widest at midlength, apex broadly rounded. Medial face of epandrium as in figure 107. In dorsal view, medial margins of epandrium shallowly concave in apical half; basal sclerite prominent.

Gonostylus. Lateral and ventrolateral teeth strong; dorsomedial tooth moderately developed. Hypandrium/gonocoxite complex in ventral view the length about 0.7 times the width, the transverse slit almost non-existent, merely a seam, broadly v-shaped, its apical angle at about 0.75 the distance from base of hypandrium to apex of gonocoxite lobes. In lateral view, exposed length of gonocoxal apodeme about 0.5 times the basal width of hypandrium; apodeme with sclerotized web ventrally.

Phallus. Paramere sheath dorsally 0.4 times the length of phallus (excluding ejaculatory apodeme). Apex of paramere sheath with ventrolateral carina, the ventral lip strongly projecting; no ventral flange. Sperm sac width in dorsal view 0.35 times the length of phallus. Ejaculatory apodeme curved ventrally in lateral view, broadly spatulate in ventral view; flat in cross-section with strong dorsal carina. Subepandrial sclerite as in figure 110. Triangular unsclerotized area in basal 0.7; spines mostly parallel-sided, blunt, densely arranged, except centrally and basolaterally on each lobe.

Female genitalia. Undissected: Setae scattered, fine, erect, pale. Tergite 8 shining black with reddish brown apical band. Sternite 8 shining reddish brown, dark brown on apical lobes; hypogynial valves black; lateral lobe setae weak. Cerci

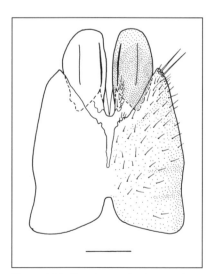

Figure 111. Female sternite 8: ventral.
Scale line = 0.3 mm.

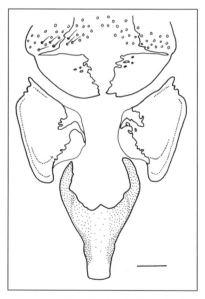

Figure 112. Female sternite 9, tergite 9 and lobes of sternite 10: ventral.
Scale line = 0.1 mm.

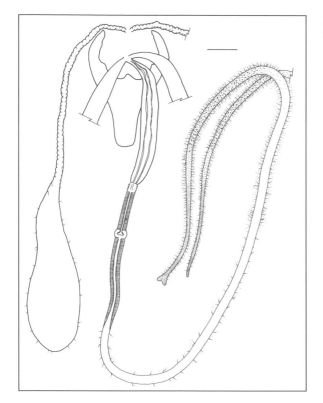

Figure 113. *L. lavignei* sp. nov. spermatheca: dorsal. Scale line = 0.1 mm.

yellow-brown with pale setae. Dissected sternite 8 (figure 111) with basal width about 0.8 times the length; undivided. Length of unsclerotized area between hypogynial valves 0.4 times as long as sternite. Hypogynial valves with medial and central carinae ventrally; a few short hairs basally.

Tergite 9 sclerites as in figure 112; sternite 9 Y-shaped, medially undivided. Tergite 10 brown/black with 6-7 black acanthophorite spines on each side. Spermathecae (figure 113) with gently curved terminal reservoirs, for most of their length only about half as wide as reservoir duct; bases of terminal reservoirs with weak wart-like protuberances on surface. Basal striated duct without fine scales; junction with basal duct with scales; basal duct moderately long.

Type Material
HOLOTYPE. (here designated) ♂ labelled: "[white rectangular label] WYO Washakie Co./8.4Mi E Ten Sleep/Leigh Creek C.G./VI-10, 1980"; "[white rectangular label] R. Lavigne/Collector. My Holotype label "HOLOTYPE/Lasiopogon ♂/lavignei Cannings/des. R.A. Cannings 2002 [red, black-bordered label]" has been attached to this specimen. Dissected genitalia in plastic vial underneath all. USNM.

PARATYPES. (3 designated) **U.S.A.: Wyoming**, Washakie Co., 8.4 mi E. of Ten Sleep, Leigh Creek C.G., 10.vi.1980, R. Lavigne (1♀, ESUW; 1♂, 1♀, USNM). My paratype labels and determination labels have been attached to these specimens.

Type Locality. U.S.A., Wyoming, Washakie Co., 8.4 mi. E. of Ten Sleep, Leigh Creek C.G.

Etymology. Named after Dr Robert Lavigne, Professor Emeritus at the University of Wyoming and collector of the type series, for his many contributions to the study of the Asilidae and his encouragement of my robber fly studies.

Distribution. Nearctic; U.S.A., mountains of western Wyoming. See map 12, p. 266.

Phylogenetic Relationships. Member of the *tetragrammus* species group and sister species to *L. quadrivittatus*.

Natural History. Habitat: Mountain forests near streams. Known flight date, 10 June.

Lasiopogon lehri sp. nov.

Diagnosis. A medium-sized to large grey and brown species. Mystax and other head bristles dark. Antennae brown, F1 long; F2+3/F1 = 0.44-0.60. Thoracic tomentum dorsally and laterally grey to brown/gold-grey with brown dorsocentral stripes and grey or brown-grey acrostichal stripes. Main thoracic and leg bristles dark; anterior dorsocentral bristles 4-5. Abdominal tergites basally with only thin brown tomentum; apical bands gold-grey, covering 0.25 to 0.4 the length of the segments. Bristles on tergite 1 dark; all lateral setae pale on tergites 1-4, mostly dark on 5-7; dorsal setulae dark. Epandrium dark chestnut/black, about 0.35-0.4 as wide as long in lateral view, the dorsal and ventral margins parallel; apex rounded; in dorsal view strongly concave medially. Female with abdominal pattern similar to that of male, the grey apical bands about 0.4-0.5 the length of the segments. Terminalia dark brown/black with light brown hairs; sternite 8 with black hypogynial valves parallel, the medial lobes ferruginous with pale brown hairs.

Description. Body length ♂ 9.2-11.3mm; ♀ 9.5-12.8mm.

Head. HW ♂ 1.94-2.32mm; ♀ 2.12-2.64mm. FW ♂ 0.38-0.48mm; ♀ 0.44-0.60mm. VW ♂ 0.92-1.08mm; ♀ 1.11-1.35mm. FW/VW = ♂ 0.38-0.52; ♀ 0.40-0.45. VD/VW = ♂ 0.11-0.15; ♀ 0.10-0.13. GH/GL = ♂ 0.31-0.37; ♀ 0.34-0.36.

Face silver-grey; vertex dark grey/brown-grey, occiput grey/brown-grey. Beard and labial hairs white, mystax bristles brown/black; all other setae brown/black. Occipital bristles rather strong (to 0.8mm), abundant, those behind the dorsomedial angle of the eye strongly curved anterolaterally; lateral and ventral ones shorter, straighter. Frontal and orbital setae abundant, to 0.6mm.

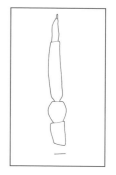

Figure 114. *L. lehri* sp. nov. antenna. Scale line = 0.1 mm.

Antennae. Brown; setae brown; F1 setae sometimes present. F1 long, more-or-less parallel-sided or slightly tapered from base. WF1/LF1 = ♂ 0.25-0.30; ♀ 0.23-0.27. LF2+3/LF1 = ♂ 0.48-0.60; ♀ 0.44-0.51.

Thorax. Prothorax grey/brown-grey, gold/brown dorsally, hairs white ventrally, brown dorsally; postponotal lobes grey/brown-grey, the lateral angle ferruginous/chestnut, hairs strong, brown. Scutum tomentum brown/grey, the dorsocentral stripes medium/dark brown, fading at some angles of view, bordered with light brown/gold. Usually, extensive gold/brown runs ventrolaterally from the postpronotal lobe to above scutellum. The acrostichal stripes grey/brown-grey, obscure. All strong bristles black, finer setae brown. Dorsocentral setae usually abundant; anteriors 4-5, longest to 0.8mm, mixed with only slightly finer setae; 5-6 posteriors. Notal setae abundant, erect, long (to 0.4mm). Postalars 2-4, with shorter setae; supra-alars 2-4 among shorter setae; presuturals 2-3, posthumerals 1, usually weak. Scutellar tomentum grey or gold-brown dorsally in darker specimens; apical scutellar bristles dark, abundant, in about 2 irregular rows, 3-5 strong ones on each side mixed with many shorter, weaker dark bristles and hairs.

Pleural tomentum grey ventrally to gold-grey, gold/brown-grey dorsally in darker specimens; anepisternum dark brown dorsally. Katatergite setae black, 7-9 among sparse, finer white setae; katepisternal setae white, often long. Anepisternal setae 8-30, dark (to 0.6mm); sometimes a few other pale, weaker ones ventrally; a patch of erect brown setae dorsally shelf. Anepimeron with a few white setae.

Legs. Base colour dark brown/black. Tomentum of coxae grey with silver highlights; tomentum on rest of legs grey. No coxal peg. Main bristles dark brown/black; pale hairs white. Femora dorsally with decumbent hairs in male and female pale basally, dark apically; pale hairs predominate in most, but not all, specimens. Longer erect hairs abundant ventrally and laterally, especially on profemur; pale basally, dark apically; longest pale ventral setae as long as, or longer than, thickness of femur. Femora with dark, abundant, dorsolateral bristles mixed with long hairs; fine, especially on profemur, usually fewer and stronger on meso- and metafemora (in females these are strong – 3-6 on mesofemur, 5-15 on metafemur). Tibiae and tarsi with dark, strong bristles typically arranged, hairs brown/black. Protibia with longest bristles about 4.0 times longer than tibial width.

Wings. Veins medium/dark brown; membrane brown in oblique view in male, less so in female. DCI = 0.30-0.46; cell M3 open. Halter yellow; knob without dark spot.

Abdomen: Male. Tergite basal colour dark brown/black. Tomentum on tergite bases thin brown/gold-brown, the dark cuticle shining through. Bands of gold-grey tomentum cover 0.25-0.50 the length of each tergite apically and broadly cover the ventrolateral areas of the tergite. Tergite 1 is thinly covered over most of its surface; about 6-7 strong dark bristles on each side of tergite. Lateral setae on tergites 1-4 white, erect, some dark dorsally; on 5-7 short, dark, the extreme ventrolateral hairs often pale. Dorsal setulae dark. Sternite tomentum gold-grey, hairs pale.

Female. Grey or gold-grey apical bands cover about 0.4-0.5 the tergite length; grey ventrolaterally. Lateral setae on tergites 1-3 white, erect, those on 4 shorter; setae on 5-7 brown. Dorsal setulae dark, more-or-less erect apically. Sternite tomentum gold-grey, hairs pale, mostly dark on 6-7 or 7.

Male genitalia. Epandrium and hypandrium/ gonocoxite complex dark chestnut/black and covered with gold-grey tomentum except on hypandrium in some specimens. Setal brush black; other setae brown/dark brown, numerous, prominent. Width of epandrium halves in lateral view about 0.40 times the length, the margins parallel, the apex rounded. Medial face of epandrium as in figure 115. In dorsal view, medial margins of epandrium strongly concave; basal sclerite weak.

Gonostylus. Gonostylus arched basally, the normally lateral face of the medial flange facing dorsally and with a prominent apicolateral lobe; medial flange joined to strong lateral tooth by an apical ridge that bears a broad, hooked lateral lobe. Hypandrium/gonocoxite complex in ventral view with length about 0.8 times the width, the transverse slit at 0.5 the length. Gonocoxal apodemes stout, in lateral view, ex-

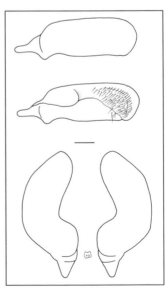

Figure 115. Epandrium (top to bottom): lateral, medial, dorsal. Scale lines = 0.2 mm.

Figure 116. *Lasiopogon lehri* sp. nov. male genitalia, hypandrium/gonocoxite complex. Clockwise from top left: dorsal, ventral, lateral and apical (scale lines = 0.3 mm); gonostylus, dorsal (scale line = 0.1 mm).

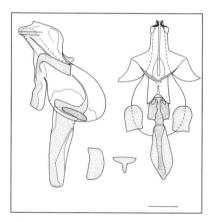

Figure 117. Phallus: lateral (left); ventral (right); lateral ejaculatory process, basal (centre left); ejaculatory apodeme, cross-section (centre right). Scale line = 0.2 mm.

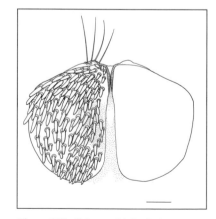

Figure 118. Subepandrial sclerite: ventral. Scale line = 0.1 mm.

posed length of apodeme about 0.5 times the basal width of hypandrium; apodeme with sclerotized web ventrally.

Phallus. Paramere sheath angled strongly ventrally, its dorsal margin 0.4 times the length of phallus (excluding ejaculatory apodeme). Apex of paramere sheath broad, the ventrolateral carina short; ventral flange absent; dorsal carina strongly arched dorsally. Sperm sac width in dorsal view 0.5 times the length of phallus. Ejaculatory apodeme long, straight in lateral view, spatulate in ventral view; flattened ventrally in cross-section, dorsal carina narrow. Subepandrial sclerite as in figure 118. Hypoproct reduced. Sclerite broader than its length, the linear unsclerotized area in basal 0.8; spines bluntly acute and densely arranged over whole sclerite.

Female genitalia. Undissected: Hairs abundant, erect, long and light to medium brown. Tergite 8 dark chestnut/black, often narrowly ferruginous apically. Sternite 8 chestnut basally; hypogynial valves parallel, black, the medial lobes ferruginous with gold/pale brown hair. Lateral lobe setae weak. Cerci black/brown with white/light brown hairs. Dissected sternite 8 (figure 119) broad, basal width about 0.9 times the length; undivided. Sclerite of hypogynial valve includes the basolateral part of the medial lobe; a carina present on medial edge of lateral part. Medial lobes broad, especially basally, about as long as, or slightly shorter than, valves. Length of unsclerotized area between medial lobes 0.4 times the sternite length.

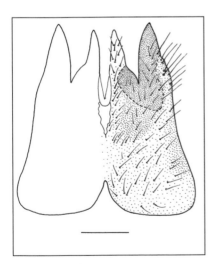

Figure 119. Female sternite 8: ventral.
Scale line = 0.3 mm.

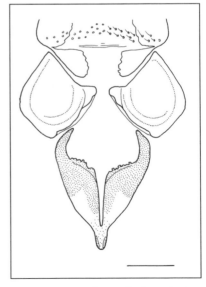

Figure 120. Female sternite 9, tergite 9 and lobes of sternite 10: ventral.
Scale line = 0.2 mm.

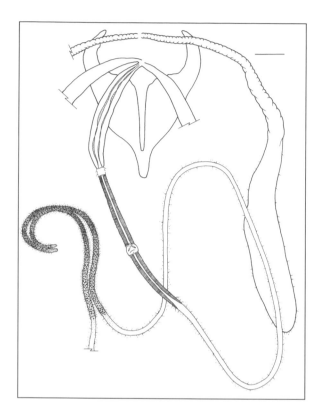

Figure 121. *Lasiopogon lehri* sp. nov. spermatheca: dorsal.
Scale line = 0.1 mm.

Tergite 9 sclerites as in figure 120; sternite 9 V-shaped, the medial area largely sclerotized but with a narrow medial slit almost dividing the sclerite. Basal lobes of sternite 10 small. Tergite 10 black, usually with 7-8 black acanthophorite spines on each side. Spermathecae (figure 121) with hooked terminal reservoirs about as thick as reservoir duct; terminal reservoirs with wart-like protuberances on surface. Both striated ducts with fine scales; junction with basal duct scaled, not golden; valve not golden; basal duct short and narrow.

Type Material.
HOLOTYPE. (here designated) ♂ labelled [in Russian]: "[rectangular white label] Primorskiy Kray/Ussuriyiskiy Reserve/Lehr 20 vi 1991". My Holotype label "HOLOTYPE/Lasiopogon ♂/lehri Cannings/des. R.A. Cannings 2002 [red, black-bordered label]" has been attached to this specimen. Dissected genitalia in plastic vial underneath all. IBPV.
PARATYPES. (41 designated). **Russia: Primorskiy Kray,** Anuchinskiy District, Anuchino, 10km W, 8.vi.1986, A. Lelej (1♂, 2♀, IBPV); Dzhigitovka, 4.vi.1985, P. Lehr (1♂, IBPV); Khasanskiy District, Tsukanovo (west of), 27.vi.1987, P. Lehr (1♀, IBPV); Khasanskiy District, Zanadvarovka, 22.vi.1987, A. Lelej (1♂, 1♀, IBPV), P. Lehr (2♀, IBPV), 23.vi.1987, A. Lelej

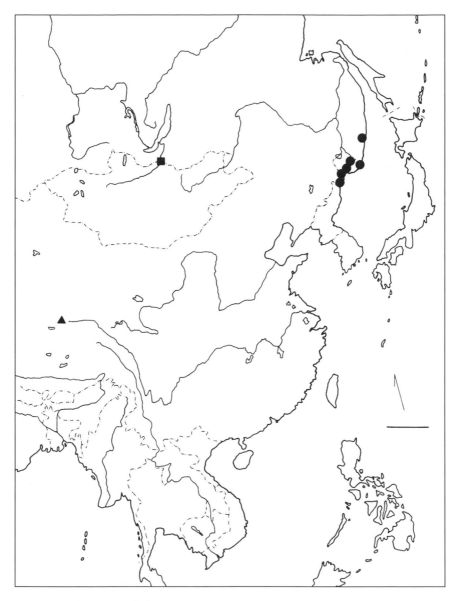

Map 5. Distribution records for
- *L. lehri*
- *L. kjachtensis*
- ▲ *L. qinghaiensis*

(1♂, IBPV; 1♂, BCPM), P. Lehr (4♂, 2♀, IBPV; 1♂, 1♀, BCPM); Milogradovo, 14.vi.1986, P. Lehr (1♀, IBPV); Ussuriyskiy District, Ussuriyskiy Reserve, 14.vi.1991, A. Lelej (1♀, IBPV), 20.vi.1991, A. Lelej (1♀, BCPM); Varvarovka (near Anuchino; reka), 6.vi.1985, P. Lehr (1♀, IBPV), 7.vi.1985, P. Lehr (8♂, 4♀, IBPV; 3♂, 3♀, BCPM).

Type Locality. Russia, Primorskiy Kray, Ussuriyiskiy Reserve

Etymology. Named for Professor Pavel Andreyivich Lehr, former Director of the Institute of Biology and Pedology, Russian Academy of Sciences, Vladivostok. In the last half of the 20th century Lehr has been the foremost systematist of the Palaearctic Asilidae and expert on Russian *Lasiopogon*; for his help and generosity in my robber fly studies.

Distribution. Palaearctic; Primorskiy Kray. Map 5, previous page.

Phylogenetic Relationships. The basal member of the *akaishii* species group.

Natural History. Stream sides in mixed forest. Dates of capture range from 4 June to 27 June.

Lasiopogon leleji sp. nov.

Diagnosis. Only one male known. A medium-sized black and grey species with dark brown legs. Antennae brown; LF2+3/LF1 = 0.63. Mystax and other head bristles dark. Scutum brown-grey with gold highlights, especially ventrolaterally and anteriorly. Dorsocentral stripes dark brown bordered by lighter brown and gold; the acrostichal stripes brown-grey. Intermediate spots faint medium brown. All strong bristles on thorax brown/black, finer setae brown. Anterior dorsocentrals 3, strong. Main bristles of legs dark brown/black. Femora dorsally with white decumbent hairs basally, dark apically. Grey-gold abdominal tergite tomentum basally very thin and disappearing at various angles of view, the black cuticle shining through. In anterolateral view apical bands of thin gold-grey tomentum cover about 0.3-0.5 of tergite length. Tergite 1 bristles pale; lateral setae white; dorsal setulae hair-like, mostly white; pale brown on 6-7.

Description. Only one male known. Body length ♂ 9.0mm.

Head. HW ♂ 1.80mm. FW ♂ 0.38mm. VW ♂ 0.71mm. VW/HW= ♂ 0.39.

FW/VW = ♂ 0.54. VD/VW = ♂ 0.21. GH/GL = ♂ 0.26.

Face gold-grey, vertex gold/brown-grey, occiput silver-grey ventrally, gold/brown-grey dorsally. Beard and labial hairs white, mystax bristles brown/black; all other setae brown/black. Occipital bristles moderately strong, to 0.6mm, those behind the dorsomedial angle of the eye strongly curved anterolaterally; lateral and ventral ones shorter, straighter. Frontal and orbital setae strong, long, the longest orbitals to 0.6mm, about as long as F1+F2+3. Ocellar bristles fine, hardly distinguished from orbitals.

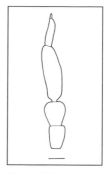

Figure 122. Antenna. Scale line = 0.1 mm.

Antennae. Brown, the base of F1 ferruginous. Setae brown, no setae on F1. F1 parallel-sided, slightly tapered apically. WF1/LF1 = 0.29; LF2+3/LF1 = 0.63.

Thorax. Prothorax gold/brown-grey with white hairs; postponotal lobes gold/brown-grey, the lateral angle ferruginous, hairs white to pale brown. Scutum brown-grey with gold highlights, especially ventrolaterally and anteriorly; variation unknown. Dorsocentral stripes dark brown bordered by lighter brown and gold; the acrostichal stripes brown-grey. Intermediate spots faint medium brown. All strong bristles brown/black, finer setae brown. Dorsocentral bristles rather strong, to 0.9mm; 3 strong anteriors with other shorter ones to 0.5-0.6mm, posteriors 3-4. Notal setae scattered over presutural area, as long as acrostichals, longest 0.2-0.3mm. Postalars 2 with shorter setae; supra-alars 2 with shorter setae; presuturals 2, 1 weak posthumeral. Scutellar tomentum gold-grey; apical scutellar bristles dark, arranged in two irregular rows, 4 strong ones on each side mixed with shorter, weaker dark bristles and hairs.

Pleural tomentum gold/brown-grey, most intense on anepisternum. Katatergite setae brown, 5-6 among finer white hairs; katepisternal setae sparse, often long. Anepisternal setae 5-6, dark (to 0.5mm); a patch of erect brown setae on dorsal shelf. Anepimeron without setae.

Legs: Base colour dark brown/black. Tomentum of coxae and other leg segments grey with gold highlights. No coxal peg. Main bristles dark brown/black. Femora dorsally with white decumbent hairs basally, dark apically. Longer, erect white/pale brown hairs ventrally (and laterally, especially on profemur); in males longest pale ventral setae as long as, or longer than, thickness of femur. Profemur with 20-30 fine long dorsolateral bristles mixed with long hairs; mesofemur with 4-5; metafemur with about 15. Tibiae and tarsi with dark, strong bristles typically arranged, hairs on tibiae and tarsi brown/black. Protibia with longest bristles about 4.5 times longer than tibial width.

Wings. Veins medium/dark brown; membrane faint medium brown in oblique view, slightly darker at main vein forks: r-m crossvein, base of cu1 cell, apex of discal cell, and forking of R4+5. DCI = 0.30; cell M3 open. Halter yellow; knob with large brown spot.

Abdomen. Male. Tergite basal colour dark brown/black, the grey-gold tomentum basally very thin and disappearing at various angles of view, the black cuticle shining through. In anterolateral view apical bands of thin gold-grey tomentum cover about 0.5 of tergite 7, 0.4 of 6, 0.3 of 4-5, and 0.5 of 2-3; tergite 1 is completely covered. Most of this grey tomentum disappears when abdomen is viewed from apex. The tomentum widens laterally and covers the length of the tergite ventrolaterally. About 10-12 fine pale bristles on each side of tergite 1. All lateral setae white, erect, rather long, shortening apically. Dorsal setulae hair-like, rather long, mostly white; pale brown on 6-7. Sternite tomentum gold-grey, hairs white.

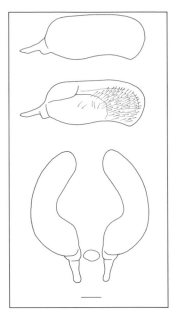

Figure 123. *L. leleji* sp. nov. epandrium (top to bottom): lateral, medial, dorsal. Scale lines = 0.2 mm.

Male genitalia. Hypandrium/gonocoxite complex brown/chestnut basally, ferruginous apically; epandrium brown basally, chestnut apically; all with thin gold-grey tomentum except on apex of gonocoxae. Setal brush 1-2 rows of very strong setae, a few very long laterally, dark brown/black; other setae brown/black, prominent. Width of epandrium halves in lateral view about 0.4 times the length, dorsal and ventral margins more-or-less parallel, the apex rounded dorsally; ventrally obtuseangular and slightly down turned. Medial face of epandrium as in figure 123. In dorsal view, medial margins of epandrium moderately concave; basal sclerite strong.

Gonostylus. Setose; medial flange a rounded lobe; dorsal flange short, thick, linked to prominent lateral tooth by apical ridge; ventrolateral tooth present. Hypandrium/gonocoxite complex in ventral view the length about 0.85 times the width, the transverse slit at 0.65 the length. In lateral view, exposed length of gonocoxal apodeme about 0.6 times basal width of hypandrium; apodeme with small sclerotized web ventrally.

Phallus. Paramere sheath dorsally 0.4 times the length of phallus (excluding ejaculatory apodeme). Ventral process short. Apex of paramere sheath with ventrolateral carina and small ventral flange. Lateral processes very short. Sperm sac width in dorsal view 0.35 times the length of phallus. Ejaculatory apodeme curved dorsally in lateral view, weakly spatulate in ventral view; triangular in cross-section, flattened ventrally with thick dorsal carina. Subepandrial sclerite as in figure 126. Broad triangular unsclerotized area in basal 0.6, the margins of

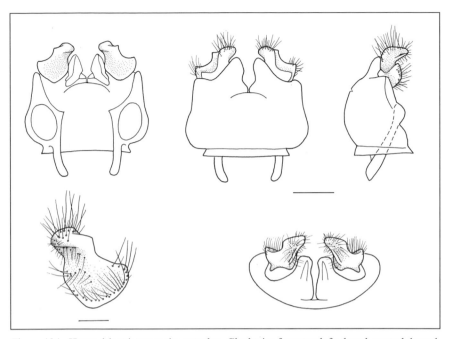

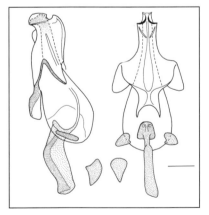

Figure 124. Hypandrium/gonocoxite complex. Clockwise from top left: dorsal, ventral, lateral and apical (scale lines = 0.3 mm); gonostylus, dorsal (scale line = 0.1 mm).

Figure 125. Phallus: lateral (left); ventral (right); lateral ejaculatory process, basal (centre left); ejaculatory apodeme, cross-section (centre right).
Scale line = 0.2 mm.

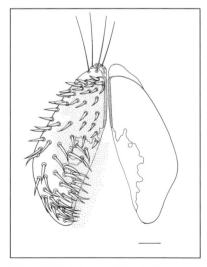

Figure 126. Subepandrial sclerite: ventral. Scale line = 0.1 mm.

the plates broadly and irregularly unsclerotized; a narrow unsclerotized portion in central 0.3; spines slender and attenuate, most dense apically and basally.
Female unknown.

Type Material.
HOLOTYPE. (here designated) ♂ labelled [in Russian]: "[rectangular white label] Primorskiy Kray/Khasanskiy Disrict/near Zanadvorovka/Lehr/22 July 1987". My Holotype label "HOLOTYPE/Lasiopogon ♂/leleji Cannings/des. R.A. Cannings 2002 [red, black-bordered label]" has been attached to this specimen. White card holding right hind leg above all labels; dissected genitalia in plastic vial underneath all. IBPV.
Known only from the holotype.

Type Locality. Russia, Primorskiy Kray, Khasanskiy District, near Zanadvorovka.

Etymology. Named for Arkady Lelej, head of the Entomology Section, Institute of Biology and Pedology, Russian Academy of Sciences, Vladivostok, for his friendship, hospitality and help with this study.

Distribution. Palaearctic; Primorskiy Kray. Known only from near Zanadvorovka in the Amba River valley, in extreme southwestern Primorskiy Kray. See map 10, p. 251.

Phylogenetic Relationships. A member of the *hinei* species group; the sister of the species pair *L. hinei* and *L. kjachtensis*.

Natural History. Date of capture of sole specimen is 22 July.

Lasiopogon marshalli sp. nov.

Diagnosis. A medium-sized grey and brown species with dark brown legs and markings; mystax and other head bristles dark. Antennae brown, F2+3 long (F2+3/F1 = 0.70-0.86). Thoracic tomentum dorsally grey-brown to gold-brown, laterally gold-grey. Dorsocentral stripes brown, strong; acrostichal stripes brown, usually with a faint brown medial stripe. Main thoracic and leg bristles dark. Anterior dorsocentral bristles 3-6. Abdominal tergites basally with dark brown tomentum; apical bands gold-grey, covering 0.2 the length of the segments in male, 0.3-0.75 the length in female; in the female the pale tomentum extends anteriorly along the midline. Bristles on tergite 1 dark, pale or mixed. Width of epandrium about 0.4 times the length, apex rounded. In dorsal view moderately concave medially. Female with black terminalia and dark hairs; strong short setae on hypogynial valves.

Description. Body length ♂ 8.5-10.2mm; ♀ 9.7-11.3mm.

Head. HW ♂ 1.86-2.18mm; ♀ 2.00-2.20mm. FW ♂ 0.36-0.42mm; ♀ 0.40-0.50mm. VW ♂ 0.82-1.05mm; ♀ 1.02-1.11mm. VW/HW = ♂ 0.41-0.51; ♀ 0.49-0.51. FW/VW = ♂ 0.38-0.49; ♀ 0.39-0.45. VD/VW = ♂ 0.11-0.13; ♀ 0.09-0.12. GH/GL = ♂ 0.23-0.24; ♀ 0.37-0.39.

Face grey/brown-grey, vertex dark grey-brown, occiput grey with brown-grey dorsally. Beard and labial hairs white, mystax bristles brown/black; all other setae brown/black. Occipital bristles rather fine, abundant, those behind the dorsomedial angle of the eye long and strongly curved anterolaterally; lateral and ventral ones shorter, straighter. Frontal and orbital setae abundant, rather fine, some as long as F1+F2+3; ocellars weak, about as long as orbitals.

Antennae. Brown to black, setae brown; F1 with or without a seta. F1 short, convex ventrally, widest in basal half or at midlength. WF1/LF1 = ♂ 0.29-.35; ♀ 0.32-0.37. LF2+3/ LF1 = ♂ 0.70-0.82; ♀ 0.70-0.86.

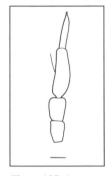

Figure 127. Antenna. Scale line = 0.1 mm.

Thorax. Prothorax brown-grey, brown dorsally; hairs white. Postpronotal lobes gold-grey to gold-brown, the lateral angle dark chestnut, hairs brown. Scutum coloration variable; tomentum dark gold-brown; the dorsocentral stripes darker brown, this darker tomentum sometimes suffused over most of scutum; intermediate spots and acrostichal stripes grey-brown, the latter sometimes more-or-less fused or, more often, separated by a thin dark medial stripe. A lighter pattern, seen most often in females, has the scutal tomentum gold-grey and gold-brown laterally, with dark brown dorsocentral stripes bordered with gold and the intermediate spots faintly brown. The acrostichal stripes are brown-grey, vague, and divided by a faint narrow brown medial stripe. All notal setae brown or black. Anterior dorsocentrals 3-6 (longest to 0.9mm), mixed with finer setae; 4-5 posteriors. Notal setae scattered, but concentrated on anterior intermediate spot and acrostichal stripes, as long as shorter dorsocentrals. Postalars 2, with shorter hairs; supra-alars 1, presuturals 2-3; often 1-2 weak posthumerals. Scutellar tomentum grey/gold-grey, usually contrasting with brown scutum; apical scutellar bristles dark, 3-4 strong ones on each side mixed with shorter, weaker dark bristles and hairs.

Pleural tomentum dark gold-grey. Katatergite setae black (a few pale in some specimens), 7-9 among finer white hairs. Anepisternal setae prominent (to

0.6mm), brown/black, 7-15, a few white ventrally; patch of erect brown hairs on dorsal shelf strong, to 0.2mm. Anepimeron with a few white setae.

Legs. Base colour dark brown/black. Tomentum of coxae grey with some gold highlights; tomentum on rest of legs grey. No coxal peg. Main bristles dark brown/black, finer setae white to black. Femora dorsally with most decumbent hairs white basally, dark apically. In females all or most hairs are white, with some apically dark, especially on profemur. Longer erect pale hairs ventrally and laterally, especially on profemur; many of these can be dark apically; in males longest white ventral setae as long as or longer than thickness of femur. Profemur with 5-9 stronger dorsolateral dark bristles; mesofemur with 2-5; metafemur with 6-15; these bristles mixed with finer bristles and setae. Tibiae and tarsi with dark, strong bristles typically arranged, hairs brown/black in male, mixed with white ones in female. Protibia with longest bristles about 3.0-4.0 times longer than tibial width.

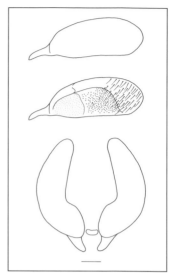

Figure 128. *L. marshalli* sp. nov. epandrium (top to bottom): lateral, medial, dorsal. Scale line = 0.2 mm.

Wings. Veins dark brown; membrane brown in oblique view. DCI = 0.30-0.48; cell M3 open. Halter yellow; knob without dark spot.

Abdomen. Male. Tergite basal colour dark brown/black. Tomentum on tergite bases dark brown, dark cuticle shining through; light brown/gold where brown meets apical bands. Bands of grey/gold-grey tomentum cover 0.2 the length of each tergite apically; ventrolateral areas are brown-grey; the apical 0.5 of tergite 1 is covered. About 4-7 strong black or pale bristles (often mixed) on each side of tergite 1. Lateral setae on tergites 1-3 white, erect, those on 4 mixed white and dark, dark on 5-7. Dorsal setulae dark, decumbent, strong. Sternite tomentum gold-grey; hairs white, dark on 4/5-7.

Female. As in male except tomentum covers tergites completely. Brown tomentum more dense; apical grey bands wider, about 0.3-0.5 the length of segment (0.5-0.75 on 7) and extending apically on the midline. Lateral setae on 1-3/4 white, dark on 4/5-7. Sternite tomentum gold-grey; hairs white, except dark on 5-7.

Male genitalia. Epandrium and hypandrium/gonocoxite complex dark brown/ black and covered with gold-grey tomentum except on parts of hypandrium. Setal

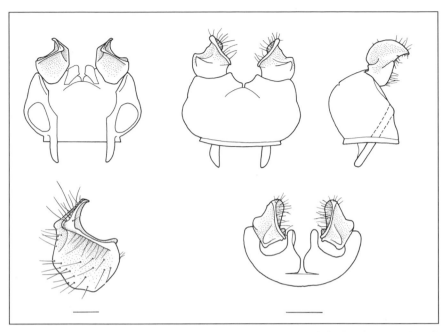

Figure 129. Hypandrium/gonocoxite complex. Clockwise from top left: dorsal, ventral, lateral and apical (scale lines = 0.3 mm); gonostylus, dorsal (scale line = 0.1 mm).

brush black; other setae dark brown/black, long. Width of epandrium halves in lateral view about 0.4 times the length, widest at about midlength; ventral margin straight, dorsal margin gently convex; apex rounded. Medial face of epandrium as in figure 128; membrane covered with numerous granular-like microsetae. In dorsal view, medial margins of epandrium moderately concave; basal sclerite strong.

Gonostylus. Medial flange low, the height about 0.3 times the dorsoventral length; in dorsal view, face of flange expanded medially. Dorsomedial tooth absent; ventrolateral tooth large. Hypandrium/gonocoxite complex in ventral view with length about 0.6 times long the width, the transverse slit at 0.80 the length; In lateral view, exposed length of gonocoxal apodeme about 0.5 times basal width of hypandrium; apodeme with sclerotized web ventrally.

Phallus. Paramere sheath dorsally 0.5 times the length of phallus (excluding ejaculatory apodeme). Apex of paramere sheath with ventrolateral carina and ventral flange. Sperm sac width in dorsal view 0.35 times the length of phallus. Ejaculatory apodeme straight in lateral view, rod-like in ventral view; triangular in cross-section, the dorsal carina thick and narrow. Subepandrial sclerite as in figure 131. Narrowly triangular unsclerotized area in basal 0.80; spines blunt, or bluntly acute; widely distributed but concentrated apically, basally and laterally.

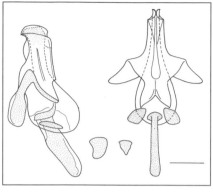

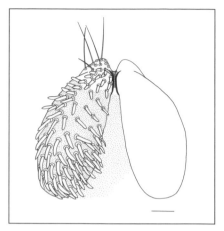

Figure 130. *L. marshalli* sp. nov. phallus: lateral (left); ventral (right); lateral ejaculatory process, basal (centre left); ejaculatory apodeme, cross-section (centre right). Scale line = 0.2 mm.

Figure 131. Subepandrial sclerite: ventral. Scale line = 0.1 mm.

Female genitalia. Undissected: Setae brown/black, erect, abundant. Tergite 8 dark brown/black, brown apically. Sternite 8 dark brown/black, lateral lobe setae strong; hypogynial valves apically brown, covered with short strong setae. Cerci black with pale setae. Dissected sternite 8 (figure 132) with basal width about 0.65 times the length; strongly divided along midline. Length of unsclerotized area between hypogynial valves 0.3 times sternite length.

Tergite 9 sclerites as in figure 133; sternite 9 V-shaped, medially undivided and with a dorsal carina. Tergite 10 brown/black with 7 black acanthophorite spines on each side. Spermathecae (figure 134) with terminal reservoirs curled in a half loop and about as wide as apical part of reservoir duct; the basal part of the duct is about half as wide. Terminal reservoirs with wart-like protuberances on surface. Narrow basal section of undifferentiated reservoir duct with a dense patch of caniculi. Basal striated duct with fine scales; junction with basal duct sclerotized, golden, scaled; valve sclerotized, golden; striated basal section of reservoir tube with fine scales, basal 0.5 without striae, bare, sclerotized; basal duct short, narrow.

Type Material.
HOLOTYPE. (here designated) ♂ labelled: "[rectangular white label] U.S.A.: VA: Giles Co./Ripplemead 2k E/Pearisburg 37°19'N/80° 44'W 27.v.99 NewR./rocky shore S.A. Marshall". My Holotype label "HOLOTYPE/Lasiopogon ♂/marshalli Cannings/des. R.A. Cannings 2002 [red, black-bordered label]" has been attached to this specimen. Dissected genitalia in plastic vial underneath all. DEBU.
PARATYPES (23 designated). **U.S.A., Virginia**, Giles Co., New River near Atherton, 17.v.1997, S.A. Marshall. (1 ♂, 1 ♀, BCPM); Giles Co., New River,

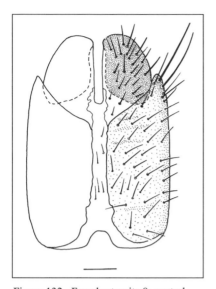

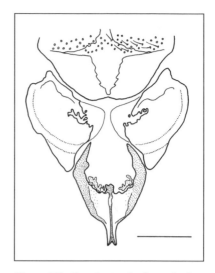

Figure 132. Female sternite 8: ventral.
Scale line = 0.2 mm.

Figure 133. Female sternite 9, tergite 9
and lobes of sternite 10: ventral.
Scale line = 0.2 mm.

Pembroke, 27.v.1999, S. Paiero (1♂, DEBU); Giles Co., Ripplemead 2km E of
Pearisburg, New River, rocky shore, 37°19'N x 80° 44'W, 27.v.1999, S.A.
Marshall (3♂, 2♀, BCPM; 2♂, 2♀, CNCI; 6♂, 2♀, DEBU; 2♂, 2♀, USNM)

Type Locality. U.S.A., Virginia, Giles Co, New River at Ripplemead
(37°19'N/80° 44'W).

Etymology. Named for my friend and colleague Dr Steve Marshall, energetic
dipterologist from the University of Guelph, who served as my supervisor for this
doctoral study and who collected the type material of the species.

Distribution. Nearctic; U.S.A., The Appalachian Mountains in Virginia; probably more widely distributed in the central regions of these mountains. See map 8,
p. 216.

Phylogenetic Relationships. Member of the *opaculus* species group and sister
species to *L. piestolophus*.

Natural History. Known only from the New River, Virginia, where specimens
were taken from the rocky river banks. Specimens examined were collected between 17 and 27 May. "They were very common on the exposed, flat, rocky
shoreline but were also seen on deadheads and isolated rocks in the water" (S.A.
Marshall *in litt.*).

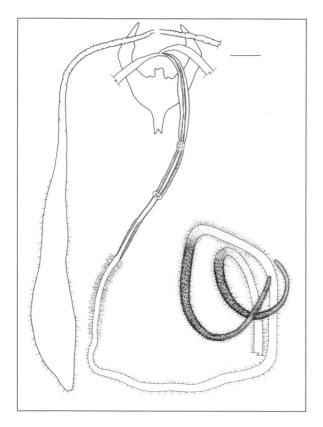

Figure 134. *L. marshalli* spermatheca: dorsal. Scale line = 0.2 mm.

Lasiopogon monticola Melander
Lasiopogon monticola Melander, 1923. *Psyche* 30: 142-143.

Diagnosis. A small to medium-sized grey/gold-brown species; mystax and other head bristles dark. Antennae brown, F2+3/F1 = 0.49-0.67. Thoracic tomentum dorsally and laterally grey/gold-grey (sometimes richer gold) with brown dorsocentral stripes and gold-grey acrostichal stripes. Main thoracic and leg bristles dark, anterior dorsocentral bristles 3-5. Abdominal tergites basally with thin brown tomentum; apical bands gold-grey/gold, covering about 0.5-0.6 the length of the segments. Bristles on tergite 1 dark. Epandrium ferruginous to black, usually shining, about 0.40-0.45 as wide as long in lateral view, slightly arched dorsally, the apex truncate. In dorsal view a strong, acute tooth basomedially. Female with black terminalia; base of sternite 8 usually ferruginous; hairs pale.

Description. Body length ♂ 5.6-11.4mm; ♀ 7.2-11.6mm.

Head. HW ♂ 1.46-2.20mm; ♀ 1.68-2.16mm. FW ♂ 0.30-0.50mm; ♀ 0.36-0.50mm. VW ♂ 0.59-0.90mm; ♀ 0.64-0.96mm. VW/HW= ♂ 0.35-0.42; ♀ 0.37-0.48. FW/VW = ♂ 0.52-0.63; ♀ 0.50-0.60; VD/VW = ♂ 0.17-0.25; ♀ 0.15-0.26. GH/GL= ♂ 0.24-0.43; ♀ 0.25-0.32.

Face and vertex grey or gold-grey to dark gold; occiput grey ventrally, gold/brown-grey dorsally. Beard and labial hairs white, mystax bristles brown/black sometimes with a few pale setae ventrally; all other setae brown/black. Occipital bristles moderately strong, those behind the dorso-medial angle of the eye long (to 0.6mm) and strongly curved anterolaterally; lateral and ventral ones shorter, straighter. Frontal and orbital setae rather sparse, to 0.4mm, as long as, or longer than, F1.

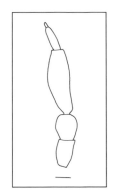

Figure 135. Antenna. Scale line = 0.1 mm.

Antennae. Brown to black, some with base of F1 chestnut. Setae brown; F1 sometimes with a seta. F1 moderately long, shorter in southern, especially Californian, specimens; variable in shape, widest in apical half, the dorsal margin straight or with a bulge in the medial third, the ventral edge gently convex. WF1/LF1 = ♂ 0.30-0.38; ♀ 0.30-0.39. LF2+3/LF1 = ♂ 0.53-0.67; ♀ 0.49-0.66.

Thorax. Prothorax grey to gold-grey/gold, hairs white; postpronotal lobes gold-grey/gold, the lateral angle ferruginous, hairs white to brown, short, sparse. Scutum tomentum variable, frequently rather evenly grey-gold with the dorsocentral stripes medium/dark brown, patchy, bordered by gold, the anterior ends not expanded except by the gold tomentum. Acrostichal stripes gold-grey and faint. Other specimens with the base tomentum grey with faint gold highlights, the gold/brown restricted to ventrolateral areas and the margins of the dorsocentral stripes; the acrostichal stripes can be almost invisible. In California, especially, many specimens are washed with gold over most of the dorsal and lateral surfaces of the mesothorax. All notal setae brown or black. Dorsocentral setae with main bristles much stronger than secondary ones. Anteriors 3-5 (to 0.8mm) mixed with finer setae; 3-4 posteriors. Notal setae sparse, to about 0.2mm; the acrostichals irregularly arranged, about the same length and density. Postalars 2-3, with very few shorter hairs; supra-alars 1-4, with very few shorter hairs; pre-suturals 2-3, posthumerals 0-1. Scutellar tomentum grey/gold-grey to gold; apical scutellar bristles dark, 3-5 strong ones on each side mixed with shorter, weaker dark bristles and hairs.

Pleural tomentum grey to gold-grey or gold. Katatergite setae brown/black, sometimes one pale, 6-8 among a few finer white hairs; katepisternal setae sparse, white. Anepisternal setae 2-4, brown, short (to 0.4mm); a patch of short erect brown hairs on dorsal margin of sclerite. Anepimeron without setae.

Legs: Base colour dark brown/black, tarsi often paler brown than basal segments. Tomentum of coxae and rest of legs grey/gold-grey to gold. No coxal peg. Setae relatively sparse, especially in Sierra Nevada specimens; main bristles dark brown/black, finer setae white to black. Femora dorsally with mainly dark decumbent hairs, often some white ones at base; in some specimens most hairs are white. Longer more erect hairs mostly ventral, mixed pale (white/gold) and dark, the pale ones predominating except apically; in males longest ventral setae as long as, or longer than, thickness of femur, except in some Sierra Nevada specimens where the ventral setae are short and sparse. Dark dorsolateral bristles fine on profemur, 10-20 (fewer and stouter in Sierra Nevada specimens); stronger on mesofemur (2-4) and metafemur (8-10). Tibiae and tarsi with dark, strong bristles typically arranged; hairs normally dark, but pale in some Sierra Nevada specimens. Protibia with longest bristles about 3.5 times longer than the tibial width.

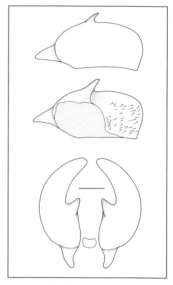

Figure 136. *L. monticola* epandrium (top to bottom): lateral, medial, dorsal. Scale lines = 0.2 mm.

Wings: Veins gold/light brown to medium brown; membrane very pale brown when viewed obliquely. DCI = 0.40-0.48; cell M3 open. Halter yellow; knob without dark spot.

Abdomen. Male. Tergite basal colour dark brown/black. Tomentum on tergite bases medium to dark brown. Bands of grey/grey-gold or gold tomentum cover about 0.5-0.6 the length of each tergite apically (tergites 6-7 are more extensively covered and segment 1 is fully covered), this tomentum often extending weakly towards the base mid-dorsally, especially on 2-3. Ventrolateral areas are gold-grey/gold, narrowly so basally. 4-5 strong dark bristles on each side of tergite 1; lateral setae all white/gold, short, sparse. Dorsal setulae mainly pale apically on tergites, dark basally. Sternite tomentum gold-grey, hairs white.

Female. Apical gold-grey bands cover about 0.5-0.6 the length of each tergite. Lateral setae pale on tergites 1-2 and the extreme ventrolateral margin of 3; short, dark and bristle-like on 3-7. In some specimens all setae are dark except for the lateral ones on 1 and the base of 2. Dorsal setulae dark, short, more-or-less erect and bristle-like. Sternites gold-grey, hairs dark on 5-7.

Male genitalia. Epandrium and hypandrium/ gonocoxite complex shining dark brown/black to ferruginous in some Sierra Nevada specimens (these specimens

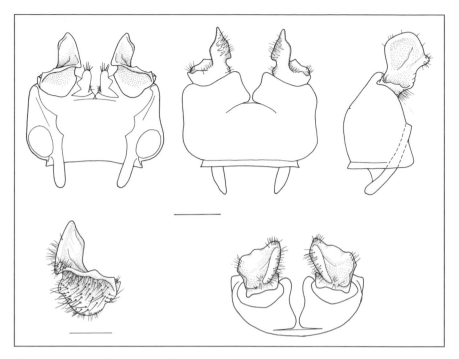

Figure 137. Hypandrium/gonocoxite complex. Clockwise from top left: dorsal, ventral, lateral and apical (scale lines = 0.3 mm); gonostylus, dorsal (scale line = 0.1 mm).

often have gold tomentum on the epandrium). Setal brush and other setae brown/black, sparse. Width of epandrium halves in lateral view about 0.4-0.45 times the length, gently convex dorsally, the ventral and dorsal margins mostly parallel; apex truncate, the ventral angle curved ventrally in most specimens. Medial face of epandrium as in figure 136. In dorsal view, medial margins convex at the base and bearing a large, acute tooth; basal sclerite prominent.

Gonostylus. Teeth weakly to moderately developed; weak ridge laterally at base of medial flange. Hypandrium/gonocoxite complex in ventral view with margins straight, the length about 0.7 times the width, the transverse slit at 0.6 the length. In lateral view, exposed length of gonocoxal apodeme about 0.6 times the basal width of hypandrium; apodeme with sclerotized web ventrally.

Phallus. Paramere sheath dorsally 0.6 times the length of phallus (excluding ejaculatory apodeme). Ventral process with low carina. Apex of paramere sheath narrow, with ventrolateral carina; ventral lip and ventral flange absent. Sperm sac width in dorsal view 0.35 times the length of phallus. Ejaculatory apodeme slightly curved ventrally in lateral view, broadly spatulate in ventral view; flattened in cross-section with strong dorsal carina. Subepandrial sclerite as in figure 139. Triangular unsclerotized area in basal 0.7; spines bluntly acute, scarce laterally.

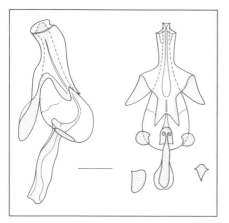

Figure 138. *L. monticola* phallus: lateral (left); ventral (right); lateral ejaculatory process, basal (left of ventral view); ejaculatory apodeme, cross-section (right of ventral view). Scale line = 0.2 mm.

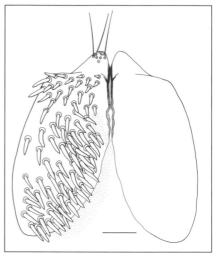

Figure 139. Subepandrial sclerite: ventral. Scale line = 0.1 mm.

Female genitalia. Undissected: Hairs pale, erect. Tergite 8 chestnut to brown/black, often with paler apical band; sternite 8 strongly keel-like; dark brown/black, often ferruginous/chestnut basally and apically on lateral lobes. Hypogynial valves black, haired; lateral lobe setae strong. Cerci brown with white/gold hairs. Dissected sternite 8 (figure 140) with basal width about 0.8 times the length; undivided along midline. Length of unsclerotized area between hypogynial valves 0.4 times sternite length. Valves with fine hairs ventrally.

Tergite 9 sclerites as in figure 141; sternite 9 narrowly V-shaped, almost divided. Tergite 10 brown/black with 6-8 black acanthophorite spines on each side. Spermathecae (figure 142) with straight terminal reservoirs about as wide as reservoir duct; terminal reservoirs with wart-like protuberances on surface. Basal striated duct without fine scales; junction with basal duct with scales; basal duct short and narrow.

Variation. *L. monticola* varies considerably over its large range, but the only striking differences that are correlated with geography are the dimensions and colour patterns of flies in the Sierra Nevada of California and Nevada when these are compared to other populations. The most obvious difference is that these southwestern specimens are small. Males examined range in length from 5.6-7.3mm (9.2-11.4mm in other areas); females 7.2-8.8mm (9.6-11.6mm in other areas). The other most noticeable characteristic is the strong golden colour of many individuals. This is not the rule, however, and some specimens are as grey as the greyest ones from farther north and east. The gold specimens are striking; males have a dark gold head, antepronotum, and scutum. The dorsocentral stripes

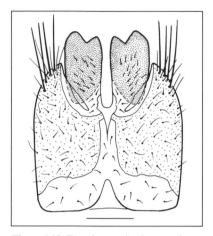

Figure 140. Female sternite 8: ventral
(notched valves are a variant).
Scale line = 0.3 mm.

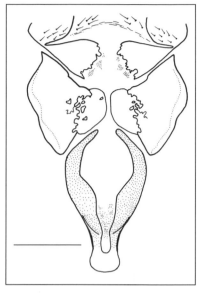

Figure 141. Female sternite 9, tergite 9
and lobes of sternite 10: ventral.
Scale line = 0.2 mm.

Figure 142.
Spermatheca: dorsal.
Scale line = 0.1 mm.

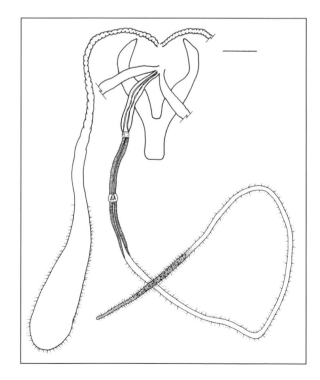

are dark brown, often obscure; the acrostichal stripes are dark gold-grey. The apical abdominal bands are golden. In some specimens the genitalia are ferruginous.

In addition, the dimensions and setation of these specimens from the Sierra Nevada often differ from those of other populations. The ratio of vertex depth to width normally is greater: 0.20-0.26 compared with 0.15-0.0.22. Antennal segment F1 is shorter and broader, WF1/LF1= 0.36-0.39 compared with 0.32-0.35. The number of bristles is reduced: presuturals 2, supra-alars 1, postalars 1, scutellars 2-3 on each side. Leg setae are reduced in number.

Type Material.
HOLOTYPE. ♂ (examined) labelled: "[rectangular beige label] Mt Adams WN./24 July 1921/AL Melander"; "[rectangular red label] TYPE/Lasiopogon/monticola/Melander"; "[rectangular white, orange and green label] AL Melander/Collection/1961"; "[rectangular beige, black-bordered label] Lasiopogon/monticola/Mel." USNM.
ALLOTYPE. ♀ (examined): same data as holotype (USNM).
PARATYPES. (22 examined). **U.S.A.: Idaho,** Moscow Mt., 3.vi.1911, A.L. Melander (1♂, CASC), 17.vi.1918, A.L. Melander (2♂, USNM), 26.v.1918, A.L. Melander (2♂, USNM), 8.vi.1921, A.L. Melander (1♂, CNCI). **Washington,** Mt. Adams, 24.vii.1921, A.L. Melander (2♂, 6♀, USNM); Mt. Rainier, 21.vii.1922, A.L. Melander/Van Trump (2♂, USNM); Mt. Rainier, Alta Vista, 22.vii.1922, A.L. Melander (1♂, USNM), 27.vii.1922, A.L. Melander (1♀, USNM), 28.vii.1922, A.L. Melander (2♂, USNM) 29.vii.1922, A.L. Melander (1♂, USNM); Mt. Rainier, Paradise Pk, viii.1917, A.L. Melander (1♀, USNM).

Other Material Examined (1112 specimens).
CANADA: British Columbia, Beaconsfield Mt., 6500', 5.viii.1984, R.J. Cannings (1♀, SMDV), Christina Lake, 3.vii.1953, J.M. Smith (1♂, CNCI); Coquihalla Hwy., Coldwater R. picnic site, 9.vii.1988, R.A. Cannings (1♂, BCPM); Creston, 12.vii.1920, W.B. Anderson (1♂, CNCI); Greenwood, Hwy.#3, 21.vi.1982, B.V. Peterson (1♂, CNCI); Hedley, Apex Mt., 6000', 11.viii.1933, A.N. Gartrell (1♀, CNCI), Hope Mts., 19.vii.1906, R.S. Sherman (1♀, SMDV; 1♀, OSUC); Keremeos, Apex Mt., 3.viii.1987 (2♀, SMDV), Keremeos, Twin Lks., 4700', 7.vii.1953, J.R. McGillis (2♂, 1♀, CNCI); Lavington, 19.v.1956, J. Grant (1♂, BCPM); Manning Prov. Park, creek at Dry Ridge Nature Trail, BCPM) 4.viii.1986, P.H. Arnaud, Jr (2♂, 1♀, CASC), Strawberry Flats, 28.vi.1990, R.A. Cannings and H. Nadel (1♀, BCPM), 20-Minute L., 28.vi.1990, R.A. Cannings and H. Nadel (2♂, 1♀, BCPM); Meager Cr. Hotsprings, 21.v.1987, P. Kroeger (1♀, SMDV); Oliver, Mt. Baldy, 1800-2304m, 14.vii.1988, R.A. Cannings (1♀, BCPM); Osoyoos, Anarchist Mt., 17.v.1980, R.A. Cannings (1♂, BCPM); Osoyoos, Mt. Baldy summit rd., 14.vii.1988, S.G. Cannings (2♂, 2♀, SMDV), 5900-6600', 24.vii.1990, S.G. Cannings (1♀, SMDV); Osoyoos, Mt. Kobau, summit, 26.vi.1988, S.G. Cannings (1♀, SMDV), Mt. Kobau, 1100m, 31.v.1991, D. Blades and C. Maier (2♂, 2♀, BCPM), 1760m, 8-12.vii.1991, D. Blades and C. Maier (1♂, 3♀, BCPM), 8.vii.1991, D. Blades and C. Maier (1♂, 1♀, BCPM); Penticton, Apex Mtn., 1900-2247m, jct. powerline x

road to summit, 15.vii.1988, R.A. Cannings (1♂, 2♀, BCPM), Apex Mtn., 13.viii.1967, J.R. Vockeroth (1♀, CNCI); Penticton, Dividend Mtn., S. side, 6350', 2.viii.1983, C.S. Guppy (3♂, 1♀, BCPM), S. side, 5500', 13.viii.1983, C.S. Guppy (1♀, BCPM); Robson (all leg. H.R. Foxlee) 18.vi.1940 (1♀, USNM), 26.v.1940 (1♀, USNM), 16.vi.1940 (1♀, USNM), 27.ix.1941 (1♀, BMNH), 7.v.1944 (3♂, 3♀, USNM), 11.v.1941 (1♂, USNM), 24.v.1947 (2♂, 2♀, CNCI), 6.vi.1947 (1♀, CNCI), 17.vi.1947 (1♂, CNCI), 19.vi.1947 (1♀, CNCI), 5.vii.1947 (1♂, CNCI), 8.vii.1947 (1♀, CNCI), 16.vii.1947 (1♀, CNCI), 31.v.1948 (1♂, CNCI), 15.vi.1948 (1♀, BCPM), 31.vi.1948 (1♂, CNCI), 2.v.1949 (1♀, SMDV), 10.v.1949 (1♂, 2♀, CNCI), 11.vi.1949 (1♂, CNCI), 12.v.1949 (4♂, 1♀, CNCI; 1♂, BCPM; 1♂, SMDV), 14.v.1949 (1♂, CNCI), 18.v.1949 (1♂, CNCI), 20.v.1949 (1♀, CNCI), 23.v.1949 (1♂, CNCI), 25.v.1949 (1♂, 1♀, CNCI), 2.vi.1949 (1♀, CNCI), 13.vi.194? (1♂, 1♀, BCPM), 31.v.1950 (2♂, CNCI), 10.v.1950 (1♂, 1♀, CNCI), 3.vi.1950 (1♂, 1♀, CNCI), 9.v.1950 (1♂, CNCI), 12.v.1950 (1♂, CNCI), 13.vi.1950 (1♀, CNCI), 15.vi.1950 (1♂, CNCI), 16.vi.1950 (1♂, 1♀, CNCI), 17.vi.1950 (1♀, BCPM), 20.vi.1950 (1♂, CNCI), 26.vi.1950 (1♂, CNCI), 5.vii.1950 (1♀, CNCI), 8.vii.1950 (1♀, CNCI), 18.vii.1950 (1♀, BCPM), 28.vii.1950 (2♀, CNCI), 24.v.1952 (1♀, CNCI), 2.vi.1952 (1♀, CNCI), 11.vi.1952 (1♀, CNCI), 27.vi.1952 (1♂, CNCI), 29.v.1953 (1♀, SMDV), 29.vii.1953 (1♂, SMDV), 3.vi.1955 (1♂, SMDV), 14.v.1956 (1♂, SMDV), 22.v.1956, 1♂, SMDV), 23.v.1956 (1♂, 1♀, SMDV), 23.v.1958 (1♂, SMDV), 26.vi.1958 (1♀, USNM), 31.v.1961 (1♀, SMDV), 5.v.1962 (1♀, SMDV), 17.v.1962 (1♀, SMDV), 8.vi.1962 (1♂, 1♀, SMDV), 14.vi.1962 (1♀, BCPM), 21.vi.1962 (1♂, SMDV), 19.vii.1962 (1♀, BCPM), 16.vi.1963 (1♀, BCPM), 10.vi.1968 (1♂, BCPM), 24.v.1969 (1♀, BCPM); Robson, Brilliant, 12.v.1944 (1♂, 1♀, USNM), 15.v.1944 (1♀, BMNH), 18.v.1944 (1♂, USNM), 19.v.1944, H.R. Foxlee (1♂, 1♀, SMDV), Rock Creek, Mt. Baldy Rd., 15.vi.1985 (1♂, 5♀, SMDV); Robson, Waldies Rd., 21.vi.1947 (1♀, CNCI), 30.vi.1947 (1♀, CNCI); Tahumming R., 14.viii.1934, J.K. Jacob (2♀, BCPM); Vernon, Becker L. Rd., 3800', 23.vi.1984, R.A. Cannings (1♂, 2♀, BCPM), Vernon, BX Range, 22.v.1984, R.A. Cannings (1♀, BCPM), 25.v.1984, R.A. Cannings (1♀, BCPM).

　　U.S.A.: California, Bluff Falls, 29.vi.1963, J. Wilcox (1♂, CASC); Calaveras Co., Big Meadow, 24.vi.1963, P.H. Arnaud, Jr (2♂, 3♀, CASC); El Dorado Co., Hwy.88 x Mormon Immigrant Trail, 13.vii.1986, L.G. Bezark (5♂, 2♀, LGBC), 20.vi.1986, L.G. Bezark (1♂, 4♀, LGBC); El Dorado Co., 12mi N. on Ice House Rd., 31.vi.1975, R.W. Brooks (1♂, UCDC); Mono Co., Sonora Pass, 9000-10000', 10.vii.1957, J.M. Burns (1♂, 1♀, EMEC); Nevada Co., Truckee, 31mi W., 13.vi.1964, M. Irwin (2♂, 2♀, UCRC); Plumas Co., Humbug Cr., Portola, 3mi NW, 5100', 18.v.1982, J.A. Powell (12♂, 12♀ [3pr in cop], EMEC); Tulare Co., Sequoia Nat. Park, Kings R. overlook, 23.vi.1948, J. Wilcox (6♂, 1♀, CASC), J. Wilcox, Jr (2♂, 3♀, CASC); Sequoia Nat. Park, Quaking Aspen campgr., 30-31.v.1979, T.D. Eichlin (1♂, 3♀, EMFC; 4♂, 1♀, LGBC), 28.vi.-5.viii.1981, J. Shelton (1♀, LGBC); Tulare Co., Wolverton, 29.vi-6.vii.1985, D.J. Burdick (1♂, 1♀, LGBC); Yosemite Nat. Park, Saddlebag L., 2.viii.1936, W.B. Herms (2♀, CASC). **Idaho,** Alpha, Long Valley, 20.v.1934, C.H. Martin (1♂, FSCA), 24.vi.1934, C.H. Martin (1♀, ODAC), 27.v.1934, D. Martin (1♂, FSCA), 10.vi.1934, C.H. Martin (1♀, FSCA), 10.vi.1934, D. Martin (1♂, FSCA), 24.vi.1934, D. Martin (1♂, 1♀, FSCA), C.H. Martin (2♂, FSCA); Blaine Co., Alice L., creek below Sawtooth Mts., 7600', 27.vii.1976, T. Griswold (1♀, EMUS); Blue Nose Mt., 8000', 17.vii.1949, C.B. Philip (1♀, CASC); Bonner Co., Grant Cr., Nordman, 6mi NW, 11.vi.1975, P.J. Landolt (1♂, 1♀, WSUC); Boundary Co., Copeland, Trout Cr., 9.vii.1968, A.R. Gittins

(1♀, ESUW); Boundary Co., Moyie R., 9mi N Moyie Springs, 22.vi.1982, R.S. Zack (1♂, 1♀, WSUC); Moyie R., Placer Cr., 3.vii.1976, D.S. Horning (5♂, 2♀, ESUW); Butte Co., Crater of the Moon Nat. Mon., 5.vii.1967, D.S. Horning, Jr (1♀, ESUW); Custer Co., Trail Cr. summit, 12.vii.1978, G.L. Jones (1♀, ESUW); Franklin Co., Cub River Canyon, 18.vii.1971, G.F. Knowlton and G.E. Bohart (1♂, 1♀, EMUS); Idaho Co., 2mi SE Burgdorff, 16.vii.1969, W.F. Barr (1♂, ESUW); Idaho Co., Frog Saddle, 7mi NNE Selway Falls, 6800', 15.vii.1979, W.J. Turner (1♀, WSUC); Idaho Co., Lolo Pass, 2.vii.1977, W.F. Barr (1♂, ESUW); Kootenai Co., Athol, 27.v.1974, W.F. Barr (1♀, ESUW), Kootenai Co., 6mi W Athol, 1.vi.1967, W.F. Barr (9♂, 8♀, ESUW), 8.vi.1967, W.F. Barr (2♂, 2♀, ESUW); Kootenai Co., Twin Lakes, 21.v.1970, W.F. Barr (6♂, 3♀, ESUW); Latah Co., Boulder Cr., 3mi S. Helmer L., 17.viii.1982, W.J. Turner (1♀, WSUC), 28.viii.1983 (1♀, WSUC), 19.viii.1983 (1♀, WSUC); Latah Co., Moscow Mt., 23.vi.1919, F.R. Cole (1♂, EMEC), 10.vi.1930, J.M. Aldrich (1♂, EMEC), 8.vi.1965, R.W. Dawson (3♂, 2♀, WSUC), 20.vii.1966, R.L. Westcott (1♀, ESUW), 4.vi.1970, W.F. Barr (1♂, 1♀, ESUW), 29.vi.1971, W. Turner (1♀, WSUC), 29.vi.1977, W.J. Turner (1♂, 1♀, WSUC); Little Round L., 10.vi.1963, D.J. Burdick (1♂, LGBC); Trout Cr., 12mi E Sandpoint, 8.vi.1967 (4♂, 2♀, ESUW); Valley Co., Warm L., 3mi S, 14.vii.1978, R.C. Biggam (1♀, ESUW); Valley Co., Ponderosa St. Park, 19.vi.1970, W.B. Garnett (2♀, WSUC); Valley Co., Upper Payette L., 4.vii.1967, E.J. Allen (1♀, ESUW). **Montana,** Flathead Co., Kalispell, 29.v.1982, D. Lester (1♂, MTEC). **Nevada,** Glenbrook, Hwy50, 5000', 16.v.1956, J. Wilcox (6♂, 6♀, CASC). **Oregon,** Baker Co., Whitman Nat. For., Anthony L., 28.vii.1954, J.H. Baker (1♂, WSUC; 1♀, CNCI); Baker Co., Wallowa Mts., 6500', Fish Cr. at Hwy 550, SEc.15, T6S, R46E, 30.vii.1971, R.L. Westcott (1♂, 3♀, ODAC); Blue Mts., Aneroid L., 7500', 23.vii.1929, H.A. Scullen (2♂, 1♀, CASC; 2♀, OSUC), 24.vii.1929, H.A. Scullen (2♂, EMEC; 1♂, OSUC; 1♂, 2♀, CASC); Blue Mts., Horseshoe L., 7500', 26.vii.1929, H.A. Scullen (1♂, OSUC); Blue Mts., Toll Gate Pass, 13.vii.1946 (1♀, EMFC), 12.vii.1950, M.T. and H.B. James (5♂, 2♀, WSUC); Clackamas Co., Timberline Lodge, 6000', 12.viii.1976, J.A. Powell (1♀, EMEC); Deschutes Co., Crater Cr. ditch, Sec.5, T18S, R9E, 31.viii.1972, R.L. Westcott (4♂, 3♀, ODAC); Deschutes Co., Dutchman Flat, 26.vii.1968, R.L. Westcott, 1♀, EMFC): Deschutes Co., McKenzie Pass, E side near summit, 1585m, 25.vii.1974, P.H. Arnaud, Jr (1♂, CASC); Deschutes Co., Mt. Batchelor, 16.vii.1970, K.J. Goeden (1♀, ODAC); Elk Cr., Elk L., 6.viii.1948, K. Fender (1♂, CASC); Grant Co., Onion Cr. Meadows, 7700', 18.vii.1936, R.E. Rieder (1♂, WSUC; 1♀, FSCA; 1♀, OSUO), H.A. Scullen (1♂, FSCA); Grant Co., Strawberry Mt., 28.vii.1946 (1♂, FSCA); Haines, 10.vii.1931, J. Nottingham (1♀, SEMC), R.H. Beamer (1♀, SEMC); Hood River, Co., Hood River Meadows, Sec.11, T3S, R9E, 22.viii.1964, K.J. Goeden (1♀, ODAC); Jefferson Co., Mt. Jefferson Wilderness Area, Rockpile L., 6200', 13.viii.1989 (R.L. Westcott (1♀, ODAC); Lane Co., Three Sisters Wilderness, Obsidian Flats, 6800', 14.viii.1988, R.L. Westcott (3♂, ODAC); Linn Co., Cascade Range, Tombstone Prairie, 25.vii.1974, R.L. Westcott (1♂, ODAC); Linn Co., Lost L., 3980', 28.vii.1970, P.H. Arnaud, Jr (1♂, CASC); Linn Co., Santiam Jct., 6mi E, 21.vii.1969, E.M. Fisher (3♂, 2♀, EMFC); Linn Co., Santiam Pass, 12.viii.1946, (1♀, FSCA), Santiam Pass, Lost Prairie, 16.vii.1946 (2♂, FSCA), 20.vii.1946 (ODAC), 26.vii.1946 (1♂, 1♀, FSCA), 1.vii.1948 (2♂, 2♀ [1 pr cop], FSCA); McKenzie Pass, 20.vi.1934, S.C. Jones (1♀, USNM); Meacham, 23.v.1947, J.E. Davis (2♂, 3♀, FSCA); Mt. Hood, 3000-6000', 5.viii.1925, C.L. Fox (3♂, 3♀, EMEC); Mt. Hood, 7000', 10.viii.1961, R.M. Bohart (1♂, 1♀, UCDC); Mt. Hood, 29.vii.1966, F.C. Harmston (2♂, 1♀, EMFC); Mt. Hood,

Cloud Cap Inn, 19.vii.1947 (1♀, EMFC); Cloud Cap Rd., 6.vii.1947, D. Martin (1♂, FSCA); Mt. Hood, N Rec. Area, 6500', 11.viii.1980, J.B. Johnson (1♂, LGBC); Mt. Hood, timberline near Gov't Camp, 20.vii.1937, E.C. Van Dyke (2♂, 4♀, CASC), 22.vii.1937, E.C. Van Dyke (2♂, 3♀, CASC), 28.vii.1937, E.C. Van Dyke (1♂, 4♀, CASC); Mt. Hood, Still Cr. Forest Camp, 3900', 19.vii.1969, E.M. and J.L. Fisher (1♂, EMFC); North Sister, White Branch Meadows, 6.viii.1935, G. Ferguson (1♀, OSUO); Pine Creek, 14 mi W of Baker, 6000', 25.vii.1968, Goeden and Westcott (1♂, ODAC); Sumpter, 9.vi.1934, C.H. Martin (4♂, FSCA); Swim, 2.vii.1942, G. Ferguson (1♂, FSCA); Timpanogas, Glacier Lake, E. Hardy (5♂, 6♀, BYUC); Umatilla Co., Tollgate Pass, Hwy 204, 9.vi.1965, K. Goeden (2♀, ODAC); Tollgate Pass, 5 mi W Tollgate, 12.vi.1968, R.L. Westcott (2♀, ODAC); Union, 15.viii.1964, E.L. Livingston (1♀, BYUC); Union Co., Ladd Canyon, 14mi S LaGrande, 4280', 6-12.vii.1976, E.J. Davis (1♂, WSUC), 17-19.vi.1976, E.J. Davis (1♂, WSUC); Union Co., Lick Cr., 26mi. SE Union, 4280', 1-3.vii.1976, E.J. Davis (1♂, WSUC), Lick Cr., 28mi SE Union, 4280', 27-30.vi.1976, E.J. Davis (1♀, WSUC), 4920', 29.vi.-5.vii.1975, E.J. Davis (1♂, WSUC), 13-16.vii.1977, E.J. Davis (1♀, WSUC); Wallowa Co., Hat Point, 29.vii.1969, K.J. Goeden (1♀, ODAC); Wallowa Nat. For., Lostine R., French For. Camp, 26.viii.1952 (1♀, CNCI). **Utah,** Cache Co., Franklin Basin, 1-14.vii.1995, W.J. Hanson (1♂, EMUS), 4-14.viii.1995, S. Keller (1♂, EMUS), 15-21.vii.1995, W. Hanson and S. Keller (4♂, EMUS); Cache Co., Green Canyon, 15-19.vi.1985, N.N. Youssef (1♂, EMUS), 18-23.vi.1985, N.N. Youssef (2♂, EMUS); Cache Co., Tony Grove Cr., 19-27.vii.1983, W.J. Hanson (1♂, EMUS), 27.vii.-2.viii.1983 (W.J. Hanson (2♂, 1♀, EMUS), 22-29.vi.1990, W.J. Hanson (1♂, EMUS), 1-3.vii.1990, W.J. Hanson (1♂, EMUS); Cache Co., Wellsville Mtns trail, E of Stewart Pass, 7200-8200', 7.vii.1993, T. Griswold (1♀, EMUS); Daniels Pass, 2mi S Wasatch Co. line, 8500', 9.vii.1961, B.H. Poole (1♂, CNCI; 1♂, BCPM); Hanna, 14.vii.1949, L.D. Beamer (2♂, 1♀, SEMC); Mt. Logan, 24.vi.1938, G.F. Knowlton (1♀, USNM), G.F. Knowlton and R.Y. Nye (1♂, 2♀, EMUS); Park City, 3.vii.1922, E.P. Van Duzee (1♂, EMEC); Parl Cr., 3.vii.1922, E.P. Van Duzee (1♀, EMEC); Provo, D.E. Johnson (2♂, 1♀, USNM); Provo Canyon, N Fork, D.E. Johnson (1♀, BYUC); Rich Co., Monte Christo, 6.vii.1968, G.F. Knowlton (1♀, EMUS), 23.vii.1975, G.F. Knowlton (1♂, EMUS), 6.vii.1976, W.J. Hanson (1♀, EMUS), 21.vii.1978, Knowlton and Hanson (1♀, EMUS); 5.viii.1980, Hanson, Clemens and Keller (1♂, BCPM); Rich Co., Logan Canyon, 26.v.1934, G.F. Knowlton (1♀, EMUS), 9.vii.1949, W.J. Hanson (1♂, 1♀, BCPM); Logan Canyon summit, 9.vii.1949, W.J. Hanson (1♂, 1♀, EMUS), 3-11.vii.1980 (1♂, 2♀, EMUS), 26.vi.-2.vii. 1982 (3♂, 2♀, EMUS), 2-8.vii.1982 (2♂, 1♀, EMUS), 8-16.vii.1982 (1♀, EMUS), 18.vi.-3.vii.1991, W.J. Hanson (4♂, 4♀, EMUS); Summit Co., Uinta Mts., Christmas Meadows, 11.vii.1996, R.W. Baumann (2♂, 2♀, BYUC); Utah Co., Aspen, Scout Falls, 11.vii.1991, H. Spafford (1♂, BYUC); Utah Co., Aspen Grove, 23.vii.1955, D.E. Johnson (1♂, BYUC); Utah Co., Mt. Timpanogos, 12.vii.1959, D.E. Johnson (1♂, BYUC); Utah Co., Nebo Bench Trailhaed, Nebo Loop Rd., 17.vii.1993, Baumann and Ochoa (2♂, 3♀, BYUC); Utah Co., Timberline Camp, 11.vii.1995, K. Hansen (1♂, 1♀, BYUC). **Washington,** Asotin, 27.vi.1932, J.M. Aldrich (1♀, EMEC); Garfield Co., Misery Spring, 28mi S Pomeroy, 27.vii.1972, W.J. Turner and W.B. Garnett (2♀, WSUC); Lake Cushman, 19.viii.1922, P.G. Putnam (1♀, WSUC); Mt. Adams, 6000', 3.vii.1935, M.C. Lane (1♂, 1♀, CASC); Mt. Adams, Bird Cr., 4-6000', 24.vii.1921, M.C. Lane (1♀, USNM), 6500', 4.viii.1973, R.L. Westcott (2♂, EMFC); Mt. Logan, 24.vi.1938, G.F. Knowlton and R.E. Nye (2♂, USNM), Mt. Rainier, 26.vii, F.M. Hull

(1♀, BCPM; 1♂, 3♀, CNCI), 19.vii.1931, R. Latta (1♂, CASC); Mt. Rainier, Alta Vista, 13.viii.1933, D. Martin (1♂, FSCA); Mt. Rainier, Currant Flat, 7.vii.1935, J. Wilcox (2♀, CASC), 25.vii.1935, Wm.W. Baker (1♀, CASC); Mt. Rainier, Ipsut Cr. Camp, 17.vii.1935, J. Wilcox (1♀, CASC); Mt. Rainier, Ohanapecosh, 14.vii.1935, Wm.W. Baker (1♀, CASC); Mt. Rainier, Paradise, 25.vii.1910, J. Wilcox (1♂, EMEC), 1.viii.1919, E.C. Van Dyke (1♂, EMEC), 17.vii.1920, E.C. Van Dyke (1♀, EMEC), 25.vii.1920, E.C. Van Dyke (3♂, EMEC), 23.vii.1935, H. Wilson (1♂, CSUC), 15.viii.1935, J. Wilcox (6♀, CASC), S.E. Crumb, Jr (3♂, 2♀, OSUO), 29.vii.1949, L.D. Beamer (1♀, SEMC), 16.viii.1957, A. and H. Dietrich (1♂, CUIC); Mt. Rainier, Shadow Lake nr Sunrise, 10.viii.1977, R.S. Zack (2♂, 1♀, WSUC); Mt. Rainier, Shadow Lake, 6200', 25.vii.1932, C.H. and D. Martin (1♂, FSCA), 31.vii.1932, C.H. and D. Martin (9♂, FSCA), 15.viii.1932, C.H. and D. Martin (1♂, FSCA); Shadow Lake, 6500', 31.vii.1932, C.H. and D. Martin (1♀, FSCA), 15.viii.1932, C.H. and D. Martin (1♂, 1♀, EMEC; 1♂, 3♀, FSCA; 2♂, SEMC), 24.viii.1932, D. Martin (1♂, FSCA), 6600', 27.viii.1933, C.H. and D. Martin (2♂, 1♀, FSCA), 6800', 31.vii.1931, C.H. and D. Martin (1♀, FSCA; 1♀, SEMC), 28.viii.1932, D. Martin (1♀, SEMC); Mt. Rainier, Sunrise, 6000', Emmons Glen View Trail, 10.viii.1977, W.J. Turner (2♂, WSUC), Sunrise, 6318', 23.vii.1932, J. Wilcox (2♂, 4♀, CASC, 1♀, EMEC; 2♀, UASM), I. Wilcox (1♀, CASC), 24.vii.1932, J. Wilcox (1♂, EMFC; 1♂, EMEC), 27.vii.1932, J. Wilcox (1♂, 1♀, AMNH; 1♂, FSCA; 4♂, 3♀, CASC; 1♂, 1♀, EDNC; 1♂, 1♀, MTEC; 2♂, UASM; 2♂, 2♀, TAMU; 1♂, 2♀, WSUC; 4♂, 1♀, EMEC; 1♀, CUIC), S.E. Crumb (1♀, BCPM; 1♂, EMFC; 2♂, FMNH; 1♂, 1♀, EMEC; 1♀, CASC), 31.vii.1932, J. Wilcox (1♂, CASC); Sunrise, 6380', 24.vii.1932, J. Wilcox (3♀, CASC; 1♂, CUCC), 27.vii.1932, Wm.W.Baker (6♀, CASC; 2♀, FMNH; 2♂, 1♀, EMEC; 1♀, EMFC), J. Wilcox (1♂, 1♀, KSUC; 1♂, BPBM; 1♂, ANSP), 31.vii.1932, S.E. Crumb (1♀, EMEC), 14.viii.1932, J. Wilcox (2♂, 1 prey of *Cyrtopogon semitarius*, EMEC; 1♀, CUCC; 1♂, EMFC; 2♂, CASC; 1♀, CUIC; 1♂, BPBM; 1♀, CSUC), 23.viii.1932, J. Wilcox (1♂, CASC), 27.viii.1932, J. Wilcox (1♂, EMEC; 1♂, 1♀, WSUC; 2♂, 1♀, CASC; 1♂, UCRC; 2♂, EMFC; 1♂, 1♀, USNM), Wm.W. Baker, (1♀, EMFC), 28.viii.1932, J. Wilcox (1♂, CSUC; 1♂, 1♀, CASC; 1♂, CUIC; 1♀, EMEC), 3.ix.1932, J. Wilcox (1♂, CASC; 1♂, UCRC), Sunrise, 6400', 29.vii.1933, J. Wilcox (4♂, CASC), 30.vii.1933, C.H. Martin (2♂, FSCA; 1♂, 1♀, CUIC), C.H. and D. Martin (2♂, EMFC; 1♂, FSCA), I. Wilcox (1♀, CASC), 17.vi.1934, J. Wilcox (3♂, 4♀, CASC), 24.vi.1935, Wm.W. Baker (1♂, CASC), 20.vii.1935, J. Wilcox, 4♂, 3♀, CASC; 1♂, USNM), A.E. Bonn (4♂, 3♀, CASC; 2♀, USNM), S.E. Crumb, Jr (1♂, 2♀, OSUO), 28.vii.1935, J. Wilcox (1♂, USNM), 5.viii.1935, Wm.W. Baker (14♂, 4♀, CASC; 1♂, EMFC; 2♀, UCRC; 1♀, USNM), I. Wilcox (8♂, 3♀, CASC), J. Wilcox (11♂, 7♀, CASC; 1♀, USNM), 6.viii.1935, J. Wilcox (2♂, 3♀, CASC), 7.viii.1935, S.E. Crumb (2♂, 3♀, CASC), I. Wilcox (1♂, CASC), J. Wilcox (1♂, CASC), 8.viii.1935, I. Wilcox (1♀, CASC), 9.viii.1935, I. Wilcox (4♀, CASC), J. Wilcox (1♂, 2♀, CASC; 1♂, USNM), 25.vii.1936, J. Wilcox (8♂, 5♀, 1 pr in cop, CASC), Wm.W. Baker (1♂, FSCA; 1♂, EMFC), 2.viii.1936, Wm. W. Baker (1♂, CASC); Mt. Rainier, Sunrise Pk, 20.vii.1936, E.C. Van Dyke (1♂, CASC), 21.vii.1936, E.C. Van Dyke (2♂, 1♀, CASC), 23.vii.1936, E.C. Van Dyke (20♂, 16♀ CASC), 24.vii.1936, E.C. Van Dyke (1♀, CASC), 25.vii.1936, E.C. Van Dyke (5♂, 1♀, CASC), 20.vii.1937, E.C. Van Dyke (1♂, CASC), Mt. Rainier, White River Camp, 4500', 28.viii.1932, I. Wilcox (1♂, CASC), 4.ix.1932, D. Martin (1♂, MSUC), J. Wilcox (3♂, CASC; 1♂, KSUC), 30.vii.1933, I. Wilcox (5♂, CASC), J. Wilcox (1♂, BCPM), 2.viii.1933, D. Martin (1♂, ESUW), 24.viii.1932, D.

Martin (1♂, CUIC), 27.viii.1933, C.H. and D. Martin (1♀, ESUW), C.H. Martin (1♂, CUIC; 1♂, FSCA), 17.vi.1934, J. Wilcox (1♂, CASC), 20.vii.1935, J. Wilcox (1♂, 1♀, CASC), A.E. Bonn (1♂, CASC), S.E. Crumb Jr (2♂, OSUO), 7.viii.1935, J. Wilcox (1♂, CASC; 1♂, USNM), 27.vii.1936, J. Wilcox (4♂, 2♀, CASC), 9-10.viii.1977, W.J. Turner (1♀, WSUC); Okanogan Co., 24.vi.1964, H.R. Dodge (1♀, WSUC); Pend Oreille Co., Bead L., 8mi NE Newport, 23.vi.1983, R.S. Zack (1♂, WSUC); Red Mtn., 8.vii.1938, K. Grey and J. Schuh (4♂, 3♀, USNM; 4♂, 5♀, WSUC; 2♂, FSCA); Salmon Meadows, 9mi NW Cononully, 4500', 3-6.vii.1975, N.E. Woodley (2♂, 2♀, WSUC); Snoqualmie Nat. For., Sheep Lake, 6.viii.1943, C.H. Martin (2♂, 1♀, FSCA; 1♀, MSUC; 1♂, CUIC); Spokane Co., Mt. Spokane, 5500', 10.vii.1984, J. Jenkins (2♂, 3♀, CSUC); Mt. Spokane, Bald Knob, 4500', 22.vi.1957, H.S. Dybas (2♂, 5♀, FMNH), 4800-5200', nr Bald Knob Campgr., 9-10.vii.1975, W.J. Turner (23♂, 32♀, 1pr in cop, WSUC; 1♀, ESUW), 21-22.vii.1975, W.J. Turner (4♀, WSUC), 25.vi.1979, W.J. Turner (1♂, 2♀, WSUC), 5200', nr Bald Knob Campgr., 6.vii.1976, W.J. Turner (4♂, WSUC), 28.vi.1977, W.J. Turner (3♂, 2♀, WSUC), 3.vi.1986, W.J. Turner (1♂, 1♀, WSUC), 10.vi.1986, W.J. Turner (2♂, 1♀, WSUC), 5500', nr Bald Knob Campgr., 10.vii.1984, W.J. Turner (2♂, 4♀, WSUC); Stevens Co., Deer Lake, 13 airmiles SE Chewelah, 29.vi.1975, M.T. James (1♀, WSUC), 30.vi.1972, M.T. James (3♂, 1♀, WSUC), 22.vi.1973, M.T. James (1♂, WSUC), 8.vii.1973, M.T. James (1♂, WSUC); Stevens Co., Little Pend Oreille Lakes, 15.vi.1951, M.T. James (2♂, 3♀, WSUC); Tipsoo Lake, 23.vii.1932, J. Wilcox (1♀, CASC), 24.vii.1932, J. Wilcox (1♂, CASC), 28.vii.1932, J. Wilcox (4♂, 1♀, CASC), 24.viii.1935, Wm.W. Baker (1♀, CASC); Yakima Co., Bear Cr., 8mi SW Tieton Res. Sta., 24-25.vi.1974, W.J. Turner (1♂, WSUC). **Wyoming,** Grand Teton Nat. Park, 14.vii.1939, D.J. and J.N. Knull (2♂, OSUC), 19.vi.1942, E. Kenaga (1♂, MICH ST); 23.vi.1938, E.C. Van Dyke (1♂, CASC); Jackson Hole, 28.vi.1938, E.C. Van Dyke (1♀, CASC); Sierra Madre Range, Battle L. Rd., 8000', 18.vii.1961, J.G. Chillcott 1♀, BCPM); Sublette Co., 39mi SE Boulder, Dutch Joe Guard Stn., 29.vi.1978, R. Lavigne (1♂, ESUW); Sublette Co., 42mi SE Boulder, 29.vi.1978, R. Lavigne (6♂, 7♀, 1pr in cop, ESUW); Sublette Co., Big Sandy Campgr., 46mi SE Boulder (cecidomyiid as prey), 29.vi.1978, R. Lavigne (5♀, ESUW); Teton Co., 6km NW Moran P.O., 19.vi.1980, K.M. O'Neill (1♂, CSUC); Teton Co., Grassy L., Targhee Nat. For., 10.viii.1967, R.J. Lavigne (1♂, ESUW), 13.viii.1967, R.J. Lavigne and W Paxton (1♀, ESUW).

Type Locality. U.S.A., Washington, Mount Adams.

Etymology. Latin *montis* = mountain; *cola* = dweller.

Distribution. Nearctic; B.C. and Montana south to California, Utah and Wyoming. Map 6, next page.

Phylogenetic Relationships. Lone member of the *monticola* group.

Natural History. Habitat: mountain forests, especially subalpine habitats. Known flight dates range from 2 May to 27 September.

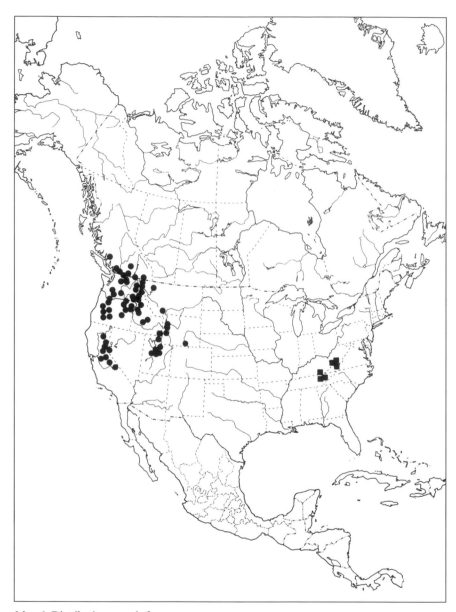

Map 6. Distribution records for
- *Lasiopogon monticola*
- ■ *L. appalachensis*

Lasiopogon oklahomensis Cole and Wilcox

Lasiopogon oklahomensis Cole and Wilcox, 1938. *Entomologica Americana* 43: 57-58.

Diagnosis. A medium-sized grey/grey-brown species with dark brown legs; mystax white, other head bristles dark. Antennae brown, F2+3/F1 = 0.56-0.69. Thoracic tomentum dorsally and laterally grey (often richer brown dorsally) with brown dorsocentral and grey or brown-grey acrostichal stripes. Main thoracic and leg bristles dark, except for pale katatergites; anterior dorsocentral bristles 2-5, strong or weak. Abdominal tergites basally with gold-brown tomentum; apical bands grey, covering at least 0.5 the length of the tergite and sometimes weakly extending anteriorly along the midline. All hairs and bristles anterior to genitalia pale; dorsal setulae in female mostly brown. Width of epandrium 0.45-0.5 the length in lateral view, the greatest width at about half the length, the apex rounded. In dorsal view strongly concave medially. Gonostylus with weak secondary medial flange. Female with brown/black terminalia and white hairs; hypogynial valves with ventral carina.

Description. Body length ♂ 8.9-9.4mm; ♀ 9.1-10.4mm.

Head. HW ♂ 1.82-2.02mm; ♀ 1.86-2.14mm. FW ♂ 0.40-0.46mm; ♀ 0.40-0.46mm. VW ♂ 0.72-0.80mm; ♀ 0.75-0.86. VW/HW = ♂ 0.40; ♀ 0.39-0.41. FW/VW = ♂ 0.55-0.58; ♀ 0.53-0.55; VD/VW = ♂ 0.19-0.21; ♀ 0.19-0.24. GH/GL = ♂ 0.32-0.35; ♀ 0.29-0.36.

Face silver or gold-silver, vertex grey with brown highlights or strongly gold-brown, occiput grey with faint brown highlights dorsally or strongly gold-brown dorsally. Beard and labial hairs white, mystax bristles white, sometimes with a few dark setae dorsally. Occipital bristles rather strong, sparse, about 5 medially the strongest, curved anterolaterally; lateral and ventral ones shorter, straighter. Frontal and orbital setae sparse, brown, sometimes a few white; the longest orbitals are about 0.4mm, about as long as F1, but shorter than the rather strong black ocellar bristles.

Antennae. Brown, some with scape+pedicel or F2+3 light brown; base of F1 usually paler. Setae brown, some specimens with a few white setae on scape; F1 and even F2+3 sometimes with a seta. F1 variable in shape, usually widest at midlength, the ventral and dorsal margins convex. WF1/LF1 = ♂ 0.29-0.31; ♀ 0.29-0.31. LF2+3/LF1 = ♂ 0.56-0.59; ♀ 0.56-0.61.

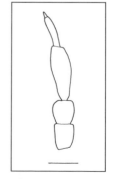

Figure 143. Antenna. Scale line = 0.2 mm.

Thorax. Prothorax silver-grey, often brown on antepronotum, with white hairs; postponotal lobes grey, the lateral angle yellow to ferruginous, hairs white to pale brown.

Scutum tomentum grey or even almost all gold-brown, at least on prescutum. Dorsocentral stripes faint gold brown when scutum is grey, obscure brown or rich gold-brown bordered with gold in the browner specimens. Acrostichal stripes range from obscure grey to dark brown-grey. All notal setae brown or black. Anterior dorsocentrals 2-5, mixed with finer setae, and ranging from all weak in some specimens to strong in others (longest to 1.0mm); 2-3 posteriors. Notal setae scattered, but concentrated on anterior intermediate spot and paramedial stripes, as long as shorter dorsocentrals. Postalars 1-2, with shorter hairs; supra-alars 1-2; presuturals 1-2, 0-1 weak posthumerals. Scutellar tomentum gold-grey; apical scutellar bristles dark, 1-2 strong ones on each side mixed with shorter, weaker dark bristles and hairs, some of which may be white.

Pleural tomentum grey to gold-grey. Katatergite setae 5-7 white/yellow, sometimes with a few brown, among finer white hairs; katepisternal setae sparse, often long. Anepisternal setae 2-4 brown, short, rather weak, in some specimens to 0.4mm, a few additional white ones ventrally; a patch of short erect brown hairs on dorsal shelf. Anepimeron without setae.

Legs: Base colour dark brown/black; some specimens with tibiae and tarsi lighter than femora. Tomentum of coxae grey with some gold highlights; tomentum on rest of legs grey. No coxal peg. Main bristles dark brown/black, finer setae white to brown. Femora dorsally with white decumbent hairs, sometimes some dark apically. Longer erect pale hairs more-or-less restricted to venter of femora; longest white ventral setae usually as long as, or longer than, thickness of femur. Profemur with 5-7 stronger dorsolateral dark bristles; mesofemur with 4-7; metafemur with 5-7. Tibiae and tarsi with dark, strong bristles typically arranged, hairs mostly white/brown on tibiae, brown on tarsi. Protibia with longest bristles about 3.0-4.0 times longer than tibial width.

Wings: Veins light to medium brown; membrane faint pale brown in oblique view. DCI = 0.36-0.44; cell M3 open. Halter yellow; knob without dark spot.

Abdomen: Male. Tergite basal colour brown. Tomentum on tergite bases gold-brown, sometimes absent on tergite 7 and faint on tergite 2, usually somewhat divided into lateral patches by weak grey extensions of the apical grey bands. These bands cover at least half of each tergite apically (segment 1 is fully covered). Ventrolateral areas are broadly grey. All bristles and hairs anterior to genitalia white. About 3-4 strong bristles on each side of tergite 1; sometimes a few are brown. Lateral setae on tergites 1-3 rather short, sparse, erect. Sternite tomentum gold-grey, hairs white.

Female. As in male except grey apical tomentum has less tendency to extend anteriorly to bases of tergites. Lateral setae on 1-3 short, sparse; setulae on ventral margins of 4-7 white. Dorsal setulae mostly brown, mixed with some white ones, especially apically. Tergite 1 bristles white/yellow, but a few sometimes brown.

Figure 144. *L. oklahomensis* epandrium (top to bottom): lateral, medial, dorsal. Scale lines = 0.2 mm.

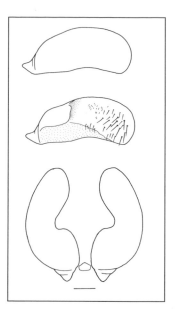

Male genitalia. Epandrium and hypandrium/ gonocoxite complex chestnut to dark brown/ black and covered with gold-grey tomentum except on parts of hypandrium in some specimens. Setal brush dark brown, other setae brown/dark brown, prominent, especially dorsally on epandrium. Width of epandrium halves in lateral view about 0.45-0.5 times the length, widest at about midlength, the apex rounded. Medial face of epandrium as in figure 144. In dorsal view, medial margins of epandrium strongly concave; basal sclerite strong.

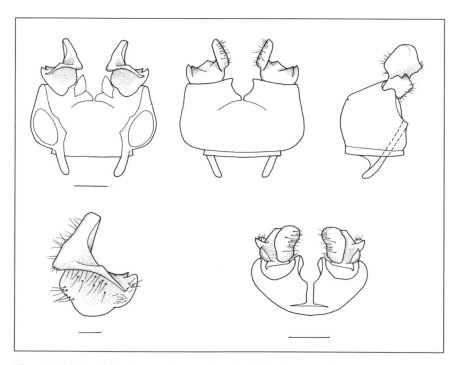

Figure 145. Hypandrium/gonocoxite complex. Clockwise from top left: dorsal, ventral, lateral and apical (scale lines = 0.3 mm); gonostylus, dorsal (scale line = 0.1 mm).

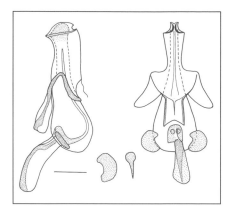

Figure 146. *L. oklahomensis* phallus: lateral (left); ventral (right); lateral ejaculatory process, basal (centre left); ejaculatory apodeme, cross-section (centre right). Scale line = 0.2 mm.

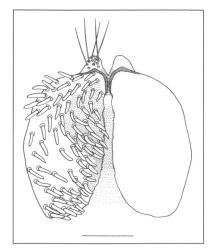

Figure 147. Subepandrial sclerite: ventral. Scale line = 0.2 mm.

Gonostylus. Medial flange strong, with weakly developed secondary flange; dorsomedial, lateral and ventrolateral teeth prominent. Hypandrium/gonocoxite complex in ventral view with length about 0.65 times the width, the transverse slit at 0.75 the length. In lateral view, exposed length of gonocoxal apodeme about 0.5 times basal width of hypandrium; apodeme with sclerotized web ventrally.

Phallus. Slender, elongate, paramere sheath dorsally 0.45 times the length of phallus (excluding ejaculatory apodeme). Apex of paramere sheath with ventrolateral carina, ventral flange absent. Sperm sac width in dorsal view 0.3 times the length of phallus. Ejaculatory apodeme with ventral margin convex in lateral view, moderately spatulate in ventral view; oval in cross-section with broad but thin dorsal carina. Subepandrial sclerite as in figure 147. Broadly triangular unsclerotized area in basal 0.25, the central 0.5 narrow with parallel sides; spines bluntly acute, sparse in central lateral area.

Female genitalia. Undissected: Hairs white, erect; Tergite 8 dark brown/black (sometimes chestnut), often paler brown apically. Sternite 8 dark brown/black, brown to ferruginous basally. Lateral lobe setae strong. Hypogynial valves with ventral carina and a few fine hairs basomedially. Cerci black with pale setae. Dissected sternite 8 (figure 148) with basal width about 0.75 times the length, undivided along midline. Length of unsclerotized area between hypogynial valves 0.4 times sternite length.

Tergite 9 sclerites as in figure 149; sternite 9 Y-shaped, medially undivided. Tergite 10 brown/black with 6-8 black acanthophorite spines on each side. Spermathecae (figure 150) with hook-shaped terminal reservoirs about as wide

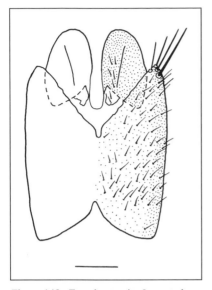

Figure 148. Female sternite 8: ventral.
Scale line = 0.2 mm.

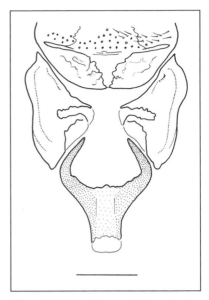

Figure 149. Female sternite 9, tergite 9
and lobes of sternite 10: ventral.
Scale line = 0.2 mm.

Figure 150.
Spermatheca: dorsal.
Scale line = 0.1 mm.

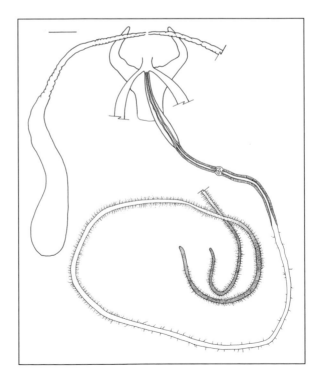

as reservoir duct; terminal reservoirs without wart-like protuberances on surface. Basal striated duct without fine scales; junction with basal duct vague, without scales; basal duct short, narrow.

Type Material.
HOLOTYPE. ♂ (examined) labelled: "[rectangular beige label] Ripley. Okla/21 iv.34/A.E. Pritchard"; "[rectangular beige label] blank; "[rectangular pink label] HOLOTYPE/Lasiopogon/oklahomensis/Cole and Wilcox". USNM.
ALLOTYPE. ♀ (examined) Same data as holotype (USNM).
PARATYPES. (18 examined). **U.S.A.: Oklahoma**, Ripley, 7.v.1934, A.E. Pritchard (1♂, USNM; 1♂, CASC). Stillwater, 21.iv.1935, C.G. Sooter (1♂, DEI; 3♂, 5♀, USNM; 1♀, EMEC; 1♀, CASC); 23.iv.1935, A.E. Pritchard (2♂, 1♀, USNM; 1♂, EMEC). Payne Co., 16.iv.1933, C.G. Sooter (1♂, USNM).

Other Material Examined (2 specimens).
U.S.A.: Oklahoma, Norman, 5.iv.1930, B. Dennis (1♀, USNM). Caddo Co., Red Rock Canyon State Park, 1 mi S Hinton, 28.iv.1965, R. and J. Matthews (1♂, LGBC).

Type Locality. U.S.A.; Oklahoma, Ripley.

Etymology. The type locality is in the state of Oklahoma.

Distribution. Nearctic; southern Great Plains of the U.S.A.; known only from Oklahoma. See map 9, p. 244.

Phylogenetic Relationships. Member of the *tetragrammus* species group; sister to clades containing *L. coconino, L. quadrivittatus, L. lavignei* and *L. woodorum, L. flammeus, L. chrysotus.*

Natural History. Habitat: riparian habitats on southern plains; known flight dates from 16 April to 4 May.

Lasiopogon opaculus Loew

Lasiopogon opaculus Loew, 1874. *Berlin Ent. Zeitschr.* 18: 367.
Lasiopogon carolinensis Cole and Wilcox, 1938. *Entomologica Americana* 43: 34-36. **New Synonymy**

Diagnosis. A medium sized grey/grey-brown species with dark brown legs and markings; mystax and other head bristles dark. Antennae brown, F2+3 rather long (F2+3/F1 = 0.60-0.80). Thoracic tomentum dorsally and laterally gold-grey (sometimes richer brown dorsally) with brown dorsocentral and acrostichal stripes. Main thoracic and leg bristles dark, except for pale katatergites; anterior dorsocentral bristles 2-4. Abdominal tergites basally with only thin brown tomentum (sometimes almost absent), the dark cuticle shining through. Apical bands gold-grey, covering 0.2 to 0.5 the length of the segments and sometimes extending anteriorly in a triangle along the midline. Bristles on tergite 1 pale. Epandrium black, the width about 0.4-0.45 the length in lateral view, the greatest width at about one-third the length, narrowing slightly to the rounded apex. In dorsal view only moderately concave medially. Female with black terminalia and dark hairs.

Description. Body length ♂ 8.7-10.5mm; ♀ 9.8-11.7mm.

Head. HW ♂ 1.80-2.00mm; ♀ 1.96-2.34mm. FW ♂ 0.36-0.40mm; ♀ 0.42-0.48mm. VW ♂ 0.86-0.96mm; ♀ 0.98-1.11mm. VW/HW = ♂ 0.48-0.50; ♀ 0.47-0.51. FW/VW = ♂ 0.40-0.42; ♀ 0.41-0.43; VD/VW = ♂ 0.13-0.17; ♀ 0.11-0.19. GH/GL = ♂ 0.29-0.34; ♀ 0.31-0.38.

Face grey or gold-grey, vertex brown-grey, occiput grey with brown highlights dorsally. Beard and labial hairs white, mystax bristles brown/black with a few white setae ventrally; all other setae brown/black. Occipital bristles rather fine, abundant, those behind the dorsomedial angle of the eye very long and strongly curved anterolaterally; lateral and ventral ones shorter, straighter. Frontal and orbital setae abundant, some as long as F1+F2+3.

Antennae. Brown to black, some with F2+3 light ferruginous. Setae brown; F1 sometimes with a seta. F1 short, slender, widest at about midlength. WF1/LF1= ♂ 0.32-0.37; ♀ 0.30-0.33. LF2+3/LF1 = ♂ 0.60-0.80; ♀ 0.65-0.79.

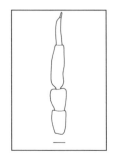

Figure 151. Antenna. Scale line = 0.1 mm.

Thorax. Prothorax grey, often brown on antepronotum, with white hairs; postponotal lobes grey, the lateral angle ferruginous, hairs white to brown. Scutum tomentum grey with brown highlights or even almost all brown. Dorsocentral stripes dark brown and, except for the widened anterior part, mostly disappearing in posterior and lateral views; acrostichal stripes brown, faint to strong. All notal setae brown or black. Anterior dorsocentrals 2-4 (longest to 0.8mm), mixed with finer setae; 4-6 posteriors. Notal setae scattered, but concentrated on anterior intermediate spot and acrostichal stripes, as long as shorter dorsocentrals. Postalars 1-3, with shorter hairs; supra-alars 1-3; presuturals 2-3, 1-3 weak posthumerals. Scutellar tomentum gold-grey; apical scutellar bristles dark, 3-6 strong ones on each side mixed with shorter, weaker dark bristles and hairs.

Pleural and scutellar tomentum gold-grey. Katatergite setae white/yellow, 6-10 among finer white hairs; katepisternal setae sparse, often long. Anepisternal setae 3-10, brown, a few moderately strong, to 0.7mm, a few white ventrally; a patch of short erect brown hairs on dorsal shelf. Anepimeron with a few long white setae.

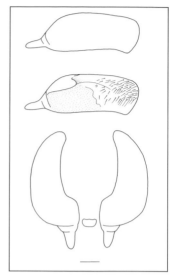

Figure 152. *L. opaculus* epandrium (top to bottom): lateral, medial, dorsal. Scale lines = 0.2 mm.

Legs. Base colour dark brown/black. Tomentum of coxae grey with some gold highlights; tomentum on rest of legs grey. No coxal peg. Main bristles dark brown/black, finer setae white to black. Femora dorsally with white decumbent hairs mainly basally, dark apically, although in some specimens most hairs are white. Longer erect pale hairs abundant ventrally and laterally, especially on profemur; many of these can be dark apically; in males longest white ventral setae longer than thickness of femur, at least on meso- and metafemora. Profemur with 5-10 stronger dorsolateral dark bristles; mesofemur with 2-6; metafemur with 3-15; these bristles often rather fine and mixed with finer setae, some basally can be straight and white. Tibiae and tarsi setose; dark, strong bristles typically arranged, hairs abundant, most brown/black. Protibia with longest bristles about 3.0-4.0 times longer than tibial width.

Wings. Veins brown to black; membrane pale brown in oblique view. DCI 0.38-0.45; cell M3 closed or narrowly open. Halter yellow; knob without dark spot.

Abdomen. Male. Tergite basal colour dark brown/black. Tomentum on tergite bases thin brown/gold-brown, usually extensively lacking so that dark cuticle shines through. Bands of grey tomentum cover 0.2 to 0.5 the length of each tergite apically (segment 1 is half to fully covered), the grey often extending thinly towards the base in a broad triangle; ventrolateral areas are narrowly gold-grey. About 4-8 strong bristles on each side of tergite 1 white, often yellow in females. Lateral setae on tergites 1-3 white, abundant, long and erect, those on 4 short and mixed with dark ones; setae on 5-7 dark. Dorsal setulae short and dark. Sternite tomentum gold-grey, hairs white.

Female. As in male except tomentum covers tergites completely. Brown tomentum more dense; apical grey bands extend more strongly in a triangle anteriorly on midline. Lateral setae on tergites 4-7 very short and dark.

Variation. Specimens from the southern parts of the range tend to be browner on the dorsum of the scutum and on the head. In males the grey apical bands on the abdominal tergites are mostly narrow (0.2-0.25 of the segment length) with little tendency to expand anteriorly along the midline; grey tomentum ventrolaterally is reduced. In females the grey bands dorsolaterally are 0.3-0.5 the length of the segment and expand anteriorly along the midline.

Male genitalia. Epandrium and hypandrium/gonocoxite complex dark brown/black and covered with grey tomentum except on parts of hypandrium in some specimens. Setal brush black; other setae dark brown/black, numerous, prominent, especially dorsally on epandrium. Epandrium long, the width in lateral view 0.4-0.45 times the length, the dorsal and ventral margins more-or-less parallel, the apex rounded dorsally, an obtuse angle ventrally. Medial face of epandrium as in figure 152. In dorsal view, medial margins of epandrium moderately concave; basal sclerite strong.

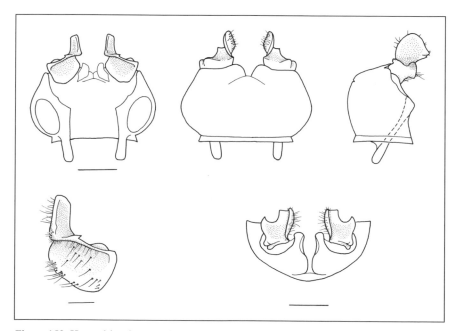

Figure 153. Hypandrium/gonocoxite complex. Clockwise from top left: dorsal, ventral, lateral and apical (scale lines = 0.3 mm); gonostylus, dorsal (scale line = 0.1 mm).

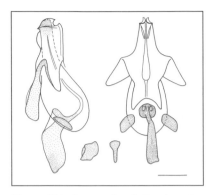

Figure 154. *L. opaculus* phallus: lateral (left); ventral (right); lateral ejaculatory process, basal (centre left); ejaculatory apodeme, cross-section (centre right). Scale line = 0.2 mm.

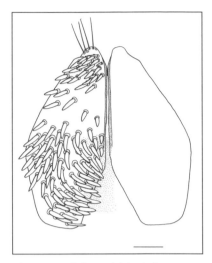

Figure 155. Subepandrial sclerite: ventral. Scale line = 0.1 mm.

Gonostylus. Small tooth at junction of dorsal flange and medial flange; dorso-medial, lateral and ventrolateral teeth strongly developed. Hypandrium/gonocox-ite complex in ventral view bulging basally, the length about 0.6 times the width, the transverse slit at 0.9 the length. In lateral view, exposed length of gonocoxal apodeme short, about 0.4 times the basal width of hypandrium; apodeme with small sclerotized web ventrally.

Phallus. Paramere sheath dorsally 0.4 times the length of phallus (excluding ejaculatory apodeme). Apex of paramere sheath with ventrolateral carina and ventral flange. Sperm sac width in dorsal view 0.35 times the length of phallus. Ejaculatory apodeme bent dorsally in lateral view, rod-like in ventral view; oval in cross-section with thick, strong dorsal carina. Subepandrial sclerite as in figure 155. Sclerite elongate, triangular unsclerotized area in basal 0.4, a narrow unsclerotized portion in central 0.5; spines bluntly acute, dense except in subapical area.

Female genitalia. Undissected: Hairs brown/black, strong, erect, abundant. Tergite 8 with a basal hump dorsally; dark brown/black, often paler brown apically. Sternite 8 dark brown/black, hypogynial valves with short, strong bristles ventrally. Lateral lobe setae strong. Cerci black with brown/black setae. Dissected sternite 8 (figure 156) elongate with basal width about 0.5 times the length, strongly divided along midline. Length of unsclerotized area between hypogynial valves 0.35 times sternite length.

Tergite 9 sclerites as in figure 157; sternite 9 V-shaped, medially undivided and with dorsal carina. Tergite 10 dark brown/black, usually with 6 black acan-

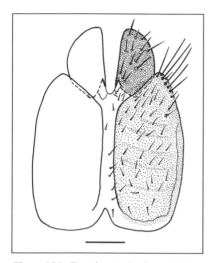

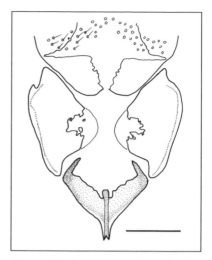

Figure 156. Female sternite 8: ventral.
Scale line = 0.2 mm.

Figure 157. Female sternite 9, tergite 9
and lobes of sternite 10: ventral.
Scale line = 0.2 mm.

thophorite spines on each side. Spermathecae (figure 158) with terminal reservoirs curled in a full loop and about as thick as reservoir duct; terminal reservoirs without wart-like protuberances on surface. Basal striated duct with fine scales; junction with basal duct sclerotized, golden, scaled; valve sclerotized, golden; basal duct moderately short, narrow.

Type Material.
LECTOTYPE. (here designated), ♀ labelled: "[rectangular beige label] Illin";
"[rectangular white label] Loew/Coll."; "[square red label] Type/12805"; "[rectangular beige label] opaculus/Lw."; "[rectangular white label] Museum of/Comparative/Zoology". My lectotype label "LECTOTYPE/Lasiopogon ♀/opaculus Loew/des. R.A. Cannings 2002 [red, black-bordered label]" has been attached to this specimen. MCZC. The holotype of *L. carolinensis* Cole and Wilcox, here synonymized with *L. opaculus*, is a male: North Carolina, Raleigh, late iv.1908, F. Sherman. OSUC.

Other Material Examined (110 specimens).
CANADA: Ontario, Lambton Co., Port Franks, Watson Property near L Lake, 6.vi.1996, J. and A. Skevington (1♀, BCPM), 8-10.vi.1996, J. Skevington (1♂, BCPM), 13.vi.1996, J. Skevington (1♂, 2♀, BCPM), 13-15.vi.1996, J. Skevington (1♂, 1♀, BCPM), 17.vi.1996, J. Skevington (1♀, BCPM); Toronto, 24.v.1888, W. Brodie (1♂, USNM), 24.v.1989, W. Brodie (1♀, USNM), 22.v.1981, Lonny Coote (1♂, DEBU); Willowdale, 24.v.1926, C. Hope (2♂, CNCI).
    U.S.A.: Georgia, Athens, 21.v.1974, A. Lavallee (1♂, LGBC); Bogart, 20.v.1972, A. Lavallee (2♂, CUCC); Clermont, 10.v.1944, R. Noblet (1♀, UGCA); Stone Mountain,

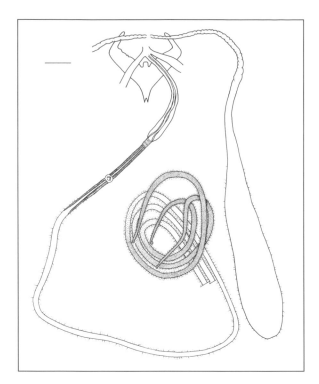

Figure 158. *L. opaculus*
spermatheca: dorsal.
Scale line = 0.1 mm.

31.iii.1946, P.W. Fattig (1♀, USNM). **Illinois,** Muncie, 1.v.1965, D.J. Burdick (1♂, LGBC); Muncie, 2mi E, 11.v.1961, G.P. Waldbauer (1♂, AGSC); Muncie, Stony Creek, 2.v.1965, D.J. Burdick (1♂, LGBC); Vermilion Co., Kickapoo St. Pk, 6.v.1966, A. Scarborough (1♂, AGSC), Kickapoo St. Pk, Middle Fork R., 8.v.1990, M.D. Baker (8♂, 1♀, MDBC; 3♂, 1♀, BCPM). **Indiana,** Chesterton, 21.vi.1916 (1♂, EMEC; 1♀, EMFC); Lafayette, v.1919, J.M. Aldrich (1♀, CASC); Putnam Co., Big Walnut Cr., 15.v.1975, P. Johnson (1♂, 1♀ in cop., ESUW). **Iowa,** Ames, 2.v.1926, J.N.T. (1♂, CUCC), 25.v.1928, G.S.W. (1♂, CNCI), 28.iv.1941, D.T. Jones (1♂, BYUC); 13.v.1949, W.L. Downes (1♀, LGBC), 14.v.1950, W.L. Downes (1♂, 1♀, LGBC), 29.v.1951, W.L. Downes (1♀, LGBC), 21.v.1956, W.L. Downes (1♀, LGBC), 5.v.1958, R.E. Johnsen (2♂, CSUC), D.E. Beck (1♂, 1♀, CASC). **Michigan,** Detroit, 17.v.1933 (1♂, MSUC; 1♀, UMMZ); Genessee Co., 4.vi.1946, R.R. Dreisbach (1♀, MSUC); Gladwin Co., 10.vi.1951, R.R. Dreisbach (1♀, MSUC), 14.vi.1953, R.R. Dreisbach (1♂, UMMZ; 1♀, FSCA), 24.v.1959, R. Dreisbach (1♂, MSUC); Ionia Co., 10.vi.1950, R.R. Dreisbach (1♂, 1♀, UMMZ); La Peer Co., 30.v.1937, R.R. Dreisbach (1♂, USNM); Midland Co., 11.vi.1936, R.R. Dreisbach (1♀, MSUC), 5.vi.1937, R.R. Dreisbach (1♀, USNM), 19.v.1939, R.R. Dreisbach (1♂, MSUC), 20.v.1942, R.R. Dreisbach (1♂, FSCA), 27.vi.1943, R.R. Dreisbach (1♂, USNM), 4.v.1944, R.R. Dreisbach (1♀, MSUC), 14.vi.1952, R.R. Dreisbach (1♀, MSUC), 7.v.1957, R. and K. Dreisbach (1♂, MSUC); Oakland Co., 15.v.1926, A.W. Andrews (1♂, 1♀, MSUC; 2♀, UMMZ); Oakland Co., Parke Davis, 23.v.1926, S. Moore (1♀, UMMZ); Saginaw Co., 1.vi.1940, C.W. Sabrosky

(1♀, USNM); St. Joseph, 30.v.1938, C.W. Sabrosky (1♂, USNM); Wayne Co., Andrews (1♂, FSCA); Wayne Co., Eloise, 11.v.1913, F.E. McCain (1♀, UMMZ). **Mississippi,** Lafayette Co., T7S. R4W. Sec.24, 24.iii.1977, T. McCraine (1♂, CUCC). **North Carolina,** Highlands, 3-5000', iv.1936, R.C. Shannon (1♂, USNM); Highlands, 3800', 7.v.1957, W.R.M. Mason (1♀, CNCI), 9.v.1957, J.R. Vockeroth (1♀, CNCI), 10.v.1957, J.R. Vockeroth (1♀, CNCI), 16.v.1957, H.C. Huckett (2♂, CNCI), 3.vi.1957, J.R. Vockeroth (1♀, CNCI); McDowell Co., Buck Creek Gap, 14.iv.1980, L.L. Pechuman (1♀, CUIC); Raleigh, late iv.1908, F. Sherman (2♂, EDNC; 1♂ *L. carolinensis* holotype, OSUC), mid iv.1921, T.S.M. (1♀, EDNC), 9.v.1924, C.S. Brimley (1♂, EDNC), 3.iv.1939, C.S. Brimley (1♂, EDNC); North Carolina (1♂ *L. carolinensis* paratype, CUIC). **Ohio,** Bainbridge, Paint Cr., 1.vi.1942 (2♂, FSCA); Cuyahogo Co., Shaker Heights, viii.1961, D.G. Furth (1♀, LGBC); Summit Co., Ira, 1.vii.1920, J.S. Hine (1♂, CASC); Clinton Co., 15.v.1962, F.J. Moore (1♀, OSUC). **South Carolina,** Anderson Co., Pendleton, Tanglewood Spring, 740', 30.iv.1987, J. Morse (1♂, 2♀, CUCC); Clemson College, 21.iv.1932, M.L. Bobb (1♀, USNM), 6.v.1939, B.M. Heniford (1♂, CUCC); Gramling, 4.iv.1939, O.L. Cartwright (1♂, CUCC; 1♂, 1♀, USNM); Oconee Co., 10mi W. Westminster, Hwy 37-48 at Little Longnose Cr., 5.v.1972, A.G. Lavallee (1♀, UGCA). **Virginia,** Petersburg, 9.v.1936, F.S. Blanton (1♂, CUIC).

Type Locality. U.S.A.; Illinois.

Taxonomic Notes. The concept of *L. opaculus* held by Cole and Wilcox (1938) embraced both this species and *L. woodorum* and both key to *L. opaculus* in their key. The name once used for the common close relative (sympatric in the southeast), *L. carolinensis* Cole and Wilcox, is unavailable because the holotype is actually a specimen of *L. opaculus* (see *L. schizopygus*). Several *L. carolinensis* paratypes also must be referred to *L. opaculus*. In addition, because *L. opaculus* was one of the first species described from the northeast, several other species (e.g., *L. slossonae, L. currani*) were often confused with it in collections and literature.

Etymology. Latin *opacus* = shady, dark; *-ulus* = little; little dark one.

Distribution. Nearctic; Southern Ontario, Michigan, Illinois and Ohio east to Virginia, south to Georgia and Mississippi. The Colorado specimen listed by Cole and Wilcox (1938) is almost certainly misidentified, but could not be located. Map 7, next page.

Phylogenetic Relationships. Member of the *opaculus* group and sister species of *L. slossonae.*

Natural History. Habitat: stream sides in deciduous woodland; dunes near lakeshores; sandy grassland habitats. Recorded from 24 March (Lafayette Co., Mississippi) to 3 June (Highlands, North Carolina) in the South, 29 April to 1 July (both Ohio) in the North. There is an undated August record, also from Ohio.

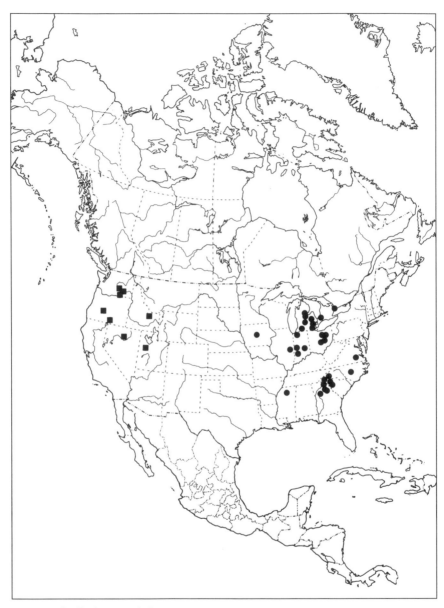

Map 7. Distribution records for
- *Lasiopogon opaculus*
- *L. chaetosus*

At Port Franks, Lambton Co., Ontario, on the shores of Lake Huron, *L. opaculus* was collected from 6-17 June 1996 in a mosaic of dune, prairie and oak woodland habitats about 500-800m inland from the lake. This habitat was bordered by a narrow sand-bottomed creek in riparian woods of Tuliptree (*Liriodendron tulipifera* L.), Swamp White Oak (*Quercus bicolor* Willd.) and Pawpaw (*Asimina triloba* (L.) Dunal). Predominant plant species in the drier habitats included Dwarf Chinquipin (*Quercus prinoides* Willd.) and Red Oak (*Quercus rubra* L.), with Little Bluestem (*Andropogon scoparius* Michx.), Rough Blazing Star (*Liatris aspera* Michx.) and scattered Red Cedars (*Juniperus virginiana* L.) in the prairie (J. Skevington, *in litt.*).

Lasiopogon phaeothysanotus sp. nov.

Diagnosis. Only one male known. A medium-sized, hirsute, dark grey species with obscure markings and almost all the fine hairs brown instead of white. Facial gibbosity prominent; mystax and other head bristles dark. Antennae brown, F2+3 short (F2+3/F1 = 0.38). Thorax dark grey with gold highlights. Dorsocentral stripes faint, dark grey/black, gold anteriorly; acrostichal stripes obscure grey with a paler grey, faint medial stripe anteriorly. All thoracic and leg bristles dark; anterior dorsocentral bristles 7-9. Abdominal tergites broadly but vaguely banded with grey apically; faintly dark brown basolaterally. Bristles on tergite 1 dark. Epandrium elongate, the width 0.45 times the length in lateral view, the greatest width at about half the length; apex truncate, the ventral angle slightly down turned. In dorsal view, medial margins of epandrium weakly concave medially. Gonostylus in dorsal view triangular, tapered apically.

Description. Only male known. Single specimen rather rubbed; descriptions of colour may not be completely reliable. Body length ♂ 10.0mm.

Head. HW ♂ 2.12mm. FW ♂ 0.48mm. VW ♂ 0.93mm. VW/HW = ♂ 0.44. FW/VW = ♂ 0.52. VD/VW = ♂ 0.10. Gibbosity height/length = ♂ 0.44.

Face grey, vertex grey with faint brown under the frontal and orbital setae; occiput grey/brown-grey. Beard and labial hairs light brown, Gibbosity prominent; mystax and all other setae brown/black. Occipital bristles abundant, those behind the dorsomedial angle of the eye strongly curved anterolaterally; longest about 0.8mm, longer than F1+f2+3; lateral and ventral ones shorter, straighter. Frontal and orbital setae strong, longest to 0.7mm, about as long as F1+F2+3.

Antennae. Brown, base of F1 dark chestnut. Setae brown/ black, no setae on F1. F1 long, the dorsal and ventral margins parallel; F2+3 short. WF1/LF1 = 0.25; LF2+3/LF1 = 0.38.

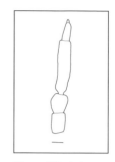

Figure 159. Antenna.
Scale line = 0.1 mm.

Thorax. Prothorax gold/brown-grey, strongly gold-grey anteriorly, hairs light brown; postponotal lobes grey, the lateral angle ferruginous, hairs sparse, strong, brown. Scutum tomentum grey/gold-grey; dorsocentral stripes faint dark grey/black, gold-brown anteriorly; acrostichal stripes grey, obscure, divided anteriorly by faint grey medial stripe. Faint gold-brown anterolaterally on scutum, below supra-alar area and on scutum above scutellum. Strong bristles abundant, black, finer setae brown. Anterior dorsocentrals 7-9, mixed with even finer setae, the most anterior ones rather long, to 0.7mm; 4-5 posteriors. Notal setae rather abundant over presutural area, as long as shorter dorsocentrals. Postalars 4-5, with shorter setae; supra-alars 5-8; presuturals 4, posthumerals 2. Scutellar tomentum grey with faint gold highlights; apical scutellar bristles dark, numerous strong ones in 2-3 irregular rows mixed with shorter, weaker dark bristles and hairs.

Pleural tomentum dark gold-grey. Katatergite setae black, 8-9 among finer brown hairs; katepisternal setae brown, sparse, only 3-5. Anepisternal setae 14-15, dark (to 0.5mm); a small patch of short brown setae on dorsal shelf. Anepimeron without setae.

Legs. Base colour dark brown/black; most of trochanters, extreme ends of femora, bases of tibiae, and venter of tarsi ferruginous. Tomentum gold-grey. No coxal peg. Setae on coxae medium/dark brown, finer setae pale. Main bristles of rest of legs dark brown/black, finer setae predominantly brown, some white. Femora dorsally with brown decumbent hairs, some white basally. Longer erect brown hairs abundant ventrally and laterally, especially on profemur; in males longest ventral setae as long as thickness of pro- and mesofemur, longer on metafemur. Profemur without strong dorsolateral dark bristles among finer ones; mesofemur with 0-6; metafemur with 20-30. Tibiae and tarsi with dark, strong bristles typically arranged, hairs on tibiae brown, abundant and almost as long as longest bristles, which are rather fine; apical bristles very strong. Protibia with longest bristles about 4.0 times longer than tibial width.

Wings. Veins medium/dark brown; membrane faint brown in oblique view. DCI = 0.35; cell M3 open. Halter dark yellow; knob without dark spot.

Abdomen. Male. Tergite basal colour dark brown/black. Faint bands of thin dark grey tomentum cover about 0.5 the length of each tergite apically (the apical 0.25 is strongest), extending mid-dorsally to the base and dividing the faint brown basal tomentum into obscure lateral patches. Segment 1 is thinly covered with grey. Ventrolateral areas have faint grey tomentum. Abundant, strong bristles on each side of tergite 1 dark. Lateral setae light/dark brown, long, weaker white hairs on ventral margins of tergites. Dorsal setulae appressed, brown, some white apically on tergites. Sternite tomentum grey, hairs brown and white.

Male genitalia. Epandrium and hypandrium/gonocoxite complex brown/black. Setal brush with long, brown/black bristles; other setae long, brown. Epandrium elongate, the width of the halves in lateral view 0.45 times the length, widest at about midlength; apex truncate, the ventral angle slightly down turned. Medial face of epandrium as in figure 160. In dorsal view, medial margins of epandrium weakly concave; basal sclerite strong.

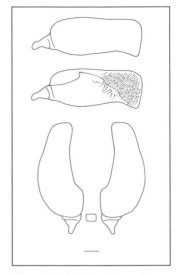

Figure 160. *L. phaeothysanotus* sp. nov. epandrium (top to bottom): lateral, medial, dorsal. Scale line = 0.2 mm.

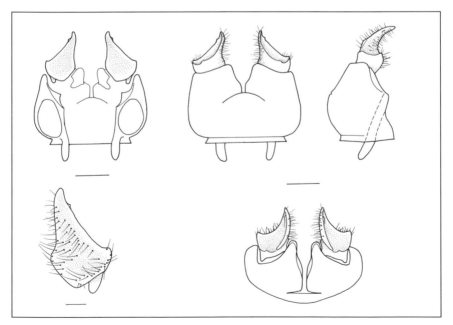

Figure 161. Hypandrium/gonocoxite complex. Clockwise from top left: dorsal, ventral, lateral and apical (scale lines = 0.3 mm); gonostylus, dorsal (scale line = 0.1 mm).

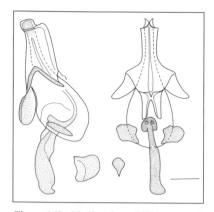

Figure 162. Phallus: lateral (left); ventral (right); lateral ejaculatory process, basal (centre left); ejaculatory apodeme, cross-section (centre right). Scale line = 0.2 mm.

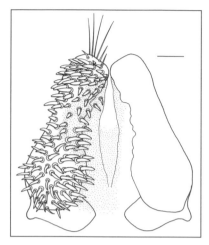

Figure 163. Subepandrial sclerite: ventral. Scale line = 0.1 mm.

Gonostylus. Gonostylus compressed laterally; the basal areas elongated apically, hiding almost all other parts of the gonostylus in dorsal view. Medial flange flattened; dorsal flange a lateral, basoapical ridge; lateral tooth present. Hypandrium/gonocoxite complex in ventral view with length about 0.7 times the width, the transverse slit at 0.65 the length. In lateral view, exposed length of gonocoxal apodeme about 0.4 times basal width of hypandrium; apodeme without sclerotized web ventrally.

Phallus. Paramere sheath dorsally 0.5 times the length of phallus (excluding ejaculatory apodeme). Apex of paramere sheath with ventrolateral carina; ventral flange absent. Sperm sac width in dorsal view 0.35 times the length of phallus. Ejaculatory apodeme curved dorsally in lateral view, weakly spatulate in ventral view; oval in cross-section with narrow dorsal carina. Subepandrial sclerite as in figure 163. Hypoproct minute, hidden behind apex of sclerite. Central unsclerotized area broad apically and occupying 0.95 of the length of the sclerite; basodorsal angles expanded; spines acute or bluntly acute, concentrated apically and basomedially.

Female unknown.

Type Material.

HOLOTYPE. (here designated) ♂ labelled [in Chinese]: "[rectangular white label] Mt. Wuxue, Hohxil/Qinghai 4800m/Academia Sinica"; "[rectangular white label] 1990.VIII.6/Collected by: Zhang, X." My Holotype label "HOLO-TYPE/*Lasiopogon* ♂/*phaeothysanotus* Cannings/des. R.A. Cannings 2002 [red, black-bordered label]" has been attached to this specimen. Dissected genitalia in plastic vial underneath all. IZAS.

Type Locality. People's Republic of China, Qinghai, Hohxil, Mt. Wuxue (approximately 35°N x 90.1°E), 4800m.

Etymology. From the Greek *phaios* = dusky, brown and *thysanotos* = fringed. *L. phaeothysanotus* is the only species in the genus known to have almost all fine hairs brown, including the fringe-like genal hairs below the eyes (the "beard").

Distribution. Palaearctic; northern Tibetan Plateau. See map 10, p. 251.

Phylogenetic Relationships. Member of the *hinei* species group. Although in the proposed phylogeny, *L. phaeothysanotus* is the sister to *L. qinghaiensis*, there are many differences between the two species and they may not be so closely related. An understanding of the true relationships in the *hinei* group must await the discovery of additional undescribed species, which surely exist in Asia.

Natural History. Habitat: high, cold desert.

Lasiopogon piestolophus sp. nov.

Diagnosis. Only one male and one female known. A small to medium-sized grey/grey-brown species with dark brown legs. Antennae brown, F2+3/F1 = 0.66. Mystax and other head bristles dark. Thoracic tomentum laterally gold-grey, dorsally brown/gold-grey with brown dorsocentral stripes and faint brown-grey acrostichal stripes. Main thoracic and leg bristles dark, except for pale katatergites; anterior dorsocentral bristles weak. Abdominal tergites basally with only thin brown tomentum, the apical bands gold-grey, very narrow, covering 0.2 the length of the segments. Bristles on tergite 1 pale. Epandrium about half as wide as long in lateral view, the greatest width at about half the length; apex rounded. In dorsal view moderately concave medially. Female abdomen with pale apical bands 0.4 the length of the tergites and not extending apically. Genitalia black with brown hair; sternite 8 medially divided, hypogynial valves with short, strong setae. Spermathecae with valves sclerotized and golden, the tube immediately apical to the valve bare and sclerotized.

Description. Body length ♂ 8.9mm; ♀ 8.4mm.

Head. HW ♂ 2.00mm; ♀ 1.96mm. FW ♂ 0.32mm; ♀ 0.38mm. VW ♂ 0.84mm; ♀ 0.92mm. FW/VW = ♂ 0.38; ♀ 0.41. VD/VW = ♂ 0.13; ♀ 0.12. GH/GL = ♂ 0.26; ♀ 0.27.

Face and vertex pale brown-grey, occiput grey with brown-grey dorsally, all with gold highlights. Beard and labial hairs white, mystax bristles brown/black; all other setae brown/black. Occipital bristles rather sparse,, those behind the dorsomedial angle of the eye strongly curved anterolaterally; lateral and ventral ones shorter, straighter. Frontal and orbital setae fine, orbitals to 0.5mm; ocellar bristles fine, hardly distinguished from orbitals.

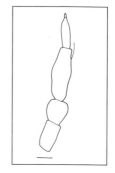

Figure 164. Antenna. Scale line = 0.1 mm.

Antennae. F1 and F2+3 missing in male. Brown; setae brown, a few white on scape in male; female with seta on one F1. F1 short, dorsal and ventral margins slightly convex, widest at midlength. WF1/LF1 = 0.34; LF2+3/LF1 = 0.66.

Thorax. Prothorax brown/gold-grey, with white hairs; postponotal lobes gold-grey, the lateral angle dark brown, hairs light brown. Scutum base tomentum brown-grey with gold highlights, the dorsocentral stripes brown bordered with gold, the intermediate spots and acrostichal stripes faint brown-grey, the latter almost invisible; a faint brown/gold medial stripe anteriorly. All strong bristles black/brown, finer setae brown. Anterior dorsocentrals 0-2, weak (longest in female to 0.5mm), mixed with even finer setae; 3-4 posteriors. Notal setae scattered over presutural area, as long as shorter dorsocentrals. Postalars 1-2, with shorter setae; supra-alars 1-2; presuturals 2. Scutellar tomentum grey with brown/gold highlights; apical scutellar bristles dark, 2-3 strong ones on each side mixed with shorter, weaker dark bristles and hairs.

Pleural tomentum gold-grey. Katatergite setae white/yellow, 7-8 among finer white hairs; katepisternal setae sparse. Anepisternal setae 3-7 dark, short (to 0.3mm); a few other pale, weaker ones ventrally; a patch of short erect brown setae on dorsal shelf. Anepimeron with a few white setae.

Legs. Base colour dark brown/black. Tomentum of coxae grey with gold highlights; tomentum on rest of legs gold-grey. No coxal peg. Main bristles dark brown/black, finer setae predominantly dark. In male, femora dorsally with dark decumbent hairs, some white ones basally; hairs all white in female. Longer erect pale hairs ventrally and on profemur, laterally also; in male longest white ventral setae longer than thickness of femur. Profemur with 2-3 strong dorsolateral dark bristles; mesofemur with 2-3; metafemur with 12-14. Tibiae and tarsi with dark, strong bristles typically arranged; hairs black/brown. Protibia with longest bristles about 3.5-4.0 times longer than tibial width.

Wings. Veins dark brown; wing membrane pale brown in oblique view. DCI = 0.35-0.40; cell M3 open. Halter yellow; knob without dark spot.

Abdomen. Male. Tergite basal colour dark brown/black. Tomentum on tergite bases thin brown/gold-brown, usually extensively lacking so that dark cuticle shines through. Bands of gold-grey tomentum cover 0.2 the length of each tergite apically; segment 1 is covered except mid-dorsally. Ventrolateral areas very narrowly covered with thin brown-grey tomentum. About 5-6 strong white/yellow bristles on each side of tergite 1. Lateral setae on tergites 1-4 white, sparse, erect, short; those on 5-7 dark. Dorsal setulae appressed, short and dark. Sternite tomentum brown-grey, hairs white on 1-5, short and dark from apex of 5 to 7.

Female. Tomentum covers tergites completely. Basal tomentum brown; brown/gold-grey apical areas covering about 0.4 the length of the tergite (0.5 on 7), slightly wider mid-dorsally, but not extending anteriorly on the midline. Grey tomentum not extending anteriorly ventrolaterally, at least on middle segments. Erect white hairs on 1-3 relatively short, brown on 4-7. Dorsal setulae dark. Sternite tomentum brown-grey, hairs white on 1-4, dark on apex of 4 to 7.

Male genitalia. Epandrium and hypandrium/ gonocoxite complex dark brown with gold-grey tomentum except on parts of hypandrium. Setal brush black; other setae brown/dark brown. Width of epandrium halves in lateral view about 0.5 times the length, widest at about midlength; ventral margin straight or slightly concave, dorsal margin gently convex, apex rounded. Medial face of epandrium with apical shelf as in figure 165; membrane covered with numerous granular-like microsetae. In dorsal view, medial margins of epandrium moderately to weakly concave; basal sclerite prominent.

Gonostylus. Medial flange narrow, the flange flattened apically, its width about 0.2-0.25 times the dorsoventral length. Ventrolateral and lateral teeth moderately developed. Hypandrium/gonocoxite complex in ventral view with length about 0.6 times the width, bulging basally, the transverse slit at 0.8 the length. In lateral view, exposed length of gonocoxal apodeme about 0.5 times basal width of hypandrium; apodeme with sclerotized web ventrally.

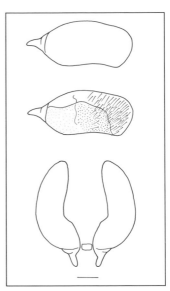

Figure 165. *L. piestolophus* sp. nov. epandrium (top to bottom): lateral, medial, dorsal. Scale lines = 0.2 mm.

Phallus. Paramere sheath dorsally 0.45 times the length of phallus (excluding ejaculatory apodeme). Apex of paramere sheath with ventrolateral carina and ventral flange. Sperm sac relatively large, the width in dorsal view 0.45 times the length of phallus and 0.75 times the width of lateral processes. Ejaculatory apodeme straight, emarginate in basal 0.3 in lateral view, moderately spatulate in ventral view; triangular in cross-section, the dorsal carina thick. Subepandrial

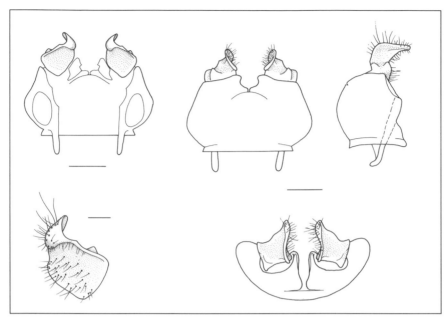

Figure 166. *L. piestolophus* sp. nov. hypandrium/gonocoxite complex. Clockwise from top left: dorsal, ventral, lateral and apical (scale lines = 0.3 mm); gonostylus, dorsal (scale line = 0.1 mm).

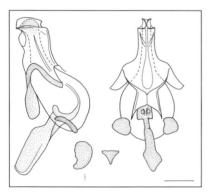

Figure 167. Phallus: lateral (left); ventral (right); lateral ejaculatory process, basal (centre left); ejaculatory apodeme, cross-section (centre right).
Scale line = 0.2 mm.

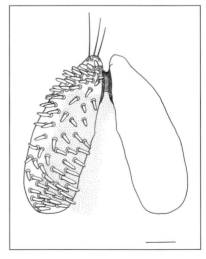

Figure 168. Subepandrial sclerite: ventral. Scale line = 0.1 mm.

sclerite as in figure 168. Narrowly triangular unsclerotized area in basal 0.75; spines blunt, widely distributed but concentrated apically and basally.

Female genitalia. Undissected: All setae brown/black. Tergite 8 dark brown/black with posterior margin pale; sternite 8 black/brown, lateral lobe setae strong. Hypogynial valves brown, ventral surface with bristles. Cerci brown with pale setae. Dissected sternite 8 (figure 169) elongate, basal width about 0.55 times the length; strongly divided along midline. Length of unsclerotized area between hypogynial valves short, about 0.25 times sternite length.

Tergite 9 sclerites as in figure 170; sternite 9 V-shaped, medially undivided and with a dorsal carina. Tergite 10 brown with 6 brown/black acanthophorite spines on each side. Spermathecae (figure 171) with terminal reservoirs curled in a complete loop and about as wide as reservoir duct; terminal reservoirs without wart-like protuberances on surface. Basal section of undifferentiated reservoir duct with a dense patch of caniculi. Basal striated duct with fine scales; junction with basal duct sclerotized, golden, scaled; valve sclerotized, golden; striated basal section of reservoir tube with fine scales, basal 0.5 without striae, bare, sclerotized; basal duct short.

Type Material.
HOLOTYPE. (here designated) ♂ labelled: "[rectangular white label] U.S.A.: ALABAMA/Conecuh County/6.2 mi E of Evergreen/Old Town Cr. at Hwy 31/17 MAR 1997/Michael Baker". My Holotype label "HOLOTYPE/

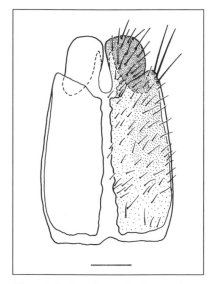

Figure 169. Female sternite 8: ventral.
Scale line = 0.2 mm.

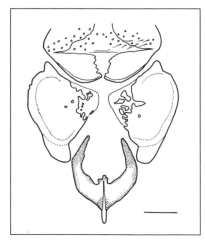

Figure 170. Female sternite 9, tergite 9 and lobes of sternite 10: ventral.
Scale line = 0.1 mm.

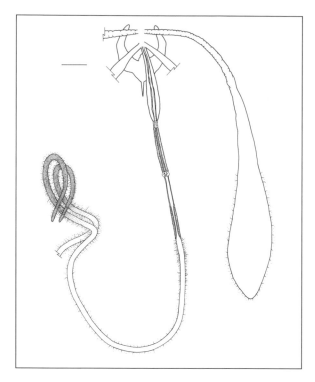

Figure 171.
L. piestolophus
spermatheca: dorsal.
Scale line = 0.1 mm.

Lasiopogon ♂/*piestolophus* Cannings/des. R.A. Cannings 2002 [red, black-bordered label]" has been attached to this specimen. Dissected genitalia in plastic vial underneath all. USNM.
PARATYPES.(1 designated) **U.S.A.: Alabama,** Conecuh Co., Evergreen, 6.2 mi E, Old Town Creek at Hwy.#31, 17.iii.1997, Michael Baker (1 ♀, USNM).

Type Locality. U.S.A., Alabama, Conecuh Co., 6.2 mi E of Evergreen, Old Town Cr. at Hwy 31.

Etymology. From the Greek *piestos* = compressed and *lophos* = ridge; a reference to the distinctive narrow medial flange on the gonostylus of *Lasiopogon piestolophus*.

Distribution. Nearctic; U.S.A.; known only from the type locality on the Alabama coastal plain. See map 9, p. 244.

Phylogenetic Relationships. A member of the *opaculus* species group and sister species to *L. marshalli*.

Natural History. An early spring species on the coastal plain of the Gulf states. Riparian habitat. Recorded flight date 17 March.

Lasiopogon qinghaiensis sp. nov.

Diagnosis. Only one male and two females known. Male and one female badly greased; description of tomentum colour based on clean female. A medium-sized, grey species with pale ferruginous tibiae and tarsi. Mystax dark dorsally, pale ventrally; other head bristles mostly dark with some pale. Antennae brown, F2+3/F1 = 0.47-0.51. Thorax grey laterally, blue-grey dorsally with stripes obscure. Scutum bristles abundant, dark; katatergites mostly pale, a few dark ones in females. Anterior dorsocentral bristles 4-6. Abdominal tergites with an apical grey band covering 0.4 the length of the segment, the basal part slightly darker brown-grey. Bristles on tergite 1 pale. Epandrium width about 0.45 times the length in lateral view, the greatest width at about half the length; apex rounded, the ventral angle slightly down turned. Epandrium in dorsal view weakly concave medially. Female with tergite 8 black, sternite 8 mostly yellow/ferruginous; hairs pale.

Description. Knowledge of variation limited; only 1♂, 2♀ known. Male greased; tomentum colour described from female. Body length ♂ 9.9mm; ♀ 9.5-10.8mm.

Head. HW ♂ 2.12mm; ♀ 2.14-2.16mm. FW ♂ 0.48mm; ♀ 0.50-0.52mm. VW ♂ 0.86mm; ♀ 0.88-0.90mm. VW/HW = ♂ 0.41; ♀ 0.41-0.42. FW/VW = ♂ 0.56; ♀ 0.57-0.58. VD/VW = ♂ 0.14; ♀ 0.13-0.14. GH/GL = ♂ 0.34; ♀ 0.36.
Face silver-grey, vertex blue-grey, occiput grey, blue-grey dorsally. Beard and labial hairs white, mystax bristles brown/black dorsally, white ventrally. Occipital bristles mostly brown/black, but some white, especially medially in females; those behind the dorsomedial angle of the eye strongly curved anterolaterally, longest about 0.7mm; lateral and ventral ones shorter, straighter. Frontal and orbital setae dark, rather strong; orbitals to 0.6mm, almost as long as F1+F2+3. Ocellar bristles hardly distinguished from orbitals in strength, but slightly longer.

Antennae. Brown; scape, pedicel, base of F1 and F2+3 lighter. Setae brown and white; seta on F1 in one specimen. F1 long, wider in apical half; WF1/LF1 = ♂ 0.24; ♀ 0.21-0.25. LF2+3/LF1 = ♂ 0.48; ♀ 0.47-0.51.

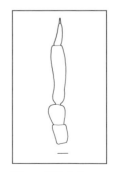

Figure 172. Antenna. Scale line = 0.1 mm.

Thorax. Prothorax grey, hairs white; postponotal lobes grey, the lateral angle ferruginous, hairs brown, strong. Scutum cuticle brown, pale brown in some areas such as postalar slope; tomentum blue-grey. Dorsocentral stripes almost absent, some very faint brown tomentum anteriorly; acrostichal stripes dark grey, obscure. Faint brown tomentum on intermediate spots and along ventrolateral margins of scutum. Notal setae abundant, prominent; all bristles black, finer setae black/brown. Anterior dorsocentrals 4-6, longest to 0.8mm), mixed with finer setae; 4-6 posteriors. Notal setae abundant over presutural area, as long as shorter dorsocentrals. Postalars 2-5, with shorter setae; supra-alars 2-5; presuturals 3; 2-3 posthumerals. Scutellar tomentum grey; apical scutellar bristles numerous, dark, forming 2-3 irregular rows with shorter and weaker dark and pale bristles and hairs intermixed.

Pleural tomentum grey. Katatergite setae white, 5-6 (9-11 with a few black ones in female) among finer white hairs; katepisternal setae white, sparse. Anepisternal setae 8-12, dark, to 0.6mm; a few other pale, weaker ones ventrally; a patch of short erect pale setae (brown in female) on dorsal shelf. Anepimeron without setae.

Legs. Base colour of coxae brown/black with ferruginous areas; trochanters ferruginous, dark brown at extremities; femora dark brown/black with ferruginous bases and apices; tibiae and tarsal segments pale ferruginous, darker apically. Tomentum grey, rather dense. No coxal peg. Main bristles mostly dark brown/black, some white ones, especially in females. Finer setae predominantly white. Femora dorsally with white decumbent hairs. Longer erect pale hairs abundant ventrally and laterally, especially on profemur; in males longest white ventral setae longer than thickness of femur. Profemur with 10-15 very fine brown dorsolateral bristles apically; mesofemur with 6-10, a few white basally; metafemur with about 20 fine white bristles (about 10 strong white ones in female). Tibiae with abundant, long, white hairs and long, fine white bristles ventrolaterally; protibia with one strong black bristle (3.5 times width of tibia) below white ones. Other dark bristles on protibia rather fine, stronger on other tibiae; apical bristles black, typically arranged. Female has stronger bristles, some of them white. Tarsi with white hairs, dark bristles.

Wings. Veins light brown with some dark brown, especially on R1, R4+5, CuA1 and CuA2; membrane very pale light brown/yellow in oblique view. DCI 0.25-0.29; cell M3 open. Halter yellow; knob without dark spot.

Abdomen. Tergite basal colour brown/black. Based on female coloration, tergites completely covered with grey tomentum; the apical 0.4 is slightly paler, the basal areas vaguely darker brown-grey. Segment 1 is grey, slightly darker middorsally at the base. Ventrolateral areas are grey. About 5-10 bristles on each side of tergite 1white. Lateral setae on tergites 1-4 white, erect, long, those on 5-7 considerably shorter. Dorsal setulae appressed, white; in female these are mixed with dark setulae, especially basally. Sternite tomentum grey, hairs white.

Male genitalia . Epandrium and hypandrium/ gonocoxite complex brown. Setal brush with long brown bristles; other setae white/light brown, numerous, strong. Width of epandrium halves in lateral view about 0.45 times the length, the widest point at about midlength; ventral margin straight, dorsal margin gently convex, the dorsal angle broadly rounded, the ventral one more angular and slightly down turned. Medial face of epandrium as in figure 173. In dorsal view, medial margins of epandrium weakly concave; basal sclerite strong.

Gonostylus compressed laterally; the basal areas elongated apically and the medial flange flattened; dorsal flange short, its lateral portion a basoapical ridge; lateral tooth present. Hypandrium/gonocoxite complex in ventral view with length about 0.7 times the width, the transverse slit at 0.7 the length. In lateral view, exposed length of gonocoxal apodeme about 0.6 times basal width of hypandrium; apodeme with sclerotized web ventrally.

Phallus. Elongate; paramere sheath dorsally 0.45 times the length of phallus (excluding ejaculatory apodeme). Ventral suture with flanges basally. Apex of paramere sheath narrow, with ventrolateral carina; ventral flange absent.

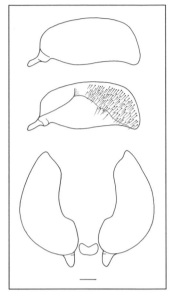

Figure 173. *L. qinghaiensis* sp. nov. epandrium (top to bottom): lateral, medial, dorsal. Scale line = 0.2 mm.

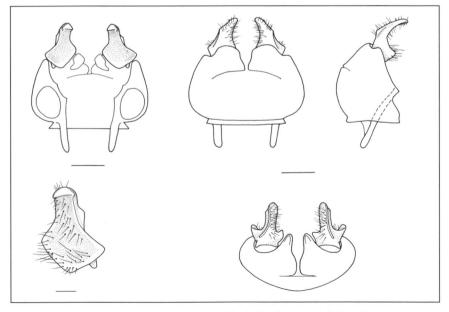

Figure 174. Hypandrium/gonocoxite complex. Clockwise from top left: dorsal, ventral, lateral and apical (scale lines = 0.3 mm); gonostylus, dorsal (scale line = 0.1 mm).

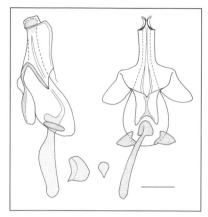

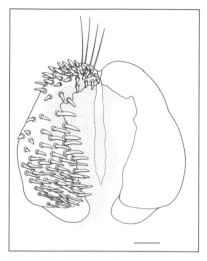

Figure 175. *L. qinghaiensis* sp. nov. phallus: lateral (left); ventral (right); lateral ejaculatory process, basal (centre left); ejaculatory apodeme, cross-section (centre right). Scale line = 0.2 mm.

Figure 176. Subepandrial sclerite: ventral. Scale line = 0.1 mm.

Sperm sac width in dorsal view 0.35 times the length of phallus. Ejaculatory apodeme with ventral margin convex in lateral view, rod-like in ventral view; oval in cross-section with reduced dorsal carina. Subepandrial sclerite as in figure 176. Hypoproct minute, hidden behind apex of sclerite. Central unsclerotized area broad apically and occupying 0.95 the length of the sclerite; basodorsal angles expanded; spines acute or bluntly acute, concentrated apically and basomedially.

Female genitalia. Undissected: Hairs white. Tergite 8 short, black with posterior margin brown. Sternite 8 brown/black basally, yellow/ferruginous apically. Hypogynial valves with short strong white hairs ventrally. Apical lobe setae white, strong. Cerci yellow-brown with pale brown setae. Dissected sternite (figure 177) with basal width about 0.65 times the length; undivided along midline. Length of unsclerotized area between hypogynial valves 0.4 times sternite length.

Tergite 9 sclerites as in figure 178; sternite 9 Y-shaped, medially undivided. Tergite 10 brown with 7-8 pale acanthophorite spines on each side. Spermathecae (figure 179) with curled terminal reservoirs about as wide as reservoir duct; terminal reservoirs without wart-like protuberances on surface. Valve clear; basal striated duct with fine scales; junction with basal duct sclerotized, golden, and with scales; basal duct short and narrow.

Type Material.
HOLOTYPE. (here designated) ♂ labelled [in Chinese]: "[rectangular white label] Xijir Ulan Hu, Hohxil/Qinghai 4800m/Academia Sinica"; "[rectangular

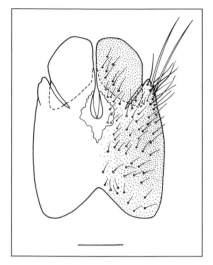

Figure 177. Female sternite 8: ventral.
Scale line = 0.3 mm.

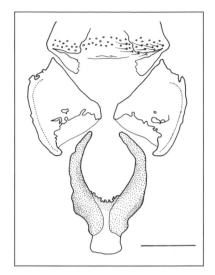

Figure 178. Female sternite 9, tergite 9
and lobes of sternite 10: ventral.
Scale line = 0.2 mm.

Figure 179.
Spermatheca: dorsal.
Scale line = 0.1 mm.

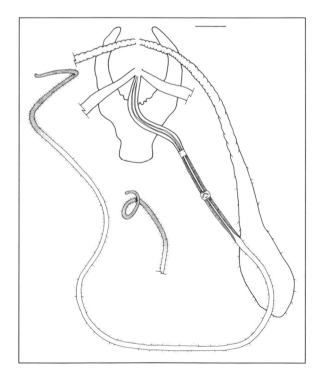

white label] 1990.VII.12/Collected by: Zhang, X." My Holotype label "HOLO-TYPE/*Lasiopogon* ♂/*qinghaiensis* Cannings/des. R.A. Cannings 2002 [red, black-bordered label]" has been attached to this specimen. Dissected genitalia in plastic vial underneath all. IZAS.
PARATYPES. (2♀ designated). Same data as male.

Type Locality. Peoples Republic of China, Qinghai, Hohxil, Xijir Ulan Hu (approximately 35.2°N x 90.4°E), 4800m.

Etymology. *L. qinghaiensis* is named for Qinghai Province in north-central China, where the species lives.

Distribution. Palaearctic; northern Tibetan Plateau. See map 5, p. 155.

Phylogenetic Relationships. Member of the *hinei* species group. Although in the proposed phylogeny, *L. qinghaiensis* is the sister to *L. phaeothysanotus*, there are many differences between the two species and they may not be so closely related. An understanding of the true relationships in the *hinei* group must await the discovery of additional undescribed species, which surely exist in Asia.

Natural History. Habitat: high, cold desert.

Lasiopogon quadrivittatus Jones

Lasiopogon quadrivittatus Jones, 1907. *Transactions of the American Entomological Society* 33: 278.
Lasiopogon aridus Cole and Wilcox, 1938. *Entomologica Americana* 43: 25-27.
New Synonymy.

Diagnosis. A medium-sized to large grey or gold-grey species variable in colour. Antennae brown, ratio F2+3/F1 = 0.49-0.64. Dorsocentral stripes usually brown, often bordered with gold, almost completely obscured in some specimens; acrostichal stripes grey to brown, sometimes obscure. Setae abundance and colour variable in northern and southern specimens as detailed below. Typically, all setae, except sometimes some on the tarsi, are white in both sexes. Anterior dorsocentral bristles 2-7, some prominent. Tergites usually bearing brown patches of tomentum basolaterally but sometimes completely grey. Epandrium in lateral view with greatest width about 0.6 times the length; broadly rounded dorsoapically, ventral corner a right-angle. Hypogynial valves with ventral carina; basal tubes of spermathecae unusually large.

Description. Body length ♂ 8.00-11.5mm; ♀ 10.1-12.9mm.

Head. HW ♂ 2.00-2.26mm; ♀ 2.16-2.70mm. FW ♂ 0.48-0.58mm; ♀ 0.50-0.54mm. VW ♂ 0.80-0.94mm; ♀ 0.88-1.06mm. VW/HW = ♂ 0.38-0.43; ♀ 0.41-0.45. FW/VW = ♂ 0.53-0.62; ♀ 0.51-0.57; VD/VW = ♂ 0.19-0.21; ♀ 0.15-0.23. GH/GL = ♂ 0.34-0.39; ♀ 0.30-0.38.

Face and vertex with silver-grey tomentum, sometimes mixed with gold. Beard and labial hairs white. In northern specimens (Colorado and north) all bristles and hairs white, sometimes slightly yellow at bases; occipital bristles variable, length to 0.6mm, most moderately curved anterolaterally, abundant, 20-30 on each side, numerous hairs intermixed; frontal and orbital setae about as long as scape+pedicel.

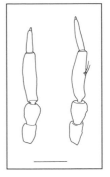

Antennae. Brown, some with basal areas paler. Setae white, dark in some specimens from the extreme south of the range; F1 sometimes with 1-2 weak setae. F1 moderately long, usually widest at about midlength, the dorsal margin straight or with a bulge in the basal third, the ventral edge gently convex. WF1/LF1 = ♂ 0.24-0.31; ♀ 0.27-0.33. LF2+3/LF1 = ♂ 0.44-0.57; ♀ 0.36-0.64.

Figure 180. Antenna. Scale line = 0.3 mm.

Thorax. Prothorax grey, hairs white; postpronotal lobes grey, the lateral angle ferruginous, hairs fine, white to brown. Scutum tomentum grey, often with gold highlights and faint brown patches on either side of suture. Dorsocentral stripes variable, light to dark brown, often bordered with gold, sometimes almost absent; acrostichal stripes grey or brown, sometimes fused with dorsocentrals or sometimes obscure. Some specimens with faint brown intermediate spots. A few specimens suffused with gold across much of scutum, especially anteriorly. Scutum rather setose, especially in males; all bristles and hairs white. Dorsocentrals variable, but mostly prominent, longest 1.0mm long, 2-7 anteriors, 2-4 posteriors. Acrostichals and notal hairs fine, short, more sparse in females. Postalars 2-5, mixed with finer hairs; supra-alars 2-3; presuturals 1-3, usually 2; posthumerals 0-1, usually 1. Scutellar tomentum grey to grey-gold; apical scutellar setae few, 2-4 on each side mixed with finer ones.

Pleural tomentum grey to gold-grey. Katatergite setae 6-8 among finer hairs; katepisternal setae normal; 3-5 strong setae (longest 0.7mm) at posterior edge of anepisternum and patch of fine hairs along dorsal margin. Anepimeron without setae.

Legs. Base colour brown or black, paler on tarsi; tomentum grey or grey-gold, often thin. No coxal peg. In northern specimens all hairs and bristles white except sometimes a few dark ones on tarsi. Femora with relatively long decumbent hairs and numerous erect ones, especially on pro-femur; fine ventral bristles common, the longest as long as or longer than thickness of femur. Apical and dorsolateral bristles strong, numer-ous; profemur in male with 5-10 dorsolateral bristles; mesofemur with 1-7; metafemur with 5-11. Bristles on tibiae and tarsi strong; protibia in male with longest bristle 4 times longer than tibial width. In southern specimens pale bristles are often golden and the number of dark bristles increases, especially on tarsi and tibiae. Finer hairs less abundant, especially in females.

Wings. Veins yellow to light brown; membrane white to pale brown in oblique view. DCI = 0.35-0.46; cell M3 open or closed. Halter white to pale yellow; knob without dark spot.

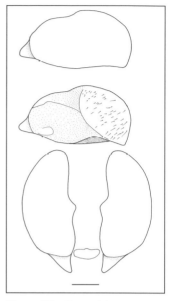

Figure 181. *L. quadrivittatus* epandrium (top to bottom): lateral, medial, dorsal. Scale lines = 0.2 mm.

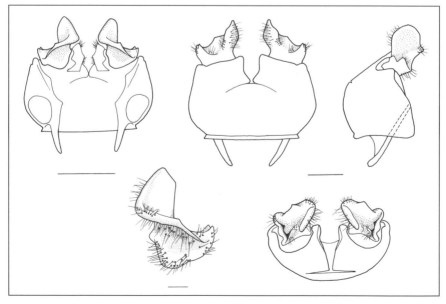

Figure 182. Hypandrium/gonocoxite complex. Clockwise from top left: dorsal, ventral, lateral and apical (scale lines = 0.3 mm); gonostylus, dorsal (scale line = 0.1 mm).

Abdomen. Tergites with grey tomentum; most specimens with dorsal paired pale brown/gold patches of tomentum covering basal 0.5 or less, these patches diminishing apically. Some specimens lack brown patches. All setae white. Tergite 1 with 6-9 major bristles on each side and many smaller setae; lateral setae on tergites fine, rather short (longer on 1-3) and directed posteriorly. Dorsal setulae hair-like, decumbent. Sternite tomentum all grey; hairs white.

Male genitalia. Epandrium and hypandrium/ gonocoxite brown to black with thin grey-gold tomentum; tomentum absent on hypandrium and sometimes at apex of epandrium. Setal brush white to gold in northern specimens, brown to black in southern ones; other setae numerous, prominent, pale in northern specimens, mostly brown in southern ones. Width of epandrium halves in lateral view about 0.55-0.6 times the length, widest at midlength; apex right-angled ventrally, broadly rounded dorsally. Medial face of epandrium as in figure 181. In dorsal view, medial margins of epandrium more-or-less parallel-sided with obtuse prominence at about midpoint; basal sclerite prominent.

Gonostylus. Dorsomedial, lateral and ventrolateral teeth well developed. Hypandrium/gonocoxite complex in ventral view with length about 0.65 times the width, the transverse slit at 0.7 the length. In lateral view, exposed length of gonocoxal apodeme about 0.6 as long as basal width of hypandrium; apodeme with strong sclerotized web ventrally.

Phallus. Paramere sheath dorsally 0.45 times the length of phallus (excluding the ejaculatory apodeme). Apex of paramere sheath broad, the ventral lip

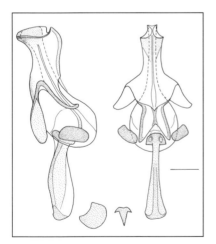

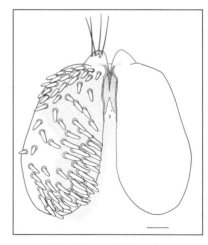

Figure 183. Phallus: lateral (left); ventral (right); lateral ejaculatory process, basal (centre left); ejaculatory apodeme, cross-section (centre right).
Scale line = 0.2 mm.

Figure 184. Subepandrial sclerite: ventral. Scale line = 0.1 mm.

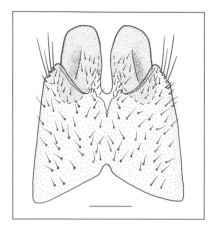

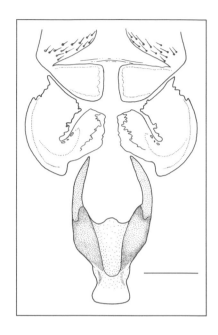

Figure 185 (above). *L. quadrivittatus* female sternite 8: ventral. Scale line = 0.3 mm.

Figure 186 (right). Female sternite 9, tergite 9 and lobes of sternite 10: ventral. Scale line = 0.2 mm.

pronounced; ventrolateral carina present, ventral flange absent. Sperm sac width in dorsal view 0.4 times the length of phallus. Ejaculatory apodeme gently curved ventrally in lateral view, moderately spatulate in ventral view; flattened in cross-section with strong dorsal carina. Subepandrial sclerite as in figure 184. Broad triangular unsclerotized area in basal 0.3 narrowing to a parallel-sided area about 0.4 the length; spines more-or-less parallel-sided, blunt; most densely arranged apically and basomedially.

Female genitalia. Undissected: All setae pale, rather abundant, erect. Tergite 8 dark brown/black with posterior margin paler. Sternite 8 dark brown/black, sometimes dark chestnut basally; lateral lobe setae moderately strong. Hypogynial valves with fine hairs basally. Cerci brown with pale setae. Dissected sternite (figure 185) with basal width about 0.8 times the length; undivided. Length of unsclerotized area between hypogynial valves 0.35 times sternite length. Hypogynial valves with ventral carina.

Tergite 9 sclerites in figure 186; sternite 9 Y-shaped, medially undivided. Tergite 10 brown/black with 6-8 brown/black acanthophorite spines on each side. Spermathecae (figure 187) with more-or-less straight terminal reservoirs about as wide as reservoir duct; terminal reservoirs with poorly developed wart-like protuberances on surface. Basal striated duct without fine scales; junction with basal duct with scales; basal duct very large, lumen 6-parted.

Figure 187. Spermatheca: dorsal. Scale line = 0.2 mm.

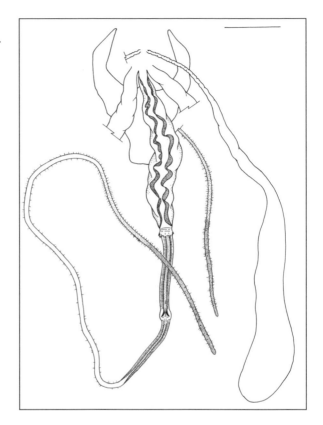

Variation. Southern specimens (Utah, New Mexico, some Colorado) have more gold infused in the tomentum of the head, and the occipital bristles are stronger, straighter and more golden. Some New Mexico specimens have black ocellar bristles. Antennal setae are dark in some specimens from the extreme south of the range. On the thorax, southern specimens are less setose and generally have darker tomentum; the notal setulae are mixed dark and light, the larger thoracic bristles largely gold or dark; 1-3 anterior dorsocentrals. On the legs, pale bristles are often golden and the number of dark bristles increases, especially on tarsi and tibiae. The finer hairs are less abundant, especially in females. In southern specimens some females have brown setulae mixed with the normal pale ones.

Type Material.
HOLOTYPE. ♀ (examined) labelled: "[rectangular white label] Halsey/6/1/06"; "[rectangular beige, black-bordered label] Lasiopogon/quadrivittatus/TYPE (Jones)" (UNSM). The holotype of *L. aridus* Cole and Wilcox, here synonymized with *L. quadrivittatus*, is a male: U.S.A., New Mexico, Jemez Springs Mts., vi.1917. OSUC.

PARATYPES. (2 examined). **U.S.A.: Nebraska,** Sioux Co., War Bonnet Canyon, 27.v.1901, M.A. Carricker (1 ♀, UNSM); Pine Ridge, badlands, mouth of Monroe Canon, 28.v. 1900, L. Bruner (1 ♀, UNSM).

Other Material Examined (224 specimens).
CANADA: Alberta, Burdette, 5mi N, 31.v.1964, S. Adisoemarto (6 ♂, UASM); Dinosaur Prov. Park, 9.vi.1964, S. Adisoemarto (6 ♂, 1 ♀, UASM); Drumheller, 18.vi.1957, Brooks and MacNay (1 ♂, CNCI); Edmonton, Emily Murphy Park, 18.vi.1963, S. Adisoemarto (3 ♂, 1 ♀, UASM), 21.vi.1963 (2 ♂, UASM), 2.vii.1963, S. Adisoemarto (1 ♂, UASM), 3.vii.1963, S. Adisoemarto (1 ♂, UASM), 4.vii.1963, S. Adisoemarto (2 ♂, UASM); Edmonton, Whitemud, 11.vi.1963 (1 ♂, UASM); Edmonton, N. Saskatchewan R. valley, 20.v.1998, John H. Acorn (4 ♂, 1 ♀, BCPM); Empress, Red Deer R., 21.v.1964, S. Adisoemarto (6 ♂, 2 ♀, UASM); Fabyan, campsite, 15.v.1964, S. Adisoemarto, (3 ♂, UASM); 16.v.1964, S. Adisoemarto, (6 ♂, UASM); Grande Prairie, 11.vi.1961, A.R. Brooks (1 ♀, CNCI); Lethbridge, 20.v.1928, G.F. Manson (1 ♂, CNCI), 30.v.1929, J.H. Pepper (1 ♂, UASM); 31.v.1929, J.H. Pepper (1 ♂, CNCI), 6.vi.1929, J.H. Pepper (2 ♂, 1 ♀: 1pr in cop, CNCI); Lethbridge, Old Man R., 3.vi.1964, S. Adisoemarto (1 ♂, UASM); Medicine Hat, 8.v.1926, F.S. Carr (1 ♂, 1 ♀ in cop, CNCI), 24.v.1933, F.S. Carr (2 ♂, CASC), 1.vi.1933, F.S. Carr (2 ♂, CASC), 10.vi.1933, F.S. Carr (1 ♂, CASC), 14.vi.1929, J.H. Pepper (1 ♂, 1 ♀, in cop, CNCI); Milk River, 49°08'Nx110°48'W, 5.vi.1965, J.R. Vockeroth (1 ♂, CNCI); N. Saskatchewan, Army Base Exp. Stn., 21.v.1964, S. Adisoemarto (1 ♂, UASM); Onefour, 6.vi.1955, A.R. Brooks (1 ♂, CNCI); Onefour, 49°06'Nx110°48'W, 7.vi.1955, J.R. Vockeroth (2 ♂, CNCI); Sandy Point, Hwy 41, 20.v.1964, S. Adisoemarto (3 ♂, UASM); Seven Persons, S. Adisoemarto (1 ♂, UASM); Taber, Old Man R., 4.vi.1964, S. Adisoemarto (1 ♂, 1 ♀, UASM); TP1/Rng4/W4Mer/Lot 1, 24.v.1982, B.F. and J.L. Carr (1 ♂, BCPM); TP14/Rng25/W4Mer/Lot 2, 15.v.1982, B.F. and J.L. Carr (1 ♂, BCPM); TP14/Rng25/W4Mer/Lot 1, 15.v.1982, B.F. and J.L. Carr (5 ♂, BCPM); TP15/Rng14/W4Mer/Lot 1, 21.v.1982, B.F. and J.L. Carr (1 ♂, BCPM); Writing-on-Stone Prov. Park, 2.vi.1964, S. Adisoemarto (5 ♂, UASM), 26.v-6.vi.1990, D. McCorquodale (1 ♀, DEBU). **British Columbia,** Attachie, 26.vi.1970, R.J. Cannings (1 ♂, SMDV); Fort Nelson, jct. Fort Nelson and Muskwa rivers, 10V 5266E 65214N, 4.viii.1997, H. Nadel (3 ♂, 1 ♀, BCPM). **Saskatchewan,** Elbow, 23.v.1960, A.R. Brooks (3 ♂, CNCI), 23.vi.1954, Brooks and Wallis, 1 ♂, CNCI), 3.vi.1960, A.R. Brooks (1 ♂, CNCI), 10.vi.1960, A.R. Brooks (1 ♂, CNCI); Estuary, 17.vi.1977, R. Hooper (1 ♂, SMNH); Great Sand Hills, 27.v.1939, A.R. Brooks (5 ♂, CNCI), 28.v.1939, A.R. Brooks (1 ♂, BCPM); Pike Lake, 14.v.1933, A.R. Brooks (1 ♂, CNCI); Winter, 12.vi.1970, R. Hooper (2 ♀, SMNH).

U.S.A.: Colorado, Alamosa Co., Blanca, 8 mi W on US Hwy 160, salt flats, 5.vi.1992, D. Leatherman (1 ♀, CRNC); El Paso Co., Foster Ranch, 5700', T15S/R65W/SEC22/NE1/4, 16.v.1976, R.J. Lavigne (10 ♂, ESUW), 1.v.1977, F.M. Brown (1 ♂, ESUW), 15.v.1977, F.M. Brown (1 ♂, ESUW), 8.v.1977, F.M. Brown (2 ♂, ESUW), 2.v.1981, F.M. Brown (1 ♂, ESUW); Foster Ranch, T15S/R65W/SEC23/N1/2, 2.v.1976, F.M. Brown (1 ♂, ESUW), 3.v.1976, F.M. Brown (1 ♂, syrphid prey, ESUW), 30.v.1976, F.M. Brown (1 ♂, ESUW); Mesa Co., Gateway, Dolores River, 3.v.1992, B. Kondratieff, W. Cranshaw and H. Knuttel (1 ♀, CSUC); Montrose Co., Jct. Route 141 and Route 90, Dry Creek San Miguel R., 3.v.1992, B. Kondratieff, W. Cranshaw and H.

Knuttel (1♀, CSUC); Rio Blanco Co., Route 139, MM59, 2.vi.1991, B. Kondratieff, and J. Welch (1♂, 1♀ in cop, CSUC); Saguache Co., Great Sand Dunes Nat. Mon., Medano Cr., 11.vi.1968, J.E. Slansky (1♂, EMFC); Great Sand Dunes Nat. Mon., picnic grounds, 13.v.1969 (4♂, EMFC); Great Sand Dunes Nat. Mon., Medano Cr., R73W/T26S/Sec26, 17.v.1977, S. Condie (1♂, ESUW), F.M. Brown (5♂, ESUW), R73W/T26S/Sec31, 9.vi.1977, F.M. Brown (1♂, ESUW). **Montana,** C.U. Lot 35 (1♂, OSUC). **Nebraska,** Sheridan Co., 2mi S Lakeside, 9.vi.1967, R.L. and J.A. Westcott (1♂, EMFC). **New Mexico,** Rio Arriba Co., Abiquiu, 14 mi N, 12.v.1992, H.E. Evans (2♂, 2♀, CRNC); Sandoval Co., Jemez Mts., 2.vi.1914 (1♀, ANSP), 5.ix.1914 (1♂, BCPM); Jemez Springs Mts., "May" (1♂, CASC), vi.1917 (1♀, CASC), vi.1917 (1♂, 1♀, OSUC [holo-type and allotype of *L. aridus* Cole and Wilcox]. **North Dakota,** Bismarck, 14.vi.1918, J.M. Aldrich (1♂, USNM; 1♂, EMEC). **Utah,** Garfield Co., Calf Cr., 5400', 24.v.1981, T. Griswold (1♂, EMUS); Gunnison Butte, Green River, 5.vii.1933, H.B. Stafford (1♀, BYUC); Moab, 25.vi.1938, G.F. Knowlton and F.C. Harmston (1♂, EMUS); San Juan Co., Indian Creek, Hwy 211, Newspaper Rock St. Pk, 5.vi.1994, P. Opler, Kondratieff and Nelson (3♂, 1♀, CSUC; 7♂, 3♀, CRNC); Newspaper Rock campgr., 5.v.1997, R.W. Baumann (2♂, 2♀, BYUC; 1♂, BCPM). **Wyoming,** no loc., 1881, Morrison, (3♂, USNM); Big Horn Co., Jct. Hwy 14 and 32, 21.v.1982, R.J. Lavigne (1♂, 1♀ in cop, ESUW); Glendo, 11.v.1960, R.J. Lavigne (4♂, 1♀, ESUW), 13.v.1960, R.J. Lavigne (5♂, 3♀: 2pr in cop, ♀ with ♂ prey, ESUW; 1♂, EMFC), 24.v.1960, R.J. Lavigne (1♀, ESUW), 1.vi.1961, R.J. Lavigne (5♂, 2♀: 1pr in cop, ♀ with 1♂ prey, ♂ prey of *Cyrtopogon*, ESUW), 7.vi.1961, R.J. Lavigne (1♂, Diptera prey, ESUW), 3.v.1962, R.J. Lavigne (1♂, 1♀, ESUW), 8.v.1972, S. Dennis (1♂, 1♀ in cop, ESUW); Glenrock, D. Johnston Pl. Stn. 7, 5.v.1974, R.J. Lavigne (1♂, ESUW), 20.v.1974, R.J. Lavigne (1♂, ESUW); Greybull, Big Horn R., 2.vi.1965, R.J. Lavigne (1♂, ESUW); Guernsey, 3.v.1962, R.J. Lavigne (1♂, therevid prey, ESUW), 9.v.1962, R.J. Lavigne (1♂, ESUW), 10.v.1962, R.J. Lavigne (1♂, 1♀: ♀ with ♂ prey, ESUW), 27.vi.1962, R.J. Lavigne (1♂, ESUW), 20.v.1963, L.J. Stevens (1♂, ESUW), 4.vi.1970, M. Ali (1♂, ESUW); Hot Springs Co., Worland, 20.7mi S Rte 789, 18.v.1982, R.J. Lavigne (1♂, ESUW); Lusk, 5.vi.1967, R.J. Lavigne (1♂, ESUW); Manderson, Big Horn R., 2.vi.1965, R.J. Lavigne (2♂, 1♀ in cop, ESUW); Natrona Co., Casper, 27mi N, 14.v.1992, R.J. Lavigne (1♂, ESUW); Niobrara Co., 30mi S of Newcastle, Cheyenne R., 26.v.1982, R.J. Lavigne (1♂, 1♀ in cop, ESUW); Park Co., 2.7mi S Powell, 21.v.1982, R.J. Lavigne (1♂, 1♀ in cop, ESUW); Pine Bluff, 7.vi.1967, R.J. Lavigne (1♂, ESUW); Sage Creek, 15.vi.1963, L.L. Wu (1♂, 3♀, EMUS); Washakie Co., Worland, 3.v.1965, R.J. Lavigne (1♂, ESUW), 1.v.1974, R.J. Lavigne (1♂, ESUW), 17.v.1984 (1♂, ESUW), Wheatland, 8.v.1972, S. Dennis (1♂, ESUW).

Type Locality. U.S.A., Nebraska, Halsey.

Taxonomic Notes. *L. quadrivittatus* is one of the most wide-ranging species in the genus, with a latitudinal range of 24°. Over much of this range its coloration and setation is rather stable, but significant variation occurs in the southern populations. The sparser, shorter, darker setation of specimens from the mountains of New Mexico is so striking that these flies are hardly recognizable as *L. quadrivittatus*; the population was described as *L. aridus* by Cole and Wilcox (1938). Nevertheless, the genitalia are identical to those of other populations of *L.*

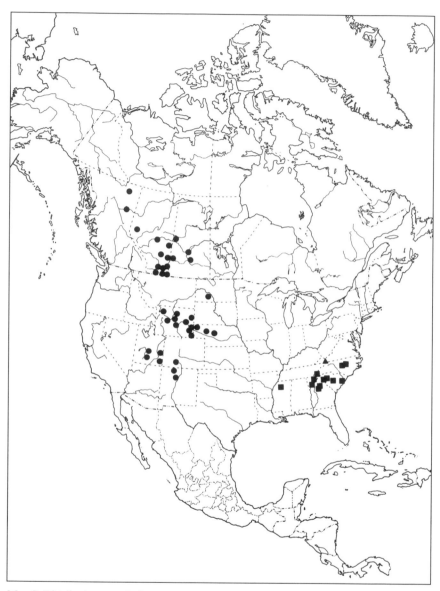

Map 8. Distribution records for
- *Lasiopogon quadrivittatus*
- *L. schizopygus*
- *L. marshalli*

quadrivittatus. Cole and Wilcox (1938) noted the surprising variation in the colour of all the major bristle groups in the four specimens they used to describe *L. aridus*. This mixture of bristle colours is found in most material from New Mexico and Utah and represents a gradation between the dark New Mexico specimens (*L. aridus*) and the pale ones to the north. *L. aridus* is here synonymized with *L. quadrivittatus*.

Etymology. *quadrivittatus* = four stripes; refers to the pair of brown dorsocentral and the pair of lighter grey/brown acrostichal stripes on the thoracic dorsum.

Distribution. Nearctic; northeastern British Columbia and northern Alberta east to North Dakota, south to Utah, New Mexico and Nebraska. In Canada, *L. quadrivitattus* is restricted to the Great Plains; south from Montana it also ranges west into the intermontane plateaus and mountains. Map 8.

Phylogenetic Relationships. Member of the *tetragrammus* species group; sister species to *L. lavignei*.

Natural History. Habitat: sandy places, often near streams or on stream banks in prairie grasslands and dry forests. Diptera such as Syrphidae and Therevidae accounted for 16 of 18 prey items in Colorado; the other two prey were small Homoptera (Lavigne 1972). Several female *L. quadrivittatus* also preyed upon conspecific males (see collection data) and one male was killed by a *Cyrtopogon* robber fly. The recorded flight period is 1 May-4 July; there is little difference in the range of collection dates from B.C./Alberta/Saskatchewan in the north and Colorado/Wyoming in the south, although most records in Canada are from June while most in the U.S. are from May.

Lasiopogon rokuroi Hradský

Lasiopogon rokuroi Hradský, 1981. *Traveaux du Musée d'Histoire Naturelle Gregor Antipa* 23: 180.

Diagnosis. A medium-sized to large species with grey thorax and shining black abdomen. Mystax and other head bristles dark. Antennae brown, F2+3/F1 = 0.48-0.65. Thoracic tomentum dorsally and laterally grey/gold-grey with brown dorsocentral stripes, grey acrostichal stripes and usually a faint gold/brown medial stripe. Main thoracic and leg bristles dark; anterior dorsocentral bristles 3-5. Abdominal tergites basally with very thin brown tomentum; apical bands gold-grey, very narrow, covering only the intersegmental membrane dorsally, widening dorsolaterally, forming triangular lateral areas of tomentum. Bristles on tergite 1 black. Epandrium dark chestnut/black, the width about 0.4 times the length in lateral view, the apex rounded and somewhat flanged. In dorsal view strongly concave medially. Female with shining black tergites much as in male, the grey apical bands wider, to about 0.2 the tergite length. Tergite 8 dark chestnut/black with brown hairs; sternite 8 with variable amounts of brown, the valves black, their medial lobes ferruginous with pale brown/gold hairs.

Description. Body length ♂ 9.1-11.5mm; ♀ 10.5-12.4mm.

Head. HW ♂ 1.90-2.04mm; ♀ 2.02-2.56mm. FW ♂ 0.36-0.40mm; ♀ 0.46-0.54mm. VW ♂ 0.93-1.23mm; ♀ 1.08-1.25mm. VW/HW = ♂ 0.53-0.63; ♀ 0.49-0.54. FW/VW = ♂ 0.33-0.39; ♀ 0.40-0.43. VD/VW = ♂ 0.10-0.13; ♀ 0.14-0.17. GH/GL = ♂ 0.29-0.32; ♀ 0.29-0.33.

Face gold-grey, vertex brown-grey, occiput grey, with gold or brown dorsally. Beard and labial hairs white (gold/ferruginous in Kyoto specimen), mystax bristles brown/black; all other setae brown/black. Occipital bristles rather fine (to 0.6mm), abundant, those behind the dorsomedial angle of the eye strongly curved anterolaterally; lateral and ventral ones shorter, straighter. Frontal and orbital setae strong, abundant, long (to 0.5mm); ocellar bristles fine, hardly distinguished from orbitals.

Antennae. Brown, sometimes with various parts (for example, apex of pedicel, base of F1, or parts of F2+3) dark ferruginous. Setae brown; some specimens with setae on F1. F1 long, narrow, more-or-less parallel-sided; WF1/LF1 = ♂ 0.21-0.25; ♀ 0.21-0.23. LF2+3/LF1 = ♂ 0.55-0.65; ♀ 0.8-0.54.

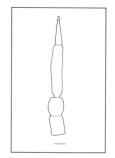

Figure 188. Antenna. Scale line = 0.1 mm.

Thorax. Prothorax grey, brown-grey on antepronotum, with white hairs ventrally, brown dorsally; postponotal lobes brown-grey, the lateral angle yellow/ferruginous, hairs brown. Scutum tomentum gold-grey, dorsocentral stripes medium brown, bordered with gold and extending to scutellum. No intermediate spots; ventrolateral areas strongly brown. Acrostichal stripes grey, sometimes obscure, the medial area usually faintly brown or gold. All strong bristles black, finer setae brown. 3-5 moderately strong anterior dorsocentral setae (to 0.7mm) mixed with slightly finer ones; 5-6 posteriors. Notal setae abundant, especially prominent in posthumeral areas, rather long (to 0.3mm), about as long as shorter dorsocentrals. Postalars 2-4 among shorter setae; supra-alars 2-4 with shorter setae; presuturals 3, 1-4 posthumerals, usually weak. Scutellar tomentum grey, sometimes strongly gold dorsally; apical scutellar bristles dark, 4-6 strong ones on each side mixed with shorter, weaker dark bristles and hairs.

Pleural tomentum grey to gold/brown-grey, darkest on anepisternum. Katatergite setae brown/black, 8-10 among finer white/brown hairs; katepisternal setae white/brown, sparse, often long. Anepisternal setae 6-20, dark (to 0.6mm), mixed with weaker ones; a large patch of erect brown setae on dorsal shelf. Anepimeron with white/brown setae.

Legs. Base colour dark brown/black; tomentum of coxae grey with some gold highlights; tomentum on rest of legs grey. No coxal peg. Main bristles dark brown/black. Femora dorsally with short hairs white basally, brown/black apically, the dark hairs predominating in males (specimen from Kyoto with almost all leg hairs brown) and the white much more extensive in females. Longer erect white/gold and dark hairs abundant ventrally and laterally, especially on profemur; the colour pattern is similar to that of the shorter hairs. Brown/black dorsolateral bristles on femora are numerous, long, and fine and are almost indistinguishable from the surrounding hairs on the profemur; these bristles are mostly stouter on other legs, about 5-15 on mesofemur, 10-20 on metafemur. In males the longest ventral setae are white or brown and longer than the width of the femur. Tibiae and tarsi have dark, strong bristles typically arranged; hairs pale brown to black, especially abundant on tibiae. Protibia with longest bristles about 5 times longer than tibial width.

Wings. Veins medium/dark brown; membrane brown in oblique view. DCI = 0.34-0.40; cell M3 open. Halter yellow; knob without a dark spot.

Abdomen. Male. Tergite basal colour shining dark brown/black, with only very thin brown tomentum, most of it basally, and most of it almost invisible. Very narrow bands of thin grey tomentum hardly cover more than the intersegmental membrane of each tergite apically; segment 1 is very thinly covered except mid-dorsally. Apical bands gradually widen dorsolaterally, expanding to cover ventrolateral areas, these gold-grey areas thus triangular, especially apically. About 4-7 black bristles on each side of tergite 1. Lateral setae on tergites 1-4 white, on 5-7 brown. In a specimen from Kyoto, all abdominal hairs are brown. Dorsal setulae dense, brown on all segments. Sternite tomentum grey, hairs white on 1-7, brown on 8.

Female. Unusually for a female *Lasiopogon*, tomentum is scarce, much as in male, except gold-grey apical bands slightly wider (0-0.2 the length of the tergite dorsally); the very thin brown tomentum sometimes more noticeable. Gold-grey ventrolateral tomentum browner basally. Erect hairs laterally on 1-4 white, but becoming brown dorsolaterally; those on 5-7 short and brown. Dorsal setulae dark. Sternites grey; hair white, brown on 7.

Male genitalia. Epandrium and hypandrium/ gonocoxite complex dark chestnut to black (lightest on hypandrium) and covered with gold-grey tomentum although often hypandrium is almost bare. Setal brush and other setae dark brown/black. Width of epandrium halves in lateral view about 0.4 times the length, slightly convex dorsally, the dorsal and ventral margins more-or-less parallel. Apex rounded,

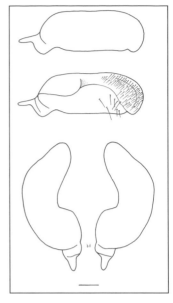

Figure 189. *L. rokuroi* epandrium (top to bottom): lateral, medial, dorsal. Scale line = 0.2 mm.

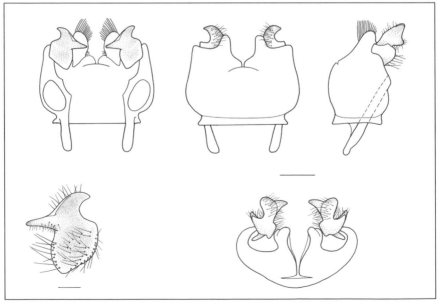

Figure 190. Hypandrium/gonocoxite complex. Clockwise from top left: dorsal, ventral, lateral and apical (scale lines = 0.3 mm); gonostylus, dorsal (scale line = 0.1 mm).

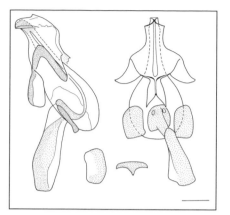

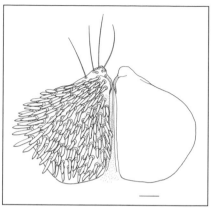

Figure 191. Phallus: lateral (left); ventral (right); lateral ejaculatory process, basal (centre left); ejaculatory apodeme, cross-section (centre right). Scale line = 0.2 mm.

Figure 192. Subepandrial sclerite: ventral. Scale line = 0.1 mm.

and slightly flanged, the ventral margin of the epandrium indented at about 0.6 its length. Medial face of epandrium as in figure 189. In dorsal view, medial margins of epandrium strongly concave; basal sclerite obscure.

Gonostylus. Medial flange expanded laterally into a curved, acute apical lobe and a narrow dorsomedial lobe; dorsal flange absent; lateral and ventrolateral teeth present. Hypandrium/gonocoxite complex in ventral view with length about 0.6 times the width, the transverse slit at 0.65 the length. In lateral view, exposed length of gonocoxal apodeme about 0.8 times basal width of hypandrium; apodeme with sclerotized web ventrally.

Phallus. Paramere sheath angled strongly ventrally, its dorsal margin 0.4 times the length of phallus (excluding ejaculatory apodeme). Apex of paramere sheath broad, the ventrolateral carina short; ventral flange absent; dorsal carina strongly arched dorsally. Sperm sac width in dorsal view 0.45 times the length of phallus. Ejaculatory apodeme long, straight in lateral view, spatulate in ventral view; flattened ventrally in cross-section, dorsal carina shallow. Subepandrial sclerite as in figure192. Sclerite broader than its length, the linear unsclerotized area in basal 0.8; spines bluntly acute and densely arranged over whole sclerite; microsetae present only on medial area.

Female genitalia. Undissected: Hairs abundant, erect, long and light to medium brown. Tergite 8 dark chestnut/black, often narrowly ferruginous apically. Sternite 8 chestnut basally; hypogynial valves parallel, narrow, the knifelike blade oriented with the edges oriented dorsoventrally; valves black, the medial lobes ferruginous with gold/pale brown hair. Lateral lobe setae weak. Cerci black/brown with hairs white/light brown. Dissected sternite (figure 193) broad, basal width about 0.9 times the length; undivided. Sclerite of hypogynial valve

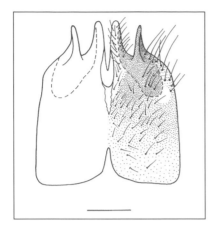

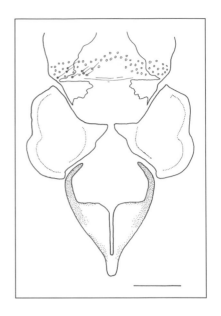

Figure 193 (above). *L. rokuroi* female
sternite 8: ventral. Scale line = 0.3 mm.

Figure 194 (right). Female sternite 9,
tergite 9 and lobes of sternite 10: ventral.
Scale line = 0.2 mm.

includes the lateral part of the medial lobe and a ventral carina is present on the
midline of lateral (valve) part. Medial lobes broad, slightly thicker and longer
than lateral valve. Length of unsclerotized area between medial lobes 0.4 times
sternite length.

Tergite 9 sclerites as in figure 194; sternite 9 V-shaped, the medial area largely sclerotized but with a narrow medial slit almost dividing the sclerite. Basal
lobes of sternite 10 small. Tergite 10 black/brown, usually with 6-7 black acanthophorite spines on each side. Spermathecae (figure 195) with hooked terminal
reservoirs about as thick as reservoir duct; terminal reservoirs with wart-like protuberances on surface. Both striated ducts with fine scales; junction with basal
duct scaled, not golden; valve not golden; basal duct short and narrow.

Type Material.
HOLOTYPE. ♂ (examined but did not copy exact label format) labelled: Japan,
Ikaho-Gumma, 30.iv.1964, Rokuro Kano. MHRC. ALLOTYPE ♀ (examined)
labelled: same data as holotype.
PARATYPES. (4 examined). Same data as holotype (2♂, 2♀, MHRC).

Other Material Examined (7 specimens).
Japan: Honshu, Kyoto, 10.v.1953 (1♂, EMFC). **Kyushu,** Kagoshima, 16.iv.1910, J.C.
Thompson (1♂, BCPM), iii.1912, J.C. Thompson (1♀, CASC); Kagoshima, Iso,
23.iii.1954, A. Nagatomi (2♂, 2♀, KUIC).

Figure 195. Spermatheca: dorsal. Scale line = 0.1 mm.

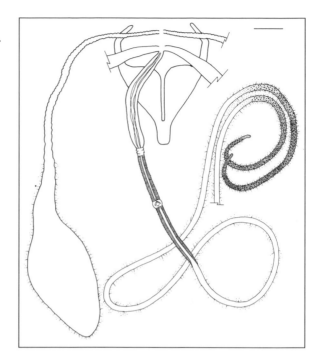

Type Locality. Japan, Honshu, Ikaho-Gumma.

Etymology. Named for the collector of the type series, Rokuro Kano (Hradský 1981).

Distribution. Palaearctic; Japan: Honshu, Kyushu. See map 3, p. 127.

Phylogenetic Relationships. Member of the *akaishii* group; sister to the species pair *L. akaishii* and *L. hasanicus*.

Natural History. Habitat: No concrete information, but probably stream sides in forests. The general area of the type locality is between 1000 and 1500 metres, although the other collection sites (vaguely defined) apparently are at lower elevations. Collections range from 30 April to 23 August; the earliest date is from the most northerly site.

Lasiopogon schizopygus sp. nov.

Lasiopogon carolinensis Cole and Wilcox, 1938. *Entomologica Americana* 43: 34-36. (Part).

Diagnosis. A small to medium-sized grey/grey-brown species with dark brown legs. Mystax and other head bristles dark. Antennae brown, F2+3 long (F2+3/F1 = 0.71-93). Thoracic colour variable, from grey and hardly marked to brown and gold, but the dark, shining abdomen is distinctive in the male. Thoracic tomentum dorsally and laterally grey/gold-grey (sometimes richer brown/gold dorsally) with brown dorsocentral stripes, grey acrostichal stripes and often a faint medial stripe. Main thoracic and leg bristles dark, except for pale katatergites; anterior dorsocentral bristles 3-5, weak. Abdominal tergites basally with only thin brown tomentum; apical bands grey, very narrow, covering 0.1 to 0.2 the length of the segments. Bristles on tergite 1 pale. Epandrium width about 0.5-0.6 times the length lateral view; apex rounded. In dorsal view weakly concave medially. Female with black/brown terminalia and light brown hairs.

Description. Body length ♂ 8.3-10.6mm; ♀ 8.5-11.5mm.

Head. HW ♂ 1.94-2.04mm; ♀ 1.80-2.22mm. FW ♂ 0.34-0.38mm; ♀ 0.34-0.44mm. VW ♂ 0.86-0.96mm; ♀ 0.90-1.08mm. FW/VW = ♂ 0.39-0.47; ♀ 0.39-0.44. VD/VW = ♂ 0.11-0.14; ♀ 0.10-0.12. GH/GL = ♂ 0.23-0.30; ♀ 0.29-0.32.

Face grey or gold-grey, vertex brown-grey, occiput grey, often with brown highlights dorsally. Beard and labial hairs white, mystax bristles brown/black, frequently with a few pale setae ventrally; all other setae brown/black. Occipital bristles rather fine, abundant, those behind the dorsomedial angle of the eye strongly curved anterolaterally; lateral and ventral ones shorter, straighter. Frontal and orbital setae abundant, long; ocellar bristles fine, hardly distinguished from orbitals.

Antennae: Brown, sometimes with various parts (for example, base of F1 or F2+3) dark ferruginous. Setae brown, sometimes a few white on scape; some specimens with setae on F1. F1 short, widest at about midlength, ventral margin convex. WF1/LF1 = ♂ 0.34-0.37; ♀ 0.25-0.42. LF2+3/LF1 = ♂ 0.75-0.93; ♀ 0.71-0.76.

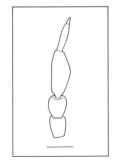

Figure 196. Antenna. Scale line = 0.2 mm.

Thorax. Prothorax grey/gold-grey, often brown on antepronotum, with white hairs; postponotal lobes grey/gold-grey, the lateral angle ferruginous, hairs white to light brown. Scutum tomentum variable. Some specimens with rather clear grey tomentum and dorsocentral stripes faint brown, the paramedial stripes grey and obscure. Others grey with gold highlights, the dorsocentral stripes light brown, the acrostichals dark grey with a faint gold-brown medial stripe. At the other extreme, the tomentum base is brown-grey to light brown, the dorsocentral stripes brown, often bordered with gold, the acrostichal stripes grey, faint to strong, with the medial area and intermediate spots and lateral areas gold-brown. All strong bristles black, finer setae brown. Anterior dorsocentrals 3-5, rather weak (longest to 0.6mm), mixed with even finer setae; 3-4 posteriors. Notal setae rather abundant over presutural area, as long as shorter dorsocentrals. Postalars 2-3, with shorter setae; supraalars 1-2; presuturals 2-3, 1-2 weak posthumerals. Scutellar tomentum grey with gold highlights; apical scutellar bristles dark, 2-4 strong ones on each side mixed with shorter, weaker dark bristles and hairs.

Pleural tomentum ranges from grey to gold. Katatergite setae white/yellow, 8-10 among finer white hairs; katepisternal setae sparse, often long. Anepisternal setae 4-8 dark (to 0.6mm); a few other pale, weaker ones ventrally; a patch of short erect brown setae on dorsal shelf. Anepimeron with a few white setae.

Legs. Base colour dark brown/black. Tomentum of coxae grey with some gold highlights; tomentum on rest of legs grey. No coxal peg. Main bristles dark brown/black, finer setae predominantly white. Femora dorsally with white decumbent hairs, often some dark apically. Longer erect pale hairs abundant ventrally and laterally, especially on profemur; in males longest white ventral setae longer than thickness of femur. Profemur with 1-5 strong dorsolateral dark bristles; mesofemur with 3-5; metafemur with 1-13. Tibiae and tarsi with dark, strong bristles typically arranged, hairs on tibiae mostly white or pale brown; tarsal hairs all white in some specimens (especially females), but usually dark. Protibia with longest bristles about 3.5-4.0 times longer than tibial width.

Wings. Veins yellow-brown to medium brown, often darkest at costa and adjacent veins; membrane usually pale brown in oblique view. DCI = ♂ 0.35-0.46; cell M3 open, sometimes only narrowly so. Halter yellow; knob without dark spot.

Abdomen. Male. Tergite basal colour dark brown/black. Tomentum on tergite bases thin brown/gold-brown, usually extensively lacking so that dark cuticle shines through. Bands of grey tomentum cover 0.1 to 0.2 the length of each tergite apically (hardly covering more than intersegmental membrane); segment 1 is thinly covered except mid-dorsally. Ventrolateral areas are narrowly covered with thin grey-brown tomentum. About 4-6 strong bristles on each side of tergite 1 white/yellow. Lateral setae on tergites 1-3 white, sparse, erect, rather short, those on 4 very short and mixed with dark ones; setae on 5-7 dark. Dorsal setulae appressed, short and dark. Sternite tomentum gold-grey, hairs white.

Female. Tomentum covers tergites completely, but the gold-grey and brown pattern is vague. Brown/grey-brown area generally in basal half or restricted to faint basolateral patches; grey or gold-grey apical areas often extending dorsally to the tergite base. Extensive grey ventrolaterally. Erect white hairs on 1-3 relatively short, 4 with very short lateral setulae mixed white and brown or all brown, brown on 5-7. Dorsal setulae on tergites 4-7 very short and dark, erect on apex of 5 and on 6-7.

Male genitalia. Epandrium and hypandrium/gonocoxite complex dark chestnut and covered with grey tomentum except on hypandrium in some specimens. Setal brush black; other setae brown/dark brown, numerous, prominent. Width of epandrium halves in lateral view about 0.5-0.6 times the length, widest at about midlength; ventral margin straight, dorsal margin gently convex, apex rounded. Medial face of epandrium as in figure 197. In dorsal view, medial margins of epandrium only weakly concave; basal sclerite small.

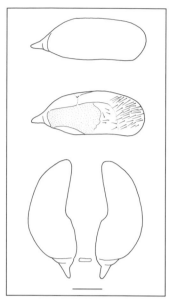

Figure 197. *L. schizopygus* sp. nov. epandrium (top to bottom): lateral, medial, dorsal. Scale line = 0.2 mm.

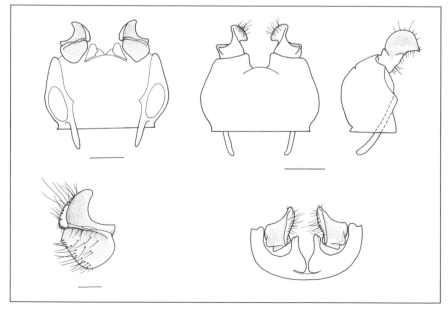

Figure 198. Hypandrium/gonocoxite complex. Clockwise from top left: dorsal, ventral, lateral and apical (scale lines = 0.3 mm); gonostylus, dorsal (scale line = 0.1 mm).

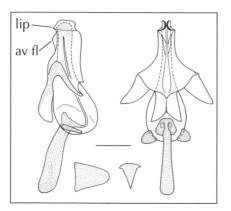

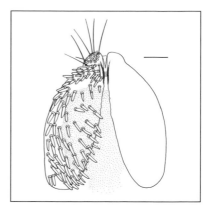

Figure 199. Phallus: lateral (left); ventral (right); lateral ejaculatory process, basal (centre left); ejaculatory apodeme, cross-section (centre right). av fl = apicoventral flange. lip = apicoventral lip. Scale line = 0.2 mm.

Figure 200. Subepandrial sclerite: ventral. Scale line = 0.1 mm.

Gonostylus. Dorsomedial tooth absent; lateral tooth small, ventrolateral tooth strong. Hypandrium/gonocoxite complex in ventral view with length about 0.65 times the width, the transverse slit at 0.8 the length. In lateral view, exposed length of gonocoxal apodeme about 0.7 as long as basal width of hypandrium; apodeme with strong sclerotized web ventrally.

Phallus. Paramere sheath dorsally 0.55 times the length of phallus (excluding ejaculatory apodeme). Apex of paramere sheath with ventrolateral carina and ventral flange. Sperm sac width in dorsal view 0.35 times the length of phallus. Ejaculatory apodeme rather straight in lateral view, weakly spatulate in ventral view; flattened in cross-section with strong dorsal carina. Subepandrial sclerite as in figure 200. Narrow triangular unsclerotized area in basal 0.85; spines blunt, sparse in central parts of sclerite halves.

Female genitalia. Undissected: Tergite 8 dark brown/black, often narrowly brown apically; hairs light to medium brown, erect. Sternite 8 chestnut to dark brown/black; lateral lobe setae moderately strong to strong. Hypogynial valves with short bristles on ventral surface. Cerci black with pale setae. Dissected sternite 8 (figure 201) elongate, basal width about 0.6 times the length; strongly divided along midline. Length of unsclerotized area between hypogynial valves 0.3 times sternite length.

Tergite 9 sclerites as in figure 202; sternite 9 V-shaped, medially undivided, with strong dorsal carina. Tergite 10 brown/black, usually with 6 black acanthophorite spines on each side. Spermathecae (figure 203) with terminal reservoirs curled in a complete loop and about as wide as reservoir duct; terminal

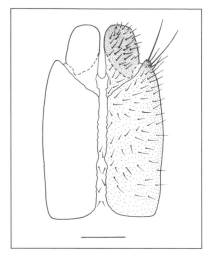

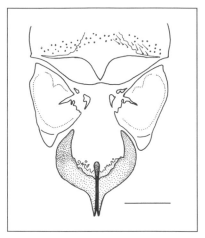

Figure 202. Female sternite 9, tergite 9 and lobes of sternite 10: ventral. Scale line = 0.2 mm.

Figure 201. *L. schizopygus* sp. nov. female sternite 8: ventral. Scale line = 0.3 mm.

reservoirs without wart-like protuberances on surface. Basal striated duct with fine scales; junction with basal duct sclerotized, golden, scaled; valve sclerotized, golden; striated basal section of reservoir tube with fine scales; basal duct short.

Variation. Specimens from the southern parts of the range tend to be browner on the dorsum of the scutum and on the head. In males the grey apical bands on the abdominal tergites are mostly narrow (0.2-0.25 of the segment length) with little tendency to expand anteriorly along the midline; grey tomentum ventrolaterally is reduced. In females the grey bands dorsolaterally are 0.3-0.5 the length of the segment and expand anteriorly along the midline.

Type Material
HOLOTYPE. (here designated) ♂ labelled: "[rectangular white label] Panthersville, Ga./3-25-45/P.W. Fattig". My Holotype label "HOLOTYPE/ *Lasiopogon* ♂/*schizopygus* Cannings/des. R.A. Cannings 2002 [red, black-bordered label]" has been attached to this specimen. Dissected genitalia in plastic vial underneath all. USNM.
PARATYPES (121 designated). **U.S.A.: Georgia,** Atlanta, 11.iv.1940, P.W. Fattig (1♂, 1♀, UGCA; 1♂, USNM), 14.iv.1940, P.W. Fattig (2♂, LGBC; 1♂, 1♀, UGCA; 1♂, 1♀, USNM), 4.v.1941, P.W. Fattig (1♀, LGBC), 13.iv.1947, P.W. Fattig (1♀, UGCA; 1♂, USNM), 27.iv.1941, P.W. Fattig (1♂, UGCA; 1♂, 1♀, USNM), 25.iv.1948 (2♂, LGBC); Bogart,, 1 20.v.1972, A. Lavallee (6♂, 5♀, UGCA; 2♂, 2♀, LGBC); Clarke Co., 27.iv.1939, G. Marsh (1♂, UGCA), 14.iv.1946, H.O. Lund (1♂, UGCA); Clarke Co., Athens, 20.v.1969, S. Howie

Figure 203. Spermatheca: dorsal. Scale line = 0.1 mm.

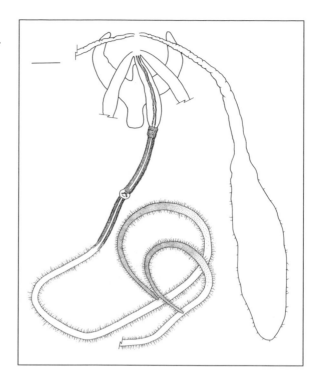

(1♀, LGBC), 10.v.1972, H. Goldberg (1♂, UGCA), 21.v.1974, A. Lavallee (1♂, LGBC; 1♀, UGCA); Oglethorpe Co., Echols Mill, 9.vi.1968, J.L. Zettler (1♂, UGCA); Panthersville, 25.iii.1945, P.W. Fattig (1♂, DEIC; 3♂, UGCA; 1♀, USNM), 28.iii.1945, P.W. Fattig (1♀, UGCA; 1♂, USNM); Rabun Co., Pine Mtn., 1400', 14.v.1957, H.C. Huckett (2♀, CNCI); Whitehall, 11.iv.1965, W.B. Sikora (2♂, UGCA). **Mississippi,** Oxford, 10.vi.1948, F.M. Hull (13♂, 4♀, CNCI). **North Carolina,** Buncombe Co., Black Mtn., 12.v.1980, L.L. Pechuman (4♂, 1♀, CUIC); Highlands, 3-5000', iv.1936, R.C. Shannon, (2♂, USNM); Highlands, 2800', 13.vi.1942, C.P. Alexander (1♂, UGCA), Highlands, 3800', 9.v.1957, W.R.M. Mason (2♀, CNCI), 10.v.1957, H.C. Huckett (1♂, CNCI); Raleigh, late iv.1908, F. Sherman (1♀ *L. carolinensis* allotype, OSUC), late iv.1920, M.R. Smith (1♂, EDNC), mid iv.1921, F. Sherman (1♂ [paratype *L. carolinensis*], OSUC), early v.1921, T.B.M. (1♂, EDNC), 15.iv.1930, C.S. Brimley (4♂, 1♀, EDNC), 17.iv.1931, C.S. Brimley (1♂, EDNC), 28.iv.1931, C.S. Brimley (3♂, 1♀, EDNC), 5.iv.1932, C.S. Brimley (1♂, EDNC), 17.v.1932, C.S. Brimley (1♀ [paratype *L. carolinensis*], CASC), 20.iv.1932, C.S. Brimley (1♂, EDNC), 21.iv.1932, C.S. Brimley (1♀, EDNC), 17.v.1932, C.S. Brimley (1♀, CASC), 9.v.1933, C.S. Brimley (1♂, EDNC), 1.v.1934, C.S. Brimley (1♂, EDNC), 24.iv.1935, C.S. Brimley (3♂, EDNC), 16.iv.1936, C.S. Brimley (7♂, 1♀, EDNC), 3.iv.1939, C.S. Brimley (1♂, EDNC); Wake Co., 2mi NW Raleigh, Crabtree Cr., 26.iv.1964, G.A. Matuza (1♂, ESUW). **South Carolina,** Clemson,

27.iv.1940, D. Dunavan (1♂, CUCC), 20.iv.1948, D. Dunavan (2♂, CUCC), 9.iv.1949, C.S. Creighton (1♀, CUCC); Clemson College, 3.v.1939, D. Dunavan (1♀, CUCC); Columbus, 4.iv.1946, O.L. Cartwright (1♂, CUCC); Gramling, 4.iv.1939, O.L. Cartwright (1♂, CUCC); Salem Co., Little River, 21.iv.1929, O.L. Cartwright (1♀, CASC; 1♀, USNM).

Type Locality. U.S.A., Georgia, Panthersville.

Taxonomic Notes. *L. schizopygus* replaces the name *L. carolinensis* Cole and Wilcox. The type series of *L. carolinensis* contains specimens of both *L. opaculus* Loew and the species Cole and Wilcox (1938) were describing. The holotype of *L. carolinensis* is a specimen of *L. opaculus*; *L. carolinensis* is thus a junior subjective synonym of *L. opaculus* and a new name must be used.

Etymology. From the Greek *schizo* = split and *pygus* = tail, rump; A reference to the strongly divided sternite 8 in the female of the species.

Distribution. Nearctic; North Carolina south to Georgia and Mississippi. See map 8, p. 216.

Phylogenetic Relationships. A member of the *opaculus* species group.

Natural History. A spring species in the Appalachian Mountains and Piedmont Plateau of the southeastern states. Common in riparian areas. Dates of capture range from 25 March to 13 June.

Lasiopogon shermani Cole and Wilcox

Lasiopogon shermani Cole and Wilcox, 1938. *Entomologica Americana* 43: 67-70.

Diagnosis. A medium-sized grey or grey-brown species with mystax and head bristles brown. Antennae brown, with F1 basally yellow and F2+3 long (F2+3/F1 = 0.60-1.00). Thoracic tomentum dorsally and laterally gold-grey (sometimes richer gold-brown dorsally) with gold-brown dorsocentral and acrostichal stripes. Thoracic and leg bristles dark; anterior dorsocentral bristles 4-5, prominent. Trochanters, bases of femora, tibiae and tarsi ferruginous. Abdominal tergites basally with brown tomentum, apical bands gold-grey, covering one-third to one-half the length of the segments. Lateral hairs white, long. Epandrium ferruginous, the width 0.35-0.5 times the length in lateral view, apex rounded, gently down turned, ventral corner angled. In dorsal view only moderately concave medially. Gonostylus with medial flange low, the height about 0.4 times the dorsoventral width; lateral tooth prominent. Female with dark hypogynial valves contrasting with brown or ferruginous base of sternite 8. Spermathecae with terminal reservoirs shallowly hooked, warty protuberances basally.

Description. Body length ♂ 8.50-10.25mm; ♀ 8.20-11.0mm.

Head. HW ♂ 1.84-2.10mm; ♀ 2.00-2.20mm. FW ♂ 0.40-0.44mm; ♀ 0.44-0.50mm. VW ♂ 0.85-0.93mm; ♀ 0.84-0.99mm. VW/HW = ♂ 0.44-0.46; ♀ 0.42-0.45. FW/VW = ♂ 0.47; ♀ 0.51-0.52. VD/VW = ♂ 0.16-0.17; ♀ 0.19-0.20. GH/GL= ♂ 0.36-0.42; ♀ 0.36-0.41.

Face silver-grey with gold highlights, vertex gold-grey; occiput grey, gold-brown dorsally. Beard and labial hairs white, mystax bristles brown with a few white setae ventrally; all other setae brown. Occipital bristles only moderately strong, the longest about 6-8 dorsomedially on each side, strongly curved anteriorly; lateral and ventral ones shorter, straighter. Frontal and orbital setae sparse, most as long as, or considerably longer than, scape+pedicel.

Antennae. Brown, F1 yellow basally; hairs brown, a few on scape white. Some specimens have setae on F1. F1 with dorsal margin straight, ventral one convex. WF1/LF1 = ♂ 0.33-0.36; ♀ 0.33-0.36. LF2+3/LF1 = ♂ 0.65-1.00; ♀ 0.60-0.65.

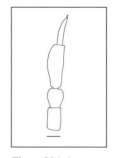

Figure 204. Antenna. Scale line = 0.1 mm.

Thorax. Prothorax silver-grey with white hairs; postponotal lobes grey to gold-grey, the lateral angle ferruginous, hairs white. Scutum tomentum in some specimens grey with extensive, gold-brown highlights, especially laterally and between the dorsocentral stripes; dorsocentral and acrostichal stripes and intermediate spots light gold-brown. In other specimens gold-brown tomentum obscures much of the grey and the stripes are darker brown. All notal setae brown or black. Dorsocentrals prominent; 4-5 presuturals (longest to 1.0mm) and 3-4 post-suturals mixed with shorter, finer setae. No definite acrostichals, but in some specimens acrostichal stripes bear irregular rows of short setae similar to the shorter dorsocentrals and the scattered notal setae. Postalars 2-3, strong, with shorter hairs; supra-alars 1-2; presuturals 2-3, posthumerals 1. Scutellar tomentum grey with extensive gold highlights in some specimens; apical scutellar bristles dark, 2-5 strong ones on each side mixed with shorter, weaker dark bristles and hairs.

Pleural tomentum grey with extensive gold highlights in some specimens. Katatergite setae 5-8 among finer white hairs, most brown, a few of the smaller ones pale; katepisternal setae sparse, long. Anepisternal setae 5-7, a few moderately strong, to 0.6mm; a patch of short erect hairs on dorsal shelf. Anepimeron without setae.

Legs. Base colour of trochanters, tibiae, tarsi, bases and apices of femora and parts of coxae light ferruginous; most of femora dark brown. Trochanter with dark spot apicoventrally. Tomentum of coxae silver with some gold highlights; tomentum on femora rather heavy, gold-grey, more sparse on tibiae and tarsi. No coxal peg. Most bristles brown or black, a few fine dorsal ones on femora pale, finer hairs white to brown. Femora dorsally with rather long white decumbent and somewhat erect hairs, some brown ones apically; long pale hairs ventrally and laterally abundant; longest white ventral setae longer than thickness of femur. Profemur with 5-8 dorsolateral dark bristles; mesofemur with 2-6; metafemur with 8-12; these bristles often rather fine, augmented by fine white setae. Tibiae and tarsi setose, dark, strong bristles typically arranged, hairs abundant, most brown, some pale. Protibia with longest bristle about 3.0-3.5 times longer than tibial width.

Wings. Veins light to medium brown; membrane pale brown in oblique view. DCI = 0.35-0.43; cell M3 open, sometimes only narrowly. Halter yellow; knob without dark spot.

Abdomen. Male. Tergite basal colour brown, ferruginous apically on segments. Tomentum on tergite bases brown with much gold in lateral and anterior views. Bands of grey tomentum cover one-third to half of each tergite apically (segment 1 is almost all covered); at the base the brown tomentum extends almost to the ventrolateral corners. About 6-8 strong bristles on each side of tergite 1 white, brown or black; usually at least three dark. Lateral setae on tergites 1-3 white, abundant, long and erect, those on posterior segments one-third to one-half as long; in some specimens some lateral hairs on posterior segments are dark. Dorsal hairs rather long, decumbent, mixed white and brown. Sternite tomentum gold-grey, hairs white.

Female. As in male, except in some specimens apical grey bands extend in a triangle anteriorly on midline. Lateral setae on tergites 4-7 short, mostly dark laterally and white along ventral margin; dorsal setulae mixed white and brown, mostly brown.

Male genitalia. Epandrium and hypandrium/ gonocoxite complex ferruginous with grey tomentum on epandrium, gonocoxites, and usually on hypandrium. Setal brush dark brown; other setae dark brown, numerous, prominent, especially dorsally on epandrium. Width of epandrium halves in lateral view 0.35-0.5 times the length, margins more-or-less parallel, usually gently arched; apex rounded, slightly down turned, ventral corner angulate. Medial face of epandrium as in figure 205. In dorsal view, medial margins of epandrium moderately concave; basal sclerite strong.

Gonostylus. Medial flange low, the height 0.4 times the dorsoventral width; dorsomedial tooth absent, dorsolateral tooth large; lateral tooth prominent. Hypandrium/gonocoxite complex in ventral view with length about 0.65 times the width, the transverse slit at 0.75 the

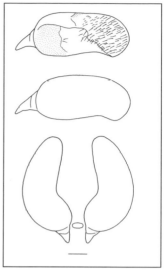

Figure 205. *L. shermani* epandrium (top to bottom): medial, lateral, dorsal. Scale line = 0.2 mm.

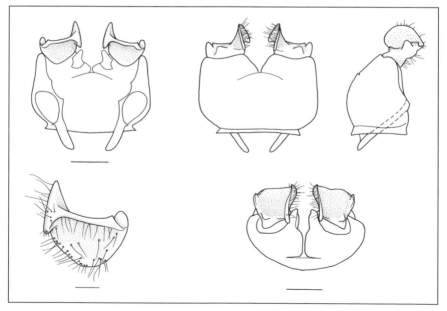

Figure 206. Hypandrium/gonocoxite complex. Clockwise from top left: dorsal, ventral, lateral and apical (scale lines = 0.3 mm); gonostylus dorsal (scale line = 0.1 mm).

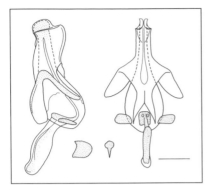

Figure 207. *L. shermani* phallus: lateral (left); ventral (right); lateral ejaculatory process, basal (centre left); ejaculatory apodeme, cross-section (centre right). Scale line = 0.2 mm.

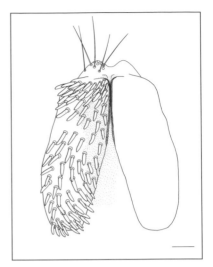

Figure 208. Subepandrial sclerite: ventral. Scale line = 0.1 mm.

length. In lateral view, exposed length of gonocoxal apodeme short, 0.4 times the basal width of hypandrium; apodeme with small sclerotized web ventrally.

Phallus. Paramere sheath dorsally 0.4 times the length of phallus (excluding ejaculatory apodeme). Ventral process long and narrow, extending almost to the base of the sperm sac. Apex of paramere sheath with ventrolateral carina and small ventral flange. Sperm sac width in dorsal view 0.35 times the length of phallus. Ejaculatory apodeme with ventral margin slightly convex in lateral view, weakly spatulate in ventral view; oval in cross-section with strong dorsal carina. Subepandrial sclerite as in figure 208. Sclerite elongate, triangular unsclerotized area in basal 0.5, a narrow unsclerotized portion in central 0.25; spines bluntly acute.

Female genitalia. Undissected: Hairs pale, erect, abundant; tergite 8 brown to black, often paler ventrally and apically. Sternite 8 brown to ferruginous with darker hypogynial valves, the valves with hairs ventrally. Lateral lobe setae moderately strong. Cerci yellow-brown with pale setae. Dissected sternite (figure 209) with basal width about 0.7 times the length, finely divided along midline. Length of unsclerotized area between hypogynial valves 0.4 times sternite length.

Tergite 9 sclerites as in figure 210; sternite 9 V-shaped, medially undivided and with weak dorsal carina. Tergite 10 brown/black, usually with 7 black acanthophorite spines on each side. Spermathecae (figure 211) with shallowly hooked terminal reservoirs about as wide as reservoir duct; terminal reservoirs with wart-like protuberances basally. Basal striated duct without fine scales; junction with basal duct without scales; basal duct short, narrow.

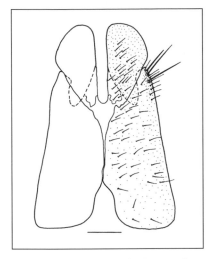

Figure 209. Female sternite 8: ventral.
Scale line = 0.2 mm.

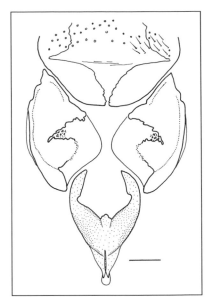

Figure 210. Female sternite 9, tergite 9
and lobes of sternite 10: ventral.
Scale line = 0.1 mm.

Figure 211. Spermatheca:
dorsal. Scale line = 0.1 mm.

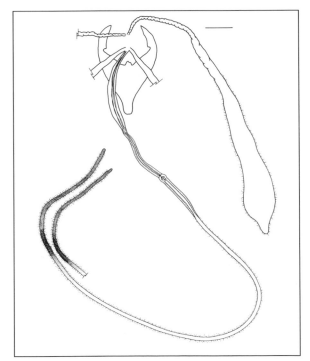

Type Material.
HOLOTYPE. ♂ (examined; allotype below on same pin) labelled: "[rectangular white label] Long Creek SC/26 Mch. 1932"; "[rectangular white label] O.L. Cartwright/Collector"; "[rectangular red label] HOLOTYPE/Lasiopogon/sher-mani/Cole and Wilcox"; "[rectangular red label] ALLOTYPE/Lasiopogon/sher-mani/Cole and Wilcox"; "[rectangular red label] Type/No. 50760/U.S.N.M.'; "[rectangular red label] U.S.N.M./Alloptype N/50760". USNM.
ALLOTYPE ♀ (examined): on same pin and with same data as holotype. See above for labelling. USNM.
PARATYPES (2 examined). **U.S.A.: North Carolina,** Raleigh, 20.iv.1932, C.S. Brimley (1♂, CASC). **South Carolina,** Salem, Little R., 21.iv.1929, O.L. Cartwright (1♂, CASC).

Other Material Examined (7 specimens).
U.S.A.: Georgia, Rabun Co., Pine Mtn., 1400', 15.v.1957, H.C. Huckett (2♀, CNCI), W.R.M. Mason (1♀, CNCI). **South Carolina,** Clemson, Clemson Univ., 14.iv.1938, J.N. Todd (1♂, CUCC), O.L. Cartwright (1♂, 1♀ in cop, CUCC), 18.v.1943, O.L. Cartwright (1♂, CUCC).

Type Locality. U.S.A., South Carolina, Long Creek.

Taxonomic Notes. I have described *L. flammeus* sp. nov. from two USNM males identified as *L. shermani* by Pritchard but which evidently were not seen by Cole and Wilcox (1938) during their revision of the genus. Because of their red legs and genitalia, these specimens key to *L. shermani* in Cole and Wilcox's key.

Etymology. "Named in honour of Prof. Franklin Sherman, who kindly allowed us to study the Clemson Agricultural College material." (Cole and Wilcox 1938).

Distribution. Nearctic; U.S.A., North and South Carolina to northern Georgia. See map 2, p. 110.

Phylogenetic Relationships. A member of the *cinereus* species group; sister species to *L. cinereus*.

Natural History. Habitat: riparian. Known flight dates range from 26 March to 18 May.

Lasiopogon slossonae Cole and Wilcox

Lasiopogon slossonae Cole and Wilcox, 1938. *Entomologica Americana* 43: 70.

Diagnosis. A medium-sized grey/grey-brown species with dark brown legs and markings; mystax and other head bristles dark. Antennae with F2+3/F1 = 0.62-0.93, but usually greater than 0.75. Thoracic tomentum dorsally and laterally gold-grey, often richer gold-brown dorsally with brown dorsocentral and acrostichal stripes. Main thoracic and leg bristles dark, except for pale katatergites; anterior dorsocentral bristles 3-5. Abdominal tergites basally with only thin brown tomentum; apical bands grey, covering 0.25 to 0.4 the length of the segments and sometimes extending anteriorly in a triangle along the midline; in female, basal brown tomentum divided into lateral patches by extension of apical pale band. Bristles on tergite 1 pale. Epandrium very long, the width about 0.4 times the length in lateral view, the dorsal margin longer than the ventral one and the apex slightly upturned. In dorsal view hardly concave medially. Female with black terminalia and dark hairs; dorsally, tergite 8 has a basal hump.

Description. Body length ♂ 8.8-11.5mm; ♀ 8.9-11.8mm.

Head. HW ♂ 1.80-2.10mm; ♀ 1.98-2.20mm. FW ♂ 0.36-0.42mm; ♀ 0.44-0.48mm. VW ♂ 0.96-1.03mm; ♀ 0.98-1.07. VW/HW = ♂ 0.48-0.54; ♀ 0.45-0.52. FW/VW = ♂ 0.40-0.41; ♀ 0.43-0.49; VD/VW = ♂ 0.13-0.17; ♀ 0.16-0.17. GH/GL = ♂ 0.31-0.38; ♀ 0.30-0.41.

Colour of head variable. Face grey to gold, vertex grey with faint brown highlights to gold-grey or golden, occiput from grey with faint gold highlights dorsally to strongly gold-brown. Beard and labial hairs white; mystax bristles dark brown/black, sometimes with a few white setae ventrally; all other setae brown/black. Occipital bristles moderately strong, abundant, those behind the dorsomedial angle of the eye long (to 0.8mm) and usually strongly curved anterolaterally; lateral and ventral ones shorter, straighter. Frontal and orbital setae abundant, some as long as F1+F2+3.

Antennae. Brown to black, some with chestnut F2+3, at base of F1, or pedicel. Setae brown; F1 sometimes with a seta. F1 short with both dorsal and ventral margins convex; range of antennal ratio large, but most specimens with ratio over 0.75; WF1/LF1 = ♂ 0.27-0.50; ♀ 0.28-0.43. LF2+3/LF1 = ♂ 0.62-0.93; ♀ 0.85-0.88.

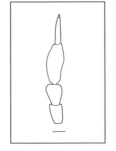

Figure 212. Antenna.
Scale line = 0.1 mm.

Thorax. Prothorax grey, gold-grey or even mostly gold dorsally, hairs white; postponotal lobes grey/gold-grey, the lateral angle chestnut/dark brown, hairs white to brown. Scutum tomentum sometimes clear grey or grey with light to medium brown tomentum between dorsocentral stripes, on intermediate spots and laterally adjacent to the pleura. Often the grey is infused with gold-brown; in the darkest specimens the base tomentum is dark gold-brown. Dorsocentral stripes dark brown, patchy, and sometimes obscure; except for the widened anterior part, mostly changing to lighter brown or gold in posterior and lateral views. Acrosichal stripes grey to brown, faint to definite. All notal setae brown or black. Anterior dorsocentrals 3-5, longest to 1.0mm and mixed with finer setae; in some specimens mostly weak and short (less than 0.6mm); 4-6 posteriors. Notal setae scattered, but concentrated on anterior intermediate spots and acrosichal stripes; these are short setulae or, when longer, are as long as shorter dorsocentrals. Postalars 2-3, with shorter setae; supra-alars 1-2; presuturals 2-3, 1-3 usually weak posthumerals. Scutellar tomentum grey or gold-grey; apical scutellar bristles dark, 3-5 strong ones on each side mixed with shorter, weaker dark bristles and hairs.

Pleural tomentum gold-grey. Katatergite setae 7-10 among finer white hairs; setae white/yellow, in some specimens with several brown or black. Katepisternal setae sparse, often long. Anepisternal setae 6-10, brown, a few moderately strong, to 0.6mm, a few white ventrally; a patch of short erect brown hairs on dorsal shelf. Anepimeron with a few long white setae.

Legs. Base colour dark brown/black. Tomentum of coxae grey with some gold highlights; tomentum on rest of legs grey. No coxal peg. Main bristles dark brown/black, finer setae white to black. Femora dorsally and laterally with mostly dark decumbent hairs, some white ones basally. Longer erect pale hairs abundant ventrally and laterally, especially on profemur; many of these can be dark apically; in males longest white ventral setae longer than thickness of femur, at least on meso- and metafemora. Profemur with 15-20 stronger dorsolateral dark bristles; mesofemur with 2-5; metafemur with 6-12; these bristles often rather fine and mixed with finer setae. Tibiae and tarsi with dark, strong bristles typically arranged, hairs brown/black. Protibia with longest bristles about 2.5-4.0 times longer than tibial width.

Wings. Veins light to dark brown; membrane pale brown in oblique view. DCI = 0.38-0.45; cell M3 closed or narrowly open. Halter yellow; knob without dark spot.

Abdomen. Male. Tergite basal colour dark brown/black. Tomentum on tergite bases thin

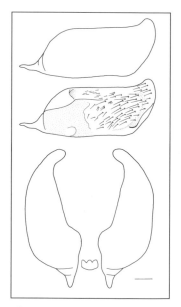

Figure 213. *L. slossonae* epandrium (top to bottom): lateral, medial, dorsal. Scale lines = 0.2 mm.

brown/gold-brown, usually extensively lacking so that dark cuticle shines through. Bands of grey tomentum cover 0.25 to 0.4 the length of each tergite apically (segment 1 is half to fully covered), the grey or grey-brown often extending thinly towards the base in a broad triangle; sometimes the tergites are more extensively washed with grey, at least on segments 2-3. Ventrolateral areas are narrowly brown-grey. About 5-8 strong bristles on each side of tergite 1 white, often yellow in females. Lateral setae on tergites 1-3 white, long and erect, those on 4 short; setae on 5-7 dark. Dorsal setulae short and dark. Sternite tomentum gold-grey, hairs white.

Female. As in male except tomentum covers tergites completely. Brown tomentum more dense; apical grey or gold-grey bands extend strongly in a triangle anteriorly on midline, separating the basal brown tomentum into lateral patches. Lateral setae on tergites 4-7 very short and dark; dorsal setulae dark. Sternite 7 usually directed strongly ventrally; hairs dark.

Male genitalia. Epandrium and hypandrium/ gonocoxite complex dark chestnut/black and covered with grey tomentum except on parts of hypandrium in some specimens. Setal brush black; other setae dark brown/black, numerous, prominent, especially dorsally on epandrium. Epandrium unusually long, about twice as long as hypandrium/gonocoxite complex and uniquely shaped: in lateral view the width 0.4 times the length, the dorsal margin longer than the ventral, the

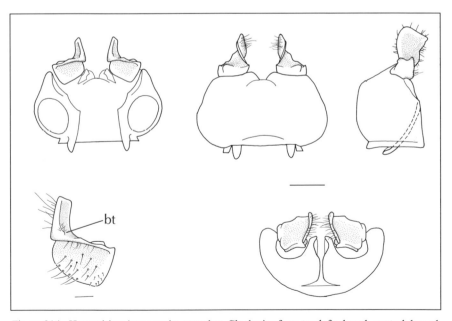

Figure 214. Hypandrium/gonocoxite complex. Clockwise from top left: dorsal, ventral, lateral and apical (scale lines = 0.3 mm); gonostylus, dorsal; bt = basal tooth of medial flange (scale line = 0.1 mm).

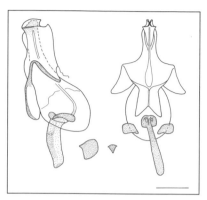

Figure 215. *L. slossonae* phallus: lateral (left); ventral (right); lateral ejaculatory process, basal (centre left); ejaculatory apodeme, cross-section (centre right). Scale line = 0.2 mm.

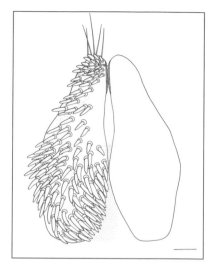

Figure 216. Subepandrial sclerite: ventral. Scale line = 0.1 mm.

apex a slightly upturned, rounded lobe. Medial face of epandrium as in figure 213. In dorsal view, medial margins of epandrium weakly concave; basal sclerite strong.

Gonostylus. Small tooth at junction of dorsal flange and medial flange; dorso-medial. lateral and ventrolateral teeth moderately developed. Hypandrium/gono-coxite complex in ventral view strongly bulging basally, the length about 0.55 times the width, the transverse slit at 0.9 the length. In lateral view, exposed length of gonocoxal apodeme very short, about 0.25 times basal width of hypandrium; apodeme with small sclerotized web ventrally.

Phallus. Paramere sheath dorsally 0.45 the length of phallus (excluding ejaculatory apodeme). Apex of paramere sheath with ventrolateral carina and ventral flange. Sperm sac width in dorsal view 0.35 the length of phallus. Ejaculatory apodeme short, curved dorsally in lateral view, rod-like in ventral view; triangular in cross-section with thick, narrow dorsal carina. Subepandrial sclerite as in figure 216. Sclerite strongly narrowed apically, triangular unsclerotized area in basal 0.3, a narrow unsclerotized portion in central 0.5; spines bluntly acute, dense except in subapical area; almost no microsetae.

Female terminalia. Undissected: Hairs brown/black, strong, erect, abundant; Tergite 8 with a distinct basal hump dorsally, dark brown/black, often ferruginous apically. Sternite 8 dark brown/black, hypogynial valves with short, strong bristles ventrally. Lateral lobe setae unusually strong. Cerci black with brown setae. Dissected sternite 8 (figure 217) elongate with basal width about 0.5 times the length, strongly divided along midline. Length of unsclerotized area between hypogynial valves 0.35 times sternite length. Lateral lobe setae strong.

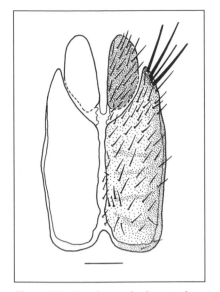

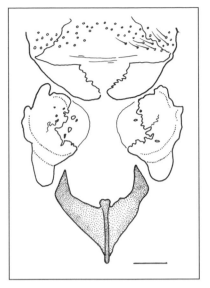

Figure 217. Female sternite 8: ventral.
Scale line = 0.2 mm.

Figure 218. Female sternite 9, tergite 9
and lobes of sternite 10: ventral. Scale
line = 0.1 mm.

Tergite 9 sclerites as in figure 218; sternite 9 V-shaped, medially undivided
and with dorsal carina. Tergite 10 brown/black, usually with 6 black acanthophorite spines on each side. Spermathecae (figure 219) with terminal reservoirs curled in a full loop and about as thick as reservoir duct; terminal reservoirs
without wart-like protuberances on surface. Basal striated duct with fine scales;
junction with basal duct sclerotized, golden, scaled; valve sclerotized, golden;
basal duct moderately long, narrow.

Type Material.
HOLOTYPE. ♂ (examined) labelled: "[rectangular white label] FRANCONIA,
N.H."; "[rectangular beige label] Collection of/Mrs. A.T. Slosson/Ac. 26226";
"[rectangular red label] HOLOTYPE/Lasiopogon/slossoni [sic]/Cole and
Wilcox". AMNH.
ALLOTYPE. ♀ (examined) labelled: "[rectangular white label] FRANCONIA,
N.H."; "[rectangular beige label] Collection of/Mrs. A.T. Slosson/Ac. 26226";
"[rectangular red label] ALLOTYPE/Lasiopogon/slossoni [sic]/Cole and
Wilcox". AMNH.
PARATYPES (7 examined). **U.S.A.: New Hampshire,** Breton Woods,
26.vi.1913, C.W. Johnson (1♀, MCZC), 28.vi.1913, C.W. Johnson (1♂,
MCZC); Franconia, Mrs. A.T. Slosson (2♂, 1♀, AMNH; 1♀, CASC).
Virginia, Glencarlyn, 21.v.1917, C.T. Greene (1♂, EMEC).

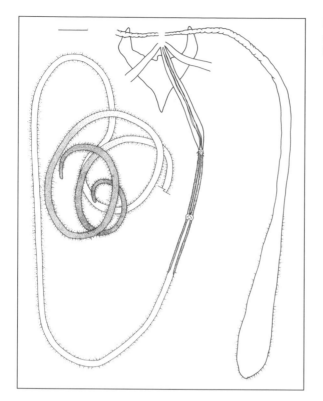

Figure 219. *L. slossonae*
spermatheca: dorsal.
Scale line = 0.1 mm.

Other Material Examined (106 specimens).
Canada: Quebec, West Brome, Yamaska R., 3.vi.1979, A. Borkent (2♀, CNCI).
    U.S.A.: Maryland, Garrett Co., 6.vi.1931, L.P. Ditman (1♀, USNM); Montgomery
Co., Colesville, 21.v.1975, A.S. Menke (1♀, LGBC), 17.v.1979, A.S. Menke (1♀,
UCDC); Montgomery Co., NW Branch Pk, 9.v.1971, A.S. Menke (1♂, 2♀, UCDC).
Massachusetts, Hampden Co., Hwy.20 3km E Mass Turnpike, Westfield R., 17.vi.1993,
H. Nadel and R.A. Cannings (1♂, 2♀, BCPM). **Michigan,** Gladwin Co., 14.vi.1953,
R.R. Dreisbach (1♀, FSCA); Saginaw Co., 1.vi.1940, C.W. Sabrosky (1♀, USNM). **New
Hampshire,** Franconia, Mrs. A.T. Slosson (1♂, USNM); Grafton Co., Pemigewasset R.,
Hwy.3, 3 km N Hwy.93, 13.vi.1993, R.A. Cannings and H. Nadel (1♂, BCPM); Grant,
AandG Academy, 24.vi.1975, R.N. Story (2♂, 2♀, DENH), 25-29.vi.1979 (DENH);
White Mtns, Osten Sacken (1♂, MCZC). **New Jersey,** Clementon, 6.v.1900 (1♀,
USNM). **New York,** vi (1♂, EMEC); Essex Co., Lake Placid, West Au Sable R. at
Hwy.86, 19.iv.1993, H. Nadel and R.A Cannings (1♂, BCPM); Hamilton Co., Indian L.,
6mi E, 43°45'30"Nx74°10'14", 22.v.1977, T.L. McCabe (1♂, NYSM); Keene Valley,
26.vi.1957, L.L. Pechuman (1♂, EMFC). **Ohio,** Ashland Co., 7.vi.1939, R.C. Osburn
(1♂, OSUC); Columbus, Alum Cr., 21.iv.1946, R.M. Goslin (1♂, OSUC; 1♂, USNM),
5.v.1946, R.M. Goslin (1♂, BMNH; 2♂, 1♀, OSUC; 1♂, MCZC), 12.v.1946, R.M.
Goslin (1♂, OSUC; 3♂, 1♀, USNM), 20.v.1946, R.M. Goslin (1♂, BMNH; 1♀, OSUC;
1♂, USNM), 2.vi.1951, R.M. Goslin (1♀, OSUC); Fairfield Co., Greenfield Twp.,
20.v.1938, R.M. Goslin (1♂, 2♀, OSUC), 21.v.1938, R.M. Goslin (2♀, OSUC),

Fairfield Co., Violet Twp., 20.v.1938, R.M. Goslin (1♂, OSUC); Hocking Co., 3.vi.1989, R.C. Osburn (1♀, OSUC); Conkle's Hollow, 4.vi.1952, R.M. Goslin (1♂, 1♀, OSUC); Hocking Co., Goodhope Twp., S16 "Neotoma", 14.v.1950, R.S. Rogers (1♂, UMMZ), E.S. Thomas (1♂, 1♀, OSUC); Licking Co., Newark, 20.v.1936, R.M. Goslin (1♀, OSUC), 1.v.1938, R.M. Goslin (2♂, 1♀, OSUC; 1♂, SEMC), 6.v.1938, R.M. Goslin (4♂, 3♀, OSUC; 2♂, 1♀, USNM), 8.v.1938, R.M. Goslin (1♀, SEMC), 29.iv.1941, R.M. Goslin (4♂, 1♀, OSUC; 1♀, USNM). **Vermont,** Essex Co., Blackbranch R. at Hwy.105, 13.vi.1993, R.A. Cannings and H. Nadel (5♂, 2♀, BCPM); Essex Co., Bloomfield, 5mi W, 28.vi.1972, H.J. Teskey (2♂, 1♀, CNCI; 1♂, 1♀, BCPM); Essex Co., Ferdinand, Rte.10, 22.vi.1971, L.L. Pechuman (1♂, CUIC), 27.vi.1972, L.L. Pechuman (1♂, CUIC), 28.vi.1972, L.L. Pechuman (1♀, CUIC), 11.vi.1973, L.L. Pechuman (2♂, 1♀, CUIC), 12.vi.1973, L.L. Pechuman (1♂, 1♀, CUIC); South Randolf, Hwy.14, 27.v.1980, B.V. Peterson (1♂, CNCI); White Mtns, Morrison (2♂, 1♀, USNM). **Virginia,** Glencarlyn, 20.v.1917, C.T. Greene (1♀, USNM), 21.v.1917, C.T. Greene (2♀, USNM), 4.v.1919, C.T. Greene (2♂, USNM). **West Virginia,** Richwood, 14mi E, Rte.39, 22.v.1966, J.B. Wallace (1♂, UGCA).

Type Locality. U.S.A., New Hampshire, Franconia.

Taxonomic Notes. Labelled as both *L. opaculus* and *L. tetragrammus* in some collections (Cole and Wilcox, 1938), although the males of the three species are not difficult to distinguish. Females of *L. slossonae* and *L. opaculus* are sometimes indistinguishable.

Etymology. Named after Mrs. A.T. Slosson, the collector of the type series.

Distribution. Nearctic; Southern Quebec south to Virginia, west to Ohio and Michigan. Map 9, next page.

Phylogenetic Relationships. Member of the *opaculus* species group; sister species of *L. opaculus*.

Natural History. Habitat: stream sides in woodlands. Recorded flight period 29 April to 29 June.

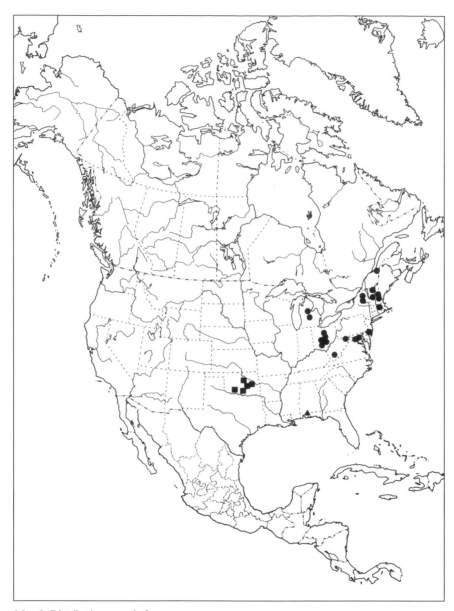

Map 9. Distribution records for
●   *Lasiopogon slossonae*
■   *L. oklahomensis*
▲   *L. piestolophus*

Lasiopogon terneicus Lehr

Lasiopogon terneicus Lehr, 1984. *Zoologicheskii Zhurnal* 63: 705.

Diagnosis. A small to medium-sized grey/grey-brown species with dark brown legs. Mystax white usually with some dark bristles dorsally; other head bristles dark. Antennae brown, partly ferruginous; F1 short and broad (F2+3/F1 = 0.57-0.73). Scutum grey with gold-brown highlights or more heavily infused with gold-brown; thorax laterally gold/brown-grey. Dorsocentral stripes dark brown, vague; acrostichal stripes grey-brown, sometimes obscure. Main thoracic and leg bristles dark; anterior dorsocentral bristles 2-5. Halter with dark spot. Abdominal tergites with apical bands grey, broad, about 0.75 the tergite length; bases gold-brown, lateral areas broadly grey. All hairs white; bristles on tergite 1 pale. Epandrium ferruginous, the width about 0.5 the length in lateral view, the greatest width at about half the length, apex broadly rounded dorsally. In dorsal view convex medially. Female with grey apical bands about 0.5 times the segment length. Tergite 8 black, apex ferruginous; sternite 8 ferruginous/brown, hairs white. Many unusual features in the male and female genitalia. Phallus with huge sperm sac and ejaculatory apodeme. Spermathecae with long, narrow basal tubes about 3 times as long as the basal striated tubes.

Description. Body length ♂ 7.8-7.9mm; ♀ 8.5-8.8mm.

Head. HW ♂ 1.72-1.76mm; ♀ 1.72-1.92mm. FW ♂ 0.38-0.40mm; ♀ 0.38-0.44mm. VW ♂ 0.65-0.68mm; ♀ 0.66-0.83mm. VW/HW = ♂ 0.39; ♀ 0.38-0.43. FW/VW = ♂ 0.59; ♀ 0.51-0.58. VD/VW = ♂ 0.16; ♀ 0.18. GH/GL= ♂ 0.25-0.26; ♀ 0.28-0.29.

Face silver-grey, vertex gold/brown-grey, occiput grey/brown-grey, intense gold-brown dorsally in some specimens. Beard and labial hairs white, mystax bristles white, with variable amounts of brown/black setae dorsally; these setae extending to the extreme dorsal margin of the gibbosity. All other setae brown/black. Occipital bristles sparse, moderately strong, rather short (longest to 0.4mm, shorter than F1+F2+3), those behind the dorsomedial angle of the eye only moderately curved anterolaterally; lateral and ventral ones shorter, straighter. Frontal and orbital setae sparse, to about 0.4mm.

Antennae. Brown; base of F1 and most of F2+3 ferruginous; tip of latter black. Setae brown, no setae on F1. F1 short and stout; WF1/LF1 = 0.33-0.42; LF2+3/LF1 = 0.57-0.73.

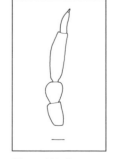

Figure 220. Antenna. Scale line = 0.1 mm.

Thorax. Prothorax grey/brown-grey, increasingly gold-brown dorsally; hairs white. Postpronotal lobes grey/gold-grey, the lateral angle yellow/dark chestnut, hairs white to light brown. Scutum tomentum ranges from grey with gold-brown highlights to intense gold/brown-grey; the gold/brown is most intense ventrolaterally and in the dorsocentral areas. Dorsocentral stripes dark brown, vague, disappearing at some angles of view; acrostichal stripes brown-grey, distinct to obscure. All strong bristles black, finer setae brown, the females more hirsute than males in the few specimens available. Anterior dorsocentrals 2-5 (longest to 0.7mm), mixed with finer setae; 3 posteriors. Notal setae sparse over presutural area, mostly shorter than shortest dorsocentrals. Postalars 1-3, with shorter setae; supra-alars 1-2 with shorter setae; presuturals 2, 0-1 weak posthumerals. Scutellar tomentum grey with gold highlights; apical scutellar bristles dark, 2-4 strong ones on each side mixed with shorter, weaker dark and pale bristles and hairs. Pleural tomentum gold/brown-grey. Katatergite setae brown/black, 6-7 among a few finer white hairs; katepisternal setae sparse. Anepisternal setae 3-5 dark (to 0.4mm); a sparse patch of erect brown setae on dorsal shelf. Anepimeron without setae.

Legs. Base colour dark brown/black; some specimens with trochanters brown/ black with chestnut apices and extreme apices of femora and bases of tibiae fer-ruginous. Tarsi with venter of segments ferruginous/brown. Tomentum of coxae grey with some gold highlights; tomentum on rest of legs grey/gold-grey. No coxal peg. Main bristles dark brown/black. Femora dorsally with white decum-bent hairs; longer erect pale hairs rather sparse ventrally (and laterally on pro-femora); in males longest white ventral setae about as long as width of femur. Profemur with 0-10 rather weak dorsolateral dark bristles; mesofemur with 0-3; metafemur with 0-6. Tibiae and tarsi with dark bristles typically arranged, hairs on tibiae mixed white and brown; tarsal hairs mostly brown. Protibia with longest bristles about 4.5 times longer than tibial width.

Wings. Veins yellow to medium brown, yellow most obvious on R1, Sc and costa near junction with Sc. Membrane usually pale brown in oblique view; faint infus-cations at vein forks, especially at r-m crossvein and at base of CuA1. DCI = 0.38-0.53; cell M3 open. Halter yellow; knob mostly covered by large dark spot.

Abdomen. Male. Tergite basal colour dark brown/black, ferruginous in some on 7. Tomentum on tergite bases gold/brown; bands of grey tomentum cover about 0.75 the length of tergites 2-5, about 0.3 of 6-7; tergite 1 mostly grey. Ventrolateral areas broadly grey. About 5-6 pale bristles on each side of tergite 1. Lateral hairs on tergites all white, sparse, erect; those on 5-7 short. Dorsal setulae white, a few brown at tergite bases. Sternite tomentum gold-grey, hairs white.

Female. Apical grey bands cover about 0.5 the length of most tergites, a little more on 2-3; tergite 1 all gold-grey. Basal tomentum gold-brown; ventrolateral areas grey/brown-grey. Lateral erect hairs rather sparse, white on 1-5, brown on 6-7. Dorsal setulae on white on tergites 1-2, mixed white and brown on 3, brown and erect on 4-7.

Figure 221. *L. terneicus* epandrium
(top to bottom): lateral, medial, dorsal.
Scale line = 0.2 mm.

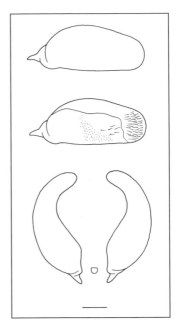

Male genitalia. Epandrium and hypandrium/
gonocoxite complex bright yellow/ferruginous
with sparse gold-grey tomentum. Setal brush
dark brown, the bristles long and straight; other
setae sparse, white/brown, white on hypandrium. Width of epandrium halves in lateral view
about 0.5 times the length, widest about midlength; ventral margin straight, dorsal margin
gently convex, apex broadly rounded dorsally,
more angular ventrally. Medial face of epandrium as in figure 221. In dorsal view, medial margins of epandrium convex centrally, concave
apically; basal sclerite small.

Gonostylus. Medial flange low, angled medially; dorsal flange undulate; lateral and ventrolateral teeth weak. Hypandrium/gonocoxite

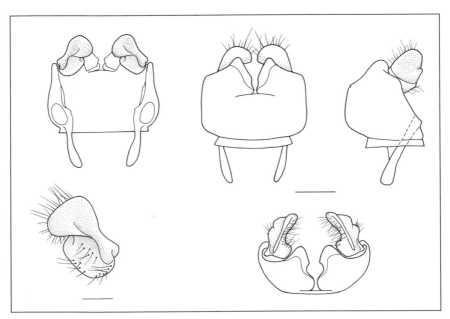

Figure 222. Hypandrium/gonocoxite complex. Clockwise from top left: dorsal, ventral, lateral
and apical (scale lines = 0.3 mm); gonostylus, dorsal; bt = basal tooth of medial flange (scale
line = 0.1 mm).

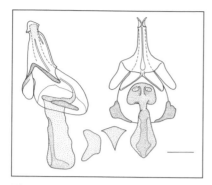

Figure 223. *L. terneicus* phallus: lateral (left); ventral (right); lateral ejaculatory process, basal (centre left); ejaculatory apodeme, cross-section (centre right). Scale line = 0.2 mm.

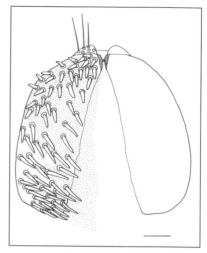

Figure 224. Subepandrial sclerite: ventral. Scale line = 0.1 mm.

complex in ventral view about as long as wide, transverse slit at 0.6 the length. Gonocoxite lobes broad, rounded and darkly sclerotized. Gonocoxal apodeme long, in lateral view, exposed length of apodeme about 0.75 the length of basal width of hypandrium; apodeme with sclerotized web ventrally.

Phallus. Paramere sheath dorsally 0.5 times the length of phallus (excluding ejaculatory apodeme). Ventral process short and extremely broad, spanning the unusually broad sperm sac. Apex of paramere sheath narrow, ventrolateral carina short; no ventral flange. Sperm sac width in dorsal view 0.5 times the length of phallus. Ejaculatory apodeme long and broad, straight in lateral view, strongly spatulate in dorsal view; flattened ventrally in cross-section with a thick, broad dorsal carina. Subepandrial sclerite as in figure 224. Broad triangular unsclerotized area in basal 0.90; spines bluntly acute, concentrated apically and basally; hypoproct setae short.

Female genitalia. Undissected: Hairs white, erect. Tergite 8 dark brown/black, narrowly ferruginous apically. Sternite 8 ferruginous/brown, hypogynial valves yellow/ferruginous with white hairs on ventral surface. Cerci brown with pale setae. Dissected sternite 8 (figure 225) with basal width about 0.7 times the length, slightly wider apically; undivided and heavily sclerotized along midline, the sternite strongly arched. Hypogynial valves elongate and narrow, bearing a lateral keel. Length of unsclerotized area between hypogynial valves narrow, 0.4 times sternite length. Lateral lobe unusually prominent and acute, 0.5 the length of the valves, the setae weak.

Tergite 9 sclerites as in figure 226; sternite 9 Y-shaped, medially divided. Tergite 10 brown/black usually with 6-7 black acanthophorite spines on each

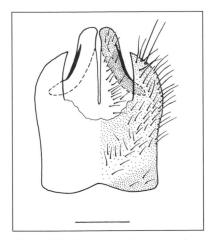

Figure 225. Female sternite 8: ventral.
Scale line = 0.3 mm.

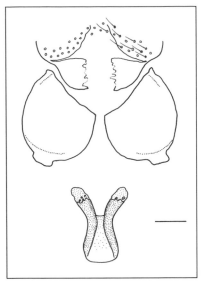

Figure 226. Female sternite 9, tergite 9
and lobes of sternite 10: ventral.
Scale line = 0.1 mm.

side. Spermathecae (figure 227) with terminal reservoirs curled 2.5 times, about as wide as very thin reservoir duct; terminal reservoirs without wart-like protuberances on surface. Basal striated section of reservoir duct and striated duct without fine scales; valve sclerotized, golden; junction with basal duct undifferentiated; basal duct very long and narrow (about 3 times as long as striated duct) and apparently striate.

Type Material.
HOLOTYPE. ♂ (examined) labelled [in Russian]: "[rectangular white label] Primorskiy Kray/Dzhigitovka River/at bridge/23.vi.1979/Lehr";"[rectangular red label] Holotypus 1981/Lasiopogon/terneicus Lehr. IBPV.
PARATYPES (2 examined). **Russia: Primorskiy Kray**, Cheremykhovaya River, 21.vi.1979, Lehr (1 ♂, 1 ♀, IBPV).

Other Material Examined (2 specimens).
Russia: Primorskiy Kray, 10km W Anuchino, 8.vi.1986, Lelej (2 ♀, BCPM).

Type Locality. Russia, Primorskiy Kray, Dzhigitovka River.

Etymology. Refers to the type locality in the Terneiskiy district of Primorskiy Kray.

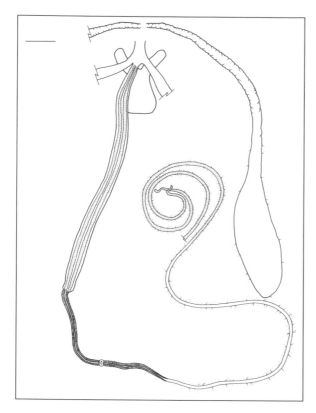

Figure 227. *L. terneicus*
spermatheca: dorsal.
Scale line = 0.1 mm.

Distribution. Palaearctic; Russia, Primorskiy Kray. Map 10.

Phylogenetic Relationships. Highly autapomorphic and difficult to place. Tentative placement in separate *terneicus* species group basal to *opaculus* group.

Natural History. Habitat: Stream sides in mixed forest. Known flight dates are 8 to 23 June.

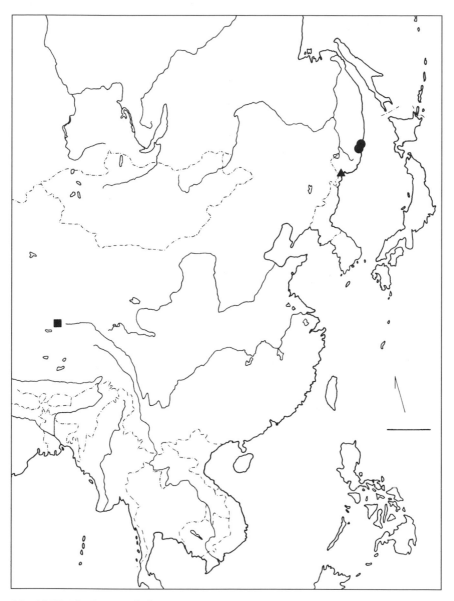

Map 10. Distribution records for
- *Lasiopogon terneicus*
- ■ *L. phaeothysanotus*
- ▲ *L. leleji*

Lasiopogon tetragrammus Loew

Lasiopogon tetragrammus Loew, 1874. *Berlin Ent. Zeitschr.* 18: 368-370.

Diagnosis. A medium-sized brown/grey-brown species with dark brown legs and markings; mystax dark, often pale ventrally; other head bristles dark. Antennae brown/black; F1 long, F2+3 short, the ratio usually about 0.5 (F2+3/F1 = 0.37-0.52). Thoracic tomentum dorsally and laterally gold-grey, usually rich gold-brown dorsally with brown dorsocentral and paramedial stripes. Main thoracic and leg bristles dark, including katatergite bristles; anterior dorsocentral bristles 4-5, rather strong. Abdominal tergites basally with thin brown tomentum; apical bands grey, covering 0.25 to 0.5 the length of the segments. In female, basal brown tomentum not divided mid-dorsally. Bristles on tergite 1 black. Epandrium ferruginous or chestnut, at least basally; usually black apically, the width about 0.35-0.45 times the length in lateral view, widening apically. In dorsal view strongly concave medially. Female with terminalia black, the apex of tergite 8 and the base of sternite 8 often ferruginous/brown; hairs white. Hypogyial valves with a ventral carina; hairs lacking.

Description. Body length ♂ 9.0-10.8mm; ♀ 9.8-11.2mm.

Head. HW ♂ 1.94-2.24mm; ♀ 2.10-2.24mm. FW ♂ 0.46-0.56mm; ♀ 0.50-5.2mm. VW ♂ 0.78-1.01mm; ♀ 0.93-0.99. VW/HW = ♂ 0.40-0.45; ♀ 0.43-0.44. FW/VW = ♂ 0.55-0.59; ♀ 0.53-0.54; VD/VW = ♂ 0.18-0.21 ♀ 0.16-0.18. GH/GL = ♂ 0.37-0.41; ♀ 0.39-0.42.

Face brown-grey to gold, vertex gold-grey, occiput gold-grey sometimes intensely gold dorsally. Beard and labial hairs white; mystax bristles dark brown/black, often with up to a third of the ventral setae white/gold; all other setae brown/black. Occipital bristles often moderately strong, rather short, those behind the dorsomedial angle of the eye the longest (to 0.4mm) and usually only moderately curved anterolaterally; lateral and ventral ones shorter, straighter. Frontal and orbital setae often moderately stout, some as long as F1 (0.4-0.5mm).

Antennae. Dark brown to black, some lighter at base of F1; setae brown/black; F1 without a seta. F1 long, widest about midlength. WF1/LF1 = ♂ 0.24-0.25; ♀ 0.20-0.29. Most specimens with antennal ratio about 0.5 (LF2+3/LF1 = ♂ 0.0.40-0.52; ♀ 0.37-0.49).

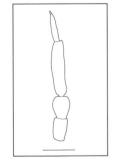

Figure 228. Antenna. Scale line = 0.3 mm.

Thorax. Prothorax gold-grey, darker dorsally, hairs white; postponotal lobes gold-grey, the lateral angle ferruginous/chestnut, hairs dark brown, short, strong. Scutum tomentum mainly grey-brown to gold-grey, often strongly gold or gold-brown ventrolaterally, and around and between the dorsocentral stripes. Dorsocentral and acrostichal stripes and intermediate spots medium to dark brown. Variation occurs. Darker specimens have the whole scutum suffused with gold-brown. In lighter ones the basic tomentum is brown-grey, the dorsocentral stripes are medium brown, the acrostichals are obscure grey, and the intermediate spots are faint brown. All notal setae brown or black. Dorsocentral bristles usually moderately strong, the anterior ones 4-5, longest to 1.0mm and mixed with finer setae; 3-5 posteriors. Notal setulae abundant, concentrated on presutural areas. Postalars 2-3, with shorter setae; supra-alars 2-3; presuturals 2-3, 1 posthumeral. Scutellar tomentum grey with gold highlights dorsally; apical scutellar bristles dark, 3-4 strong ones on each side mixed with shorter, weaker dark bristles and hairs, some of which can be white.

Pleural tomentum gold-grey, often strongly golden. Katatergite setae 7-8 among finer white hairs; black, in some specimens with several shorter, paler ones. Katepisternal setae sparse. Anepisternal setae 3-6, brown, short, to 0.4mm, often a few white hairs ventrally; a sparse patch of short erect brown hairs on dorsal shelf. Anepimeron without setae.

Legs. Base colour dark brown/black. Tomentum of coxae grey with some gold highlights; tomentum on rest of legs gold-grey. No coxal peg. Main bristles dark brown/black, finer setae white to black. Femora dorsally and laterally with mostly white decumbent hairs, sometimes dark ones apically. Longer erect pale hairs sparse ventrally; in males longest white ventral setae shorter or about as long as the thickness of the femur, usually longer on the metafemur. Profemur with 6-12 stronger dorsolateral dark bristles; mesofemur with 1-6; metafemur with 7-12. Tibiae and tarsi with dark, strong bristles typically arranged, hairs brown/black. Protibia with longest bristles about 3.0-4.0 times longer than tibial width.

Wings. Veins light to dark brown; membrane very pale brown in oblique view. DCI = 0.31-0.40, usually about 0.37; cell M3 open. Halter yellow; knob without dark spot.

Abdomen. Male. Tergite basal colour dark brown/black. Tomentum on tergite bases thin brown/gold-brown, the dark cuticle sometimes shining through. Bands of grey tomentum cover 0.25 to 0.5 the length of each tergite apically (segment 1 is half to fully covered). Ventrolateral areas are broadly gold-grey. About 3-8 strong bristles on each side of tergite 1 black, sometimes 1-2 pale. Lateral setae on tergites 1-3 white, rather short and sparse; white 4-7, sometimes some dark dorsolaterally. Dorsal setulae short and mostly white; some dark on basal brown areas and in some specimens 6-7 extensively dark haired. Sternite tomentum gold-grey, hairs white, sometimes dark on 7.

Female. As in male except tomentum covers tergites completely. Brown tomentum more dense, often mixed with gold and faint grey. Apical grey or gold-grey bands average somewhat narrower than in male, about 0.3 times the length of the segment, not extending anteriorly on midline as in related species. Lateral setae on tergites 1-3 short, sparse, white; all other setae, including dorsal setulae, dark.

Male genitalia. Epandrium and hypandrium/gonocoxite complex ferruginous or chestnut, especially basally, usually dark brown/black apically; covered with gold-grey tomentum except on parts of hypandrium in some specimens. Setal brush black; other setae dark brown/black. Width of epandrium in lateral view 0.35-0.45 times the length, widest at about 0.6-0.7 the distance from base, the apex broadly rounded dorsally, the ventral corner obtuse. Medial face of epandrium as in figure 229. In dorsal view, medial margins of epandrium strongly concave; basal sclerite prominent.

Gonostylus. Strong dorsomedial tooth; well developed lateral and ventrolateral teeth; path of setae at junction of dorsal and medial flanges. Hypandrium/gonocoxite complex in ventral view with the length about 0.6 times the width,

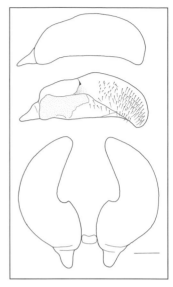

Figure 229. *L. tetragrammus* epandrium (top to bottom): lateral, medial, dorsal. Scale lines = 0.2 mm.

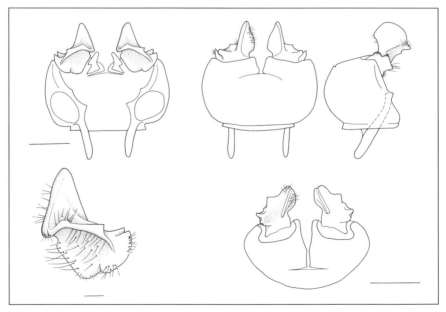

Figure 230. Hypandrium/gonocoxite complex. Clockwise from top left: dorsal, ventral, lateral and apical (scale lines = 0.5 mm); gonostylus, dorsal; bt = basal tooth of medial flange (scale line = 0.1 mm).

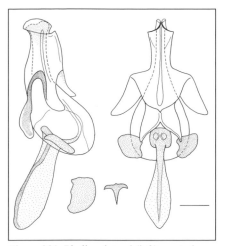

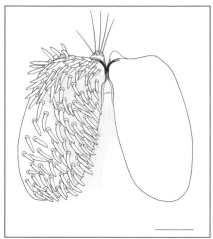

Figure 231. Phallus: lateral (left); ventral (right); lateral ejaculatory process, basal (centre left); ejaculatory apodeme, cross-section (centre right). Scale line = 0.2 mm.

Figure 232. Subepandrial sclerite: ventral. Scale line = 0.2 mm.

the transverse slit at 0.8 the length. In lateral view, exposed length of gonocoxal apodeme about 0.65 times the basal width of hypandrium; apodeme with small sclerotized web ventrally.

Phallus. Paramere sheath dorsally 0.4 times the length of phallus (excluding ejaculatory apodeme). Apex of paramere sheath with ventrolateral carina; ventral flange absent. Sperm sac width in dorsal view 0.4 times the length of phallus. Ejaculatory apodeme long; straight in lateral view, spatulate in ventral view; flattened in cross-section with thin, broad dorsal carina.

Subepandrial sclerite as in figure 232. Broad triangular unsclerotized area in basal 0.8; spines blunt, scarce laterally in basal half.

Female genitalia. Undissected: Hairs white, erect. Tergite 8 dark brown/black, often brown or ferruginous apically. Sternite 8 dark brown/black, often ferruginous or brown basally. Hypogynial valves with ventral carina and without hairs. Lateral lobe setae moderately strong. Cerci black with pale setae. Dissected sternite 8 (figure 233) broad, the basal width about 0.8 times the length; undivided but weak along midline. Length of unsclerotized area between hypogynial valves 0.4 times sternite length.

Tergite 9 sclerites as in figure 234; sternite 9 Y-shaped, medially undivided. Tergite 10 brown/black, usually with 6-7 black acanthophorite spines on each side. Spermathecae (figure 235) with straight or gently curved terminal reservoirs about as wide as reservoir duct; terminal reservoirs without wart-like protuberances on surface. Basal striated duct without fine scales; junction with basal duct without scales; basal duct moderately long.

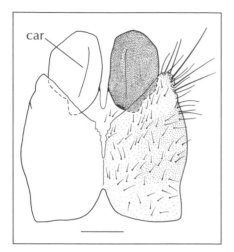

Figure 233. *L. tetragrammus* female sternite 8: ventral. car = carina of hypogynial valve. Scale line = 0.3 mm.

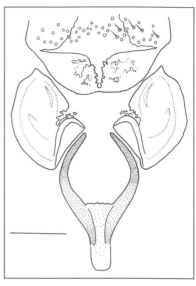

Figure 234. Female sternite 9, tergite 9 and lobes of sternite 10: ventral. Scale line = 0.2 mm.

Figure 235. Spermatheca: dorsal. Scale line = 0.2 mm.

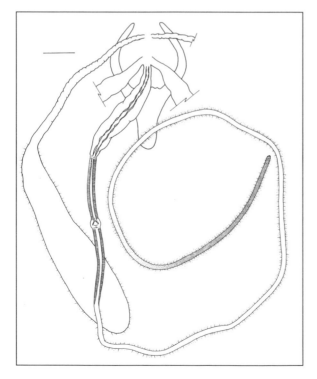

Type Material.
LECTOTYPE. (here designated), ♀ labelled: "[rectangular white label] Can.]";
"[rectangular white label] Loew/Coll."; "[square red label] Type/12803"; "[rectangular white label] tetragrammus/Lw."; "[rectangular white label] Museum of/Comparative/Zoology". My lectotype label "LECTOTYPE/Lasiopogon ♀/tetragrammus Loew/des. R.A. Cannings 2002 [red, black-bordered label]" has been attached to this specimen. MCZC.

Other Material Examined (127 specimens).
CANADA: Ontario, Lambton Co., Ipperwash, 25.v.1930 (1 ♀, CNCI); Ipperwash Army Base, 10.vi.1995, J. Skevington; Lambton Co., Pinery Prov. Pk, Burley Campgr., wet meadow, 4-6.vi.1995, J. Skevington and D. Caloren (1♂, BCPM), 15-19.vi.1995, J. Skevington (1♂, BCPM), 27.vi-1.vii.1995, J. Skevington (1♂, BCPM); Simcoe Co., Wasaga Beach, 21.v.1932, A.W.A. Brown (1♂, ROME); Wasaga Beach, Wasaga R., 8.vi.1992, R.W. Burgess (1♀, BCPM); Wasaga Beach, John St. off River Rd. E, 4.vi.1993, R.A. Cannings and H. Nadel (18♂, 22♀, BCPM; 3♂, 3♀, CNCI). **Quebec,** James Bay, Eastmain R., 1-14.viii.1934, R.C. McDonald (1♂, CNCI); Lanoraie, 24.v.1933 (1♀, EMFC; 2♀, FSCA), 26.v.1933 (1♂, AMNH; 1♂, LEMQ), 27.v.1933 (1♀, AMNH; 1♀, FSCA), 14.vi.1986, F. Liard (1♀, CNCI).

U.S.A.: **Connecticut,** Avon, Avon Old Farms, 25.vi.1929, C.H. Curran (5♂, AMNH; 1♂, 1♀, CASC; 1♀, MCZC); Cromwell, 17.v.1939, M.P. Zappe (1♀, CAES); Farmington, 16.v.1933, M.P. Zappe (1♂, CAES); Portland, 8.vi.1933, M.P. Zappe (1♂, CAES), J.P. Johnson (1♀, CAES). **Massachusetts,** Amherst, 12.vi.1963 (1♂, LGBC); Hampton Co., Hwy20 3km E Mass Turnpike, Westfield R., 17.vi.1993, H. Nadel and R.A. Cannings (1♂, BCPM); West Springfield, 13.vi.1916, H.E. Smith (1♂, MCZC; 2♀, USNM), 22.v.1916, H.E. Smith (1♀, USNM), 27.v.1916, H.E. Smith (1♂, 3♀, MCZC); Sunderland, 6.vi.1921 (2♂, 2♀, USNM; 1♀, AMNH; 1♂, KSUC); 11.vi.1939, Blanton and Bromley (1♀, 1♂, EMFC; 2♂, 1♀, CUIC). **Michigan,** Grayling, 18.vi.1953, Eliz. Thomas and E.S. Thomas (1♂, OSUC); Iosco Co., State Game Refuge, 26.v.1934, F.M. Gaige (2♀, UMMZ); Newaygo Co., 9.v.1959, R. and K. Dreisbach (1♀, UMMZ); Oscoda Co., T25N R3E Sec3, 25.v.1969, J.P. Donahue (1♂, 1♀, MSUC); Presque Isle Co., T33N R2E Sec33, 20.vi.1968, N.T. Baker (2♂, MSUC). **New Hampshire,** Grafton Co., Hwy3 3km N Hwy93, Pemigewasset R., 13.vi.1993, H. Nadel and R.A. Cannings (17♂, 11♀, BCPM); North Conway, 13.vi.1946, C.R. Frost (1♂, 1♀ in cop., USNM). **New York,** Clinton Co., Black Brook, 21.vi.1915 (1♀, CNCI); Peru, 8.vi.1916, C.R. Crosby and W.T.M. Forbes (1♀, CASC), 10.vi.1916, C.R. Crosby and W.T.M. Forbes (1♂, CNCI; 2♀, CUIC).

Type Locality. Canada, Quebec?

Taxonomic Notes. Described from a single female in the Loew Collection, apparently collected in Quebec by Provancher (Cole and Wilcox 1938). The male was described by Cole and Wilcox (1938) from a specimen from Peru, NY in the Cornell University collection.

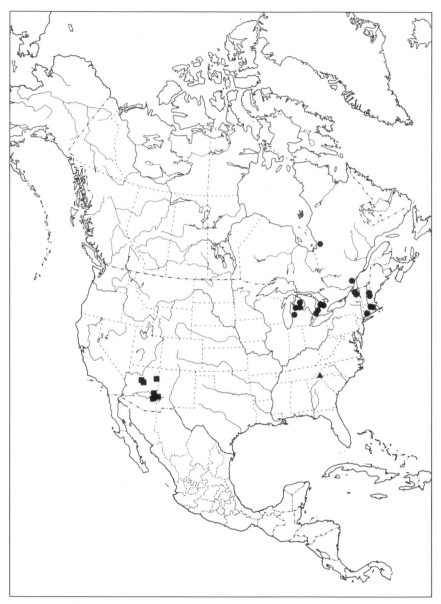

Map 11. Distribution records for
- *Lasiopogon tetragrammus*
■ *L. coconino*
▲ *L. flammeus*

Etymology. *Tetra* = four; *grammus* (*gramme*) = line; refers to the four dorsal thoracic stripes – the pair of acrostichal stripes and the two more prominent dorsocentral ones.

Distribution. Nearctic; Ontario and Quebec south to Connecticut and New York west to Michigan. The most northerly ranging of the species in the *opaculus* species group, the only *Lasiopogon* species entering the boreal forest in eastern North America. Map 11.

Phylogenetic Relationships. Member of the *tetragrammus* species group; basal to the rest of the group.

Natural History. Habitat: beach dunes; stream sides in woodland; sandy pine barrens. Records range from 16 May to 1 July in most of the range; a northern outlier near James Bay, Ontario (about 52°N) was collected in early August.

At Ipperwash, Ontario *L. tetragrammus* flew on small dunes directly behind a very large shore dune about 100m from the edge of Lake Huron. The dunes are sparsely clothed with Sand Cherry (*Prunus pumila* L.) and grasses such as *Ammophila* and bluestems (*Andropogon* spp.). Other asilids in the area include *Stichopogon trifasciatus* (Say) and *S. argentatus (Say) (J. Skevington, in litt.).* Large numbers were flying and landing on the stones and boulders at the edge of the Pemigewasset River, Grafton Co., New Hampshire on 13 June 1993.At Wasaga Beach, Simcoe County, Ontario on the beach along Lake Huron and adjacent sandy road edges hundreds flew at noon on 4 June 1993. Specimens examined had killed small and large Trichoptera and Tipulidae.

Lasiopogon woodorum sp. nov.

Diagnosis. A medium-sized grey/grey-brown species with brown legs and markings; mystax and other head bristles dark. Antennae brown, F2+3/F1 = 0.51-0.63. Thoracic tomentum dorsally varying from gold-grey to gold-brown with brown dorsocentral and acrostichal stripes grey or brown. Main thoracic and leg bristles dark; anterior dorsocentral bristles 3-4, often weak. Abdominal tergites basally with brown tomentum; apical bands grey, covering 0.5 to 0.6 the length of the segments and sometimes extending anteriorly in a triangle along the midline. Bristles on tergite 1 pale. Epandrium brown, often ferruginous basally, about 0.45 as wide as long in lateral view; in dorsal view strongly concave medially. Gonostylus with secondary medial flange and small tooth on dorsal flange near junction with medial flange; dorsomedial, lateral and ventrolateral teeth prominent. Female with black terminalia, the hairs pale.

Description. Body length ♂ 9.0-9.5mm; ♀ 8.5-11.5mm.

Head. HW ♂ 1.94-2.10mm; ♀ 1.94-2.38mm. FW ♂ 0.36-0.46; ♀ 0.44-0.50mm. VW ♂ 0.82-0.90mm; ♀ 0.84-0.98mm. VW/HW = ♂ 0.42-0.43; ♀ 0.41-0.43. FW/VW = ♂ 0.48-0.51; ♀ 0.49-0.52. VD/VW = ♂ 0.17-0.24; ♀ 0.18-0.22. GH/GL = ♂ 0.31-0.33; ♀ 0.35-0.39.

Face grey or gold-grey, vertex gold/brown-grey, occiput grey with brown highlights dorsally. Beard and labial hairs white, mystax bristles brown/black usually with a few pale setae ventrally; all other setae brown/black. Occipital bristles rather short (less than 0.5mm), strong, sparse and curved anterolaterally, about 10 strong ones medially; lateral and ventral ones shorter, straighter. Frontal and orbital setae long as F1.

Antennae. Brown; setae brown; F1 with setae in some specimens. WF1/LF1 = ♂ 0.30-0.33; ♀ 0.28-0.30. LF2+3/LF1 = ♂ 0.60-0.63; ♀ 0.46-0.53.

Thorax. Prothorax grey, often gold-brown on antepronotum, with white hairs; postponotal lobes gold-grey, the lateral angle ferruginous, hairs brown. Scutum tomentum variable: some specimens grey with faint brown highlights and darker grey acrostichal stripes; others gold-brown with brown acrostichal stripes, often all stripes merging somewhat with background tomentum. Acrostichal stripes often obscure. Dorsocentral stripes brown; intermediate spots medium brown. All notal setae brown or black. Anterior dorsocentrals 3-4 (longest to 0.7mm), mixed with a few finer setae; 2-3 posteriors. Notal setae usually short, sparsely scattered, mostly anteriorly, including the paramedial

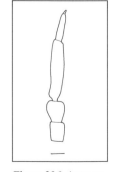

Figure 236. Antenna. Scale line = 0.1 mm.

stripes; about as long as shortest dorsocentral setae. Postalars 2-3, with shorter hairs; supra-alars 1; presuturals 2, 0-1 weak posthumerals. Scutellar tomentum gold-grey; apical bristles dark, 2-3 strong ones (usually 2) on each side mixed with a few shorter, weaker dark setae.

Pleural and scutellar tomentum grey with gold highlights. Katatergite setae dark, usually with one or two pale; 5-7 among finer white hairs. Katepisternal setae sparse. Anepisternal setae 2-5 short, brown, to 0.4mm; a patch of short erect brown hairs on dorsal shelf. Anepimeron without setae.

Legs. Base colour of femora black, tibiae dark brown, tarsi brown. Tomentum of coxae grey with some gold highlights; tomentum on rest of legs grey, sparse. No coxal peg. Main bristles dark brown/black, finer setae white to black. Femora dorsally with decumbent hairs mainly dark, some white basally; longer erect pale hairs mostly restricted to ventral areas. In males longest white ventral setae longer than thickness of femur, at least on meso- and metafemora. Profemur with 4-6 stronger dorsolateral dark bristles; mesofemur with 1-4; metafemur with 4-7. Tibiae and tarsi with dark, strong bristles typically arranged, hairs abundant, most brown/black. Protibia with longest bristles about 3.0 times longer than tibial width.

Wings. Veins light to medium brown; membrane pale brown in oblique view. DCI = 0.35-0.41; cell M3 open. Halter yellow; knob without dark spot.

Abdomen. Male. Tergite basal colour brown, some specimens with apical segments partly ferruginous. Tomentum on tergite bases light to medium brown. Bands of grey tomentum cover 0.0.5-0.6 the length of each tergite apically (segment 1 is half to fully covered), the grey usually extending towards the base in a broad triangle, dividing the brown area into two large spots. Ventrolateral areas broadly grey, narrowing basally. About 4-5 strong bristles on each side of tergite 1 white, often yellow in females. All lateral setae white, longest on tergites 1-3, erect, sparse. Dorsal setulae short and mostly white, some dark setae basally on tergites. Sternite tomentum gold-grey, hairs white.

Female. As in male. Sometimes a few of the bristles on tergite 1 light brown. Apical grey bands extending strongly or weakly in a triangle anteriorly on midline. Lateral setae on tergites 4-7 very short and dark. Dorsal setulae mostly dark, some specimens with white setulae apically on basal segments.

Male genitalia. Epandrium and hypandrium/gonocoxite complex dark brown, ferruginous basally and covered with gold-grey tomentum except on hypandrium. Setal brush dark brown; other setae dark brown/black, numerous, prominent, especially dorsally on epandrium. Width of epandrium halves in lateral view about 0.45 times the length, widest at about 0.6 the distance from the base; strongly convex dorsally, broadly rounded apically, the apicoventral corner broadly angled. Medial face of epandrium as in figure 237. In dorsal view, medial margins of epandrium strongly concave; basal sclerite strong.

Gonostylus. Secondary medial flange present; small tooth on dorsal flange near junction with medial flange; dorsomedial, lateral and ventrolateral teeth prominent. Hypandrium/ gonocoxite complex in ventral view with length about 0.5 times the width, the transverse slit at 0.8 the length. In lateral view, exposed length of gonocoxal apodeme about 0.5 times the basal width of hypandrium; apodeme with small sclerotized web ventrally.

Phallus. Paramere sheath dorsally 0.4 times the length of phallus (excluding ejaculatory apodeme). Ventral process with low carina. Apex of paramere sheath with ventrolateral carina and without ventral flange. Sperm sac width in dorsal view 0.4 times the length of phallus. Ejaculatory apodeme with slightly convex ventral margin in lateral view, moderately spatulate in ventral view; triangular in cross-

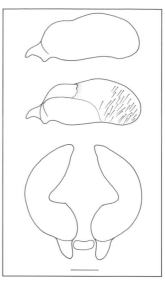

Figure 237. *L. woodorum* sp. nov. male genitalia, epandrium (top to bottom): lateral, medial, dorsal. Scale line = 0.3 mm.

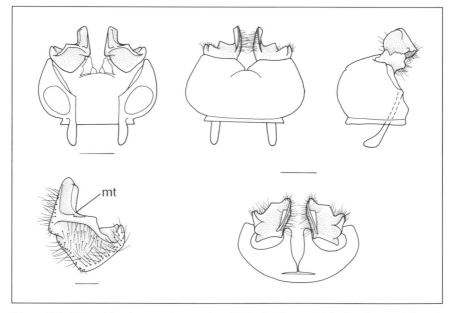

Figure 238. Hypandrium/gonocoxite complex. Clockwise from top left: dorsal, ventral, lateral and apical (scale lines = 0.3 mm); gonostylus, dorsal; mt = medial tooth of dorsal flange (scale line = 0.1 mm).

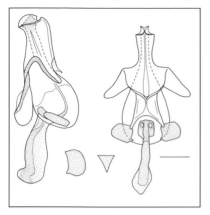

Figure 239. Phallus: lateral (left); ventral (right); lateral ejaculatory process, basal (centre left); ejaculatory apodeme, cross-section (centre right). Scale line = 0.2 mm.

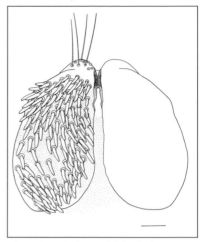

Figure 240. Subepandrial sclerite: ventral. Scale line = 0.1 mm.

section, the ventral surface flat, the dorsal carina thick and narrow. Subepandrial sclerite as in figure 240. Broad triangular unsclerotized area in basal 0.25, a narrow unsclerotized portion in central 0.50; spines blunt.

Female genitalia. Undissected: Hairs white, erect; tergite and sternite 8 dark brown/black, often brown/ferruginous apically on tergite 8 and basally on sternite. Ventral surface of hypogynial valves with medial and lateral carinae and short hairs basally. Lateral lobe setae stout and strong. Cerci brown/black with light brown setae. Dissected sternite 8 (figure 241) with basal width about 0.7 times the length; undivided. Length of unsclerotized area between hypogynial valves 0.4 times sternite length.

Tergite 9 sclerites as in figure 242; sternite Y-shaped, medially undivided. Tergite 10 brown/black, usually with 5-6 black acanthophorite spines on each side. Spermathecae (figure 243) with straight or slightly curved terminal reservoirs about as thick as reservoir duct; terminal reservoirs without wart-like protuberances on surface. Basal striated duct without fine scales; junction with basal duct with scales; basal duct short.

Type Material.
HOLOTYPE. (here designated) ♂ labelled: "[rectangular beige label] Sugar Grove/O 10.vi.28"; "rectangular beige label] Lasiopogon/opaculus/Loew/det. J. Wilcox"; "[rectangular white label] SW Bromley/Collection/1955"; My Holotype label"HOLOTYPE/Lasiopogon ♂/woodorum Cannings/des. R.A. Cannings 2002 [red, black-bordered label]" has been attached to this specimen. USNM.

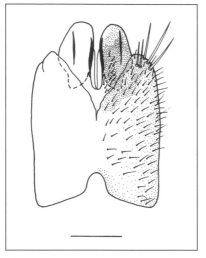

Figure 241. *L. woodorum* sp. nov. female
sternite 8: ventral. Scale line = 0.3 mm.

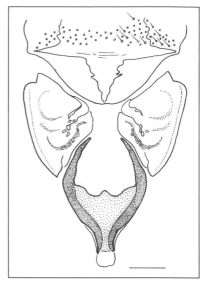

Figure 242. Female sternite 9, tergite 9
and lobes of sternite 10: ventral.
Scale line = 0.2 mm.

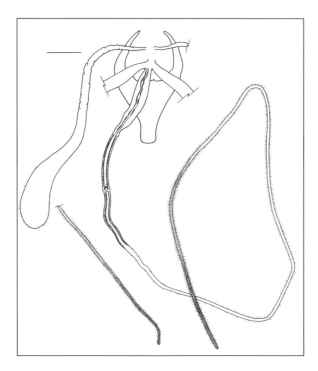

Figure 243. Spermatheca:
dorsal. Scale line = 0.2 mm.

PARATYPES (66 designated). **U.S.A.: Delaware**, Water Gap, vi, Mrs. A.T. Slosson (1 ♀, AMNH). **Indiana**, Montgomery Co., Shades St. Pk, 20.v.1990, M.D. Baker (1 ♀, MDBC). **Maryland**, Plummers Island, 7.v.1913, R.C. Shannon (1 ♀, USNM), 5.v.1914, R.C. Shannon (1 ♂, USNM), 10.v.1914, R.C. Shannon (1 ♂, USNM), 1.v.1915, R.C. Shannon (1 ♀, USNM), 6.v.1916, J.C. Crawford (1 ♂, USNM). **Ohio**, Bainbridge, Paint Cr., 1.vi.1942 (1 ♂, EMFC); Cincinnati, 22.v.1900, Dury (1 ♂, 1 ♀, OSUC), 9-16-vi.1901 (1 ♂, OSUC); Columbus, J.S. Hine (1 ♂, OSUC); Fairfield Co., 12.v.1954, R.E. Woodruff (1 ♂, FSCA); Hocking Co., Goodhope Twp., 13.v.1938, R.M. Goslin (1 ♂, BMNH), 27.v.1951, R.M. Goslin (3 ♂, OSUC), 30.v.1951 (2 ♂, OSUC), Goodhope Twp., "Neotoma", 12.v.1954 (1 ♂, 2 ♀, OSUC), "Neotoma", 14.v.1950, E.S. Thomas (2 ♂, OSUC), 20.v.1951, E.S. Thomas (2 ♂, 3 ♀, OSUC); Hocking Co., J.S. Hine (2 ♂, OSUC); Ira, J.S. Hine (1 ♂, OSUC); Lake Co., Willoughby, 18.v.1965, J.C. Pallister (2 ♂, AMNH); Licking Co., Newark, 27.v.1936, R.M. Goslin (1 ♀, OSUC), 1.v.1938, R.M. Goslin (1 ♂, 1 ♀, OSUC), 6.v.1938, R.M. Goslin (1 ♀, BMNH; 3 ♂, 1 ♀, OSUC; 1 ♀, MCZC; 1 ♂, USNM), 8.v.1938, R.M. Goslin (1 ♀, BMNH; 7 ♂, 1 ♀, OSUC; 1 ♂, SEMC), 29.iv.1941, R.M. Goslin (1 ♂, 3 ♀, OSUC); Loudonville, 6.v.1915 (1 ♂, CASC); Sugar Grove, 10-vi.1928 (1 ♂, OSUC; 1 ♂, 1 ♀, USNM); Warren Co., 23.iv.1962, F.J. Moore (1 ♂, OSUC). **Virginia**, Great Falls, 2.v.1917, C.T. Greene (1 ♂, USNM), 12.v (1 ♂, MCZC), 30.vi (1 ♂, MCZC).

Type Locality. U.S.A., Ohio, Sugar Grove.

Taxonomic Notes. Specimens described here as *L. woodorum* were subsumed in Cole and Wilcox's (1938) concept of *L. opaculus*.

Etymology. Named in honour of Dr D. Monty Wood, eminent dipterologist, and Grace C. Wood, who have long been generous friends and mentors.

Distribution. Nearctic; U.S.A., Virginia and Maryland west to Ohio. Map 12, next page.

Phylogenetic Relationships. Member of the *tetragrammus* species group; in an unresolved trichotomy with *L. chrysotus* and *L. flammeus*.

Natural History. Habitat: stream sides. Recorded dates: 23 April to 16 June.

Map 12. Distribution records for
- *Lasiopogon woodorum*
- ■ *L. lavignei*
- ▲ *L. chrysotus*

PHYLOGENY OF *LASIOPOGON*

PHYLOGENETIC INTERPRETATION OF THE STICHOPOGONINAE AND THE MONOPHYLY OF *LASIOPOGON*

This brief examination of the relationships among the genera of the Stichopogoninae, as it is now conceived (Artigas and Papavero 1990, Hull 1962), attempts to clarify the phylogenetic position of *Lasiopogon*. Eight of the ten genera that I place in the subfamily have been examined; material of *Afghanopogon* and *Stackelberginia* has not been studied. I have not studied the biogeography of the subfamily.

Character Analysis

Below, I summarize the 19 characters and their states used in the phylogenetic analysis of the genera of the Stichopogoninae. The character number corresponds to the one on the character matrix (Table 1, next page), in the cladogram (figure 244) and in the discussions concerning evolutionary relationships. For each character, I describe the inferred plesiomorphic and apomorphic states (1 for binary characters, 2, 3,... for multistate ones). Weight values assigned to character states are indicated. I did not use weights in the cladistic analysis, but indicating them below reflects my confidence in the character's utility to define robust clades (see p. 20). Genera in the subfamily Stenopogoninae, in particular *Stenopogon* Loew and *Bathypogon* Loew, are used as outgroups. Character 7 was unordered in the analysis.

1. Epandrium form (horizontally clasping). Two states: plesiomorphic – entire or incompletely divided and not clasping; apomorphic 1 – completely divided and halves horizontally clasping (++). This is the transformation from a plesiomorphic stenopogoninine epandrium to that of *Lasiopogon*. These states are discussed further on pages 272-3.

2. Epandrium form (vertically clasping). Two states: plesiomorphic – entire or incompletely divided and not clasping; apomorphic 1 – epandrium entire,

267

hood-like, vertically clasping (+++). This is the transformation from a plesiomorphic stenopogoninine epandrium to that of *Stichopogon* and its close relatives. These states are discussed further on pages 272-3.

3. Epandrial apodemes. Two states: plesiomorphic – absent; apomorphic – present (+++). In the genera that I have examined, epandrial apodemes are present only in *Lasiopogon* and are indicative of monophyly. Presumably they are associated with the horizontally clasping function of the independent epandrium halves, although they are absent in *Bathypogon*, where the epandrium is generally less strongly sclerotized compared with that of *Lasiopogon*. In some species of *Stichopogon* and *Lissoteles*, at least, the hood-like epandrium has basal apodemes. I assume that these are not homologous with the apodemes of *Lasiopogon*, but rather have developed from the base of an entire epandrium to aid in the vertical clasping mode found in these genera.

4. Fusion of hypandrium and gonocoxites. Three states: plesiomorphic – hypandrium and gonocoxites separate; apomorphic 1– hypandrium/gonocoxite fusion complex with medial horizontal slit and vertical slit (++); apomorphic 2 – gonocoxites fused but no horizontal slit present (+). Multistate character treated as ordered. See page 272 below for further discussion of this character.

5. Gonocoxal apodemes. Two states: plesiomorphic – not exserted from base of hypandrium/gonocoxites or very shallow lobes; apomorphic – strongly exserted (+). In the Stichopogoninae the apomorphic state is homoplasious, appearing in some species of *Stichopogon* and *Rhadinus*, at least. Sinclair et al. (1994)

Table 1. Character state matrix for genera of the Stichopogoninae. Character numbers correspond to character numbers in the text. 0 - plesiomorphic state; 1-2 - apomorphic state; ? - missing data.

Characters	1	2	3	4	5	6	7	8	9	1 0	1 1	1 2	1 3	1 1	1 5	1 6	1 7	1 8	1 9
Stenopogon	0	0	0	0	0	0	0	0	1	0	0	0	0	0	0	0	0	1	0
Bathypogon	1	0	0	1	0	0	0	0	0	0	0	0	0	1	0	0	0	1	0
Lasiopogon	1	0	1	1	1	1	0	1	0	0	1	1	0	0	1	0	0	0	1
Argyropogon	0	1	0	1	0	0	0	2	0	0	2	0	0	0	1	1	0	0	1
Rhadinus	0	1	0	1	1	1	0	0	0	0	1	0	1	1	1	1	1	0	1
Eremodromus	0	1	0	?	0	0	?	?	?	?	?	0	1	1	1	1	1	0	1
Townsendia	0	1	0	2	0	0	0	2	1	0	2	0	0	0	1	1	1	0	1
Lissoteles	0	1	0	2	0	0	1	2	1	0	2	0	0	0	1	1	1	0	1
Stichopogon	0	1	0	1	0	0	1	2	1	1	2	0	0	0	1	1	1	0	1
Clinopogon	0	1	0	2	0	0	0	2	1	1	2	0	0	0	1	1	1	0	1

noted that elongate apodemes have evolved independently a number of times in the Tabanomorpha, Stratiomyidae and Asiloidea.

6. Gonocoxite setal brush. Two states: plesiomorphic – absent; apomorphic – present (++). A dense patch of strong bristles clothes the medioapical area of the gonocoxites in *Lasiopogon* species. I consider this character indicative of monophyly, although a similar but less organized setal concentration is present in some *Rhadinus* species.

7. Subepandrial sclerite shape. Two states: plesiomorphic – plate-like or shelf-like; apomorphic – sclerite reduced to comb-like rods (+++). In the genera of Stenopogoninae and Stichopogoninae examined, most show the plesiomorphic state: a plate sclerotized in various intensities – either fully sclerotized and heart-shaped (e.g., *Tillobroma*) or laterally thicker (e.g., *Stenopogon, Bathypogon, Townsendia, Rhadinus, Argyropogon*). In *Stichopogon* and *Lissoteles* the apomorphic condition, rods and comb-like structures, replaces the simple plate.

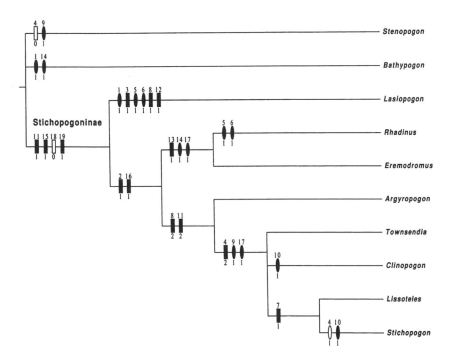

Figure 244. Cladistic relationships of the Subfamily Stichopogoninae. Characters are plotted on branches as rectangles (characters without homoplasy) or ovals (characters with homoplasy); open symbols are reversals. The numbers above the symbols correspond to the character numbers in the text; the numbers below represent the character states described in the character list: 0 = plesiomorphic state; 1, 2, etc. = apomorphic states.

8. Subepandrial sclerite vestiture. Three states: plesiomorphic – sclerite with fine setae; apomorphic 1 – setae modified into stout, blunt or attenuate spines (+++); apomorphic 2 – setae absent (++). The spines in *Lasiopogon* (apomorphic 1) are strongly indicative of monophyly. I consider the absence of setae (apomorphic 2) as an independently derived loss of the plesiomorphic state. The multiple states are treated as unordered.

9. Form of cerci. Two states: plesiomorphic – pad-like, lightly sclerotized, haired; apomorphic – sclerotized strongly and without hairs (+). The apomorphic condition shows considerable homoplasy; it is common in the Stichopogoninae and is widespread in the Stenopogoninae, but the plesiomorphic state occurs in *Bathypogon*.

10. Fusion of cerci. Two states: plesiomorphic – separate; apomorphic – fused (++). Fused cerci have arisen a number of times in the Stenopogoninae (e.g., *Cyrtopogon* Loew) and other subfamilies (e.g., *Lestomyia* Williston in the Dasypogoninae).

11. Lateral ejaculatory processes. Three states: plesiomorphic – large; apomorphic 1 – small (+); apomorphic 2 – absent (++). Processes have been lost in all genera of the Stichopogoninae examined except *Lasiopogon* and *Rhadnius*, and the state in *Eremodromus* is unknown. There is considerable homoplasy in the distribution of the apomorphic 1 state in the outgroups. This multistate character is analysed as ordered.

12. Rotation of genitalia. Two states: plesiomorphic – unrotated; apomorphic – rotated (++). See pages 38-39. The presence of the apomorphic state is variable within the Asilidae.

13. Pulvilli. Two states: plesiomorphic – present; apomorphic – absent (++). The apomorphic condition is found in *Rhadinus* and *Eremodromus*; there is no homoplasy in the ingroup. Pulvilli are lost in several other asilid lineages including the subfamilies Leptogastrinae and Stenopogoninae (e.g., *Ablautus* Loew).

14. Cell M3. Two states: plesiomorphic – open; apomorphic – closed and stalked (++). The apomorphic state is widespread in the Asilidae, including the Stenopogoninae, and it occurs in *Bathypogon*, one of the outgroups. It is especially prevalent in the Asilinae, considered to be the most derived subfamily.

15. Shape of frons. Two states: plesiomorphic – parallel-sided; apomorphic – expanded dorsally (++). Outside the Stichopogoninae, a few genera of Stenopogoninae, including *Willistonina* Back and *Hermannomyia* Oldroyd have an expanded frons. *Cerotainia* Schiner in the Atomosiini (Laphriinae) also shows this character state. I assume that these conditions are independently derived in the different lineages.

16. Structure of face. Two states: plesiomorphic – gibbous and mystax extensive with long setae; apomorphic – flattened, mystax reduced (++). The apomorphic state occurs in various asilid lineages.

17. Setation of scutum. Two states: plesiomorphic – notal bristles strong; apomorphic – notal bristles reduced (+).

18. Structure of prosternum. Two states: plesiomorphic – complete, fused laterally with proepisternum; apomorphic – reduced, bordered by membrane (+). Clements (1951) and Hull (1962) considered a complete prosternum to be plesiomorphic. Clements believed there was too much variation in the character for it to be diagnostic at the tribal level, as promoted by Hardy (1948).

19. Gonocoxite apex. Two states: plesiomorphic – with process(es); apomorphic – without process(es) (+). Many genera in the Stenopogoninae and other subfamilies, and also in the Therevidae and Mydidae, have one or more processes near the apex of the gonocoxite. Although there is no homoplasy in the ingroup, I have given this a low weight because I am unsure of the homology of the processes.

Relationships Among the Genera of the Stichopogoninae and the Monophyly of the Subfamily

Hennig86 analysis produced nine equally most parsimonious trees with a length of 31 steps, a consistency index of 70 and a retention index of 71. These trees have a basal trichotomy. But the linking of *Stenopogon* and *Bathypogon* (character 18) in a dichotomy with the Stichopogoninae is also equally parsimonius. This is a more pleasing topology, and one that mirrors the presently accepted classification (*Bathypogon* in the Stenopogoninae), but I have chosen the trichotomy (figure 244) because I believe it better illustrates the ambiguous position of *Bathypogon* with respect to the structure of the male genitalia. The fused hypandrium/gonocoxites and the separate, horizontally clasping epandrium halves (with basal sclerite in some species) are strong links to *Lasiopogon*. I prefer to leave the trichotomy unresolved until more detailed analyses are done.

All trees shared the same topology with respect to the basal clades of the Stichopogoninae. The subfamily remained monophyletic with *Lasiopogon* the sister group to the rest of the lineage. The only variations among the trees were in the relationships among *Townsendia, Clinopogon, Lissoteles* and *Stichopogon*. Figure 244 is the tree with the best support for this ambiguous clade (non-homoplasious character 4). It is also the tree that simultaneously uses, without resorting to homoplasy, the only high-weight character (7) defining relationships within the clade.

The sister group to *Lasiopogon* contains the rest of the genera of the Stichopogoninae. This clade is a well-supported monophyletic group characterized by the hood-like epandrium (2) and the flat face and reduced mystax (16). Within this group the monophyly of the clade of *Rhadinus* plus *Eremodromus* is supported by the loss of pulvilli (13) and the occurrence of a closed M3 cell in the wing (14).

The clade of *Argyropogon* plus the group of genera containing *Townsendia, Clinopogon, Lissoteles* and *Stichopogon* shares the lack of setae on the subepandrial sclerites (8) and the loss of lateral ejaculatory processes (11). The monophyly of the trichotomy consisting of *Townsendia, Clinopogon* and *Lissoteles* plus *Stichopogon* is only weakly supported by the apomorphy of the totally fused hypandrium and gonocoxites (4), a state that is reversed in *Stichopogon*. The homoplasious apomorphic states of characters 9 (sclerotized cerci) and 17 (reduction of notal setae) are also supportive, especially character 9, which is not found elsewhere in the ingroup, but only in *Stenopogon*, one of the outgroups. The monophyly of *Lissoteles* plus *Stichopogon* is strongly supported by the possession of comb-like subepandrial sclerites (7).

The monophyly of the Stichopogoninae is indicated by several synapomorphies. The dorsally expanded frons (15) occurs rarely in non-stichopogonines; I assume the state is independently derived in *Willistonina* Back. I consider the expanded frons a synapomorphy of the genera of Stichopogoninae.

The coalesced hypandrium and gonocoxites (4) is characteristic of the subfamily. These structures are inextricably fused into a bowl-shaped unit. The approximate extent of the original sclerites is indicated by the presence of a ventromedial slit running at right angles to the midline of the structure (*Lasiopogon, Rhadinus, Eremodromus*, most *Stichopogon),* by a reduction in sclerotization in this position (some *Stichopogon*), by a significant gap there (*Lissoteles*) or by complete fusion (*Clinopogon*). *Bathypogon* in the Stenopogoninae also has coalesced hypandrium and gonocoxites. Coalesced hypandrium and gonocoxites are found in other asilid lineages (e.g., Laphriinae, Laphystiinae, Trigonomiminae), but it is not clear if these fusions are homologous with that seen in the lineages treated here. The reduced or lost lateral ejaculatory processes (11) link the genera in the subfamily, although I consider this weak support for monophyly. Similarly, the loss of processes on the gonocoxite apex (19) is weakly supportive. Unlike the members of the Stenopogoninae, the Stichopogoninae retain the complete prosternum (18), which is a reversal to the ancestral state.

The plesiomorphic condition of the asilid epandrium is unclear, although Sinclair et al. (1994) confirmed that in the ground plan of the Brachycera the epandrium is an entire sclerite. Yeates (1994) considered this state to be plesiomorphic for the Muscomorpha (*sensu* Woodley 1989) and considered the divided epandrium of *Stenopogon* to be apomorphic. On the other hand, Papavero (1973a) stated that "free epandrial halves" were plesiomorphic in the Asilidae. Karl (1959) believed a partly divided epandrium was in the ground plan of the Asilidae and that entire sclerites (e.g., *Stichopogon*) as well as completely divided ones (such as in *Lasiopogon*) were apomorphic. Certainly, epandria split medially to various degrees are rather common in asilid taxa considered plesiomorphic, as are flattened epandria divided completely in two. Nevertheless, an epandrium articulating in membrane and clasping horizontally (1) is rare in subfamilies other than the Asilinae. This situation must be considered apomorphic. But it is a different picture than that found in the rest of the Stichopogoninae.

In all genera in the subfamily except *Lasiopogon* the epandrium is an undivided sclerite arched over the rest of the terminalia like a hood, forming a vertically clasping structure with the hypandrium/gonocoxite complex (2). This form of epandrium is found nowhere else in the Asilidae; it is clearly an apomorphy compared with the flap-like, rather rigid, often lobed or divided epandrium of the Dasypogoninae, Stenopogoninae and other related groups. The clade defined by the hoodlike epandrium is a stronger monophyletic group without the inclusion of *Lasiopogon*.

Lasiopogon sits uncomfortably in the Stichopogoninae. Its general appearance – prominent face and, for the most part, setose body, sets it apart from the other genera. More significantly, the divided, clasping epandrium is anomalous in the group. The present analysis places it as the sister group to the rest of the subfamily. Given this relationship, *Bathypogon*, currently placed in the Stenopogininae, is a candidate for the sister group to the Stichopogoninae. Such solutions require a more thorough phylogenetic analysis developing a satisfactory higher classification for the Asilidae. Unfortunately, such a task is not within the scope of this study.

The Monophyly of *Lasiopogon*

The following apomorphies support the monophyly of *Lasiopogon*:

1. Completely divided epandrium halves strongly concave ventrally in cross-section, horizontally clasping (found elsewhere only in *Bathypogon*) (1).

2. Well-developed epandrial apodemes (3).

3. Gonocoxal apodemes strongly exserted from the base of the fused hypandrium/gonocoxites (5). In no other genera that I have examined (except in a few species of *Rhadinus* and *Stichopogon* [Stichopogoninae]) do these apodemes extend beyond the base of the hypandrium as more than small, low lobes.

4. Dense brush of strong, posteriorly directed bristles medially on the gonocoxite lobes (found elsewhere only in *Rhadinus*, where it is poorly developed) (6).

5. Subepandrial sclerite strongly sclerotized and covered with stout, short, ventrally projecting striate spines (8).

6. At the level of the Stichopogoninae, the genitalic rotation found in *Lasiopogon* is a generic apomorphy (12).

7. Metacoxal peg, when present, slender, weak and acute (except in the aberrant *L*. unc-7sp. nov.).

Table 2. Character state matrix for proposed species groups of *Lasiopogon*. Character numbers correspond to character numbers in the text. 0 - plesiomorphic state; 1-5 - apomorphic state.

Characters	12345	67891 0	11111 12345	11112 67890	22222 12345	22223 67890	33333 12345	33334 67890	44444 12345	444 678
Ancestor	00000	00000	00000	00000	00000	00000	00000	00000	00000	000
quadrivittatus	01000	00001	01210	11110	14511	00010	10110	10110	00001	131
woodorum	01000	00001	01210	11110	14511	00010	10110	10110	00001	131
opaculus	01010	00001	01210	11110	14501	00120	11000	10110	00012	111
slossonae	01010	00001	01210	11110	14501	00120	11000	10110	00012	111
akaishii	01011	01001	01211	21120	14602	10101	00110	20110	00002	111
hasanicus	01011	01001	01211	21120	14602	00101	00110	20110	00002	111
hinei	11001	00001	01211	21011	14501	00110	10010	20110	00002	131
kjachtensis	11001	01001	01211	21011	14501	00110	10010	20110	00002	111
actius	01011	00001	01220	11110	14400	10010	10010	11110	00002	131
martinorum	01011	00001	01220	11110	14400	10010	10010	11110	00001	131
californicus	01011	00001	01220	11110	14400	00000	10010	11110	00001	131
drabicola	01011	00001	01220	11110	14400	00000	10010	11110	00001	131
currani	11100	01110	01200	11010	00300	00000	00010	31111	01302	121
fumipennis	11100	01110	01200	10010	00300	00000	00010	32111	01301	121
trivittatus	01000	10020	01200	10010	00000	00000	10010	41010	00000	000
septentrionalis	11000	10020	01200	10010	00000	00000	10010	41010	00000	000
canus	00000	11110	10000	10010	00000	00000	00001	31000	01100	000
pugeti	00000	11110	10000	10010	00000	00000	00001	31000	01100	000
aldrichii	01000	10100	00000	10010	01100	00000	00010	31010	01100	000
yukonensis	01000	00100	00000	10010	01100	00000	00010	31010	01100	000
testaceus	01100	00110	02200	10010	02000	00000	00010	31010	01202	010
tes-3 sp. nov.	01100	01110	02200	10010	02000	00000	00010	32010	01201	010
bellardii	10001	11110	00000	10010	00000	01000	00011	31010	01200	000
montanus	10001	11110	00000	10010	00000	01000	00001	31010	01200	000
apenninus	00000	11100	00100	10010	03000	01000	00011	31010	01200	000
grajus	00000	11110	00100	10010	03000	01000	00011	31010	01210	000
cinctus	11000	11110	03100	10010	00200	01000	00001	40010	11200	000
cin-3 sp. nov.	11000	11110	03100	10010	00200	01000	00001	40010	11200	000

PHYLOGENY OF THE SPECIES GROUPS

Character Analysis

The analysis of *Lasiopogon* species groups presented here is tentative and incomplete. There are few good external characters dividing the genus into monophyletic groups; most of these groups are based on genitalic characters. Two exemplars from each of several postulated monophyletic groups were included in the analysis of the species group relationships. Because of time constraints, I did not study many Palaearctic species in detail and did not include them in the phylogeny. Further analysis will likely result in the establishment of additional groups for some of these species in and around the *cinctus* clade. With a few exceptions, Nearctic species have been assigned species groups. Appendix 1 contains a list of the species groups and species of *Lasiopogon* as presently understood. The 48 characters and their states used in the phylogenetic analysis of the species groups of *Lasiopogon* are summarized in table 2. Each character is numbered; these numbers correspond to those on the character matrix (table 2), on the cladograms (figures 252-55) and in the discussions concerning evolutionary relationships. For each character, the inferred plesiomorphic and apomorphic (1 for binary characters, 2, 3, etc. for multistate ones) states are described. Weight values (see p. 20) assigned to character states are indicated; weights were not used in the analysis. Characters 12, 22, 23, 36, 43 and 47 were unordered in the analysis.

Thorax and Appendages

1. Halter knob. Two states: plesiomorphic – dark spot absent; apomorphic – dark spot present (+). The halter knob has a large dark spot in several *Lasiopogon* lineages. In the *fumipennis* group, it is a strong mark that occurs in all specimens of all species and is a synapomorphy of the group. It may be independently derived compared to its appearance in other groups where all species do not have the mark and where it may be less definite when it occurs. In some groups it can link sister taxa (e.g., *L. prima, L. septentrionalis*), but in others its presence is less reliable, even in a particular species (*cinctus, montanus* groups) (Cannings 1996).

2. Coxal peg. Two states: plesiomorphic – present; apomorphic – absent (+). The coxal peg is a small projection on the anterior face of the hind coxa present in the Therevidae plus some taxa of Apsilocephalidae and Bombyliidae; its presence in various tabanomorph families is not considered homologous by Yeates (1994). Although Yeates did not note it, many asilid genera also have a coxal peg, similar in shape and position to that of the Therevidae. I have not searched extensively for the character in the family, but it occurs in at least *Stichopogon* and *Lasiopogon* (figure 6) and in the stenopogoninine genera *Hypenetes, Tillobroma, Bathypogon, Willistonina* and *Cyrtopogon*. In *Lasiopogon* it is subject to considerable homoplasy. It is restricted to several species groups in the *cinctus* clade, most of which are from the West Palaearctic, and to the mainly Nearctic *terricola* group in the *bivittatus* clade. Except in *L.* unc-7 sp. nov., the peg in *Lasiopogon* is smaller, thinner and more acute than the stout, rounded one found in other genera. Despite its homoplasious distribution in the genus, it is often useful in determining the monophyly of species groups.

3. Colour of tibiae and external genitalia. Two states: plesiomorphic – brown/black or mostly so; apomorphic – reddish (+). The apomorphic state is scattered throughout the genus; it is found in all but one species of the *testaceus* group and is indicative of the group's monophyly.

4. Anepimeron setae. Two states: plesiomorphic – absent; apomorphic – present (+). The apomorphic state is homoplasious in the more derived clades, the anepimeron bearing a few fine hairs in the *californicus, bivittatus, opaculus* and *akaishii* species groups.

5. Scutellar bristles. Two states: plesiomorphic – arranged in a single row; apomorphic – arranged in two or more irregular rows (+). The apomorphic state is subject to homoplasy as well as to errors in interpretation owing to intermediate states in several species examined.

6. Medial stripe on scutum. Two states: plesiomorphic – absent; apomorphic – present (+). In the plesiomorphic state there is a pair of acrostichal stripes straddling the midline of the scutum in addition to the broader dorsocentral stripes. A dark stripe along the midline (medial stripe) replaces the acrostichal stripes in the apomorphic condition. This occurs in much of the *cinctus* clade and a majority of species groups in the basal part of the *bivittatus* clade. It is also found in a large proportion of specimens of *L. cinereus* in the *opaculus* clade, where it appears to be independently derived.

Epandrium

7. Basal sclerite. Two states: plesiomorphic – present as a definite sclerite (figure 245i); apomorphic – vague or absent (++) (figure 245h). The basal sclerite is small, rectangular or oval, and lies in the subepandrial membrane between the bases of the epandrial halves. It has not been described before. It probably strengthens the articulation at the base of the clasping epandrium in many species where the epandrium halves are completely separated and horizontally clasping; I have found it only in *Lasiopogon* and *Bathypogon*, the only two members of the Stenopogoninae/Stichopogoninae that show this epandrium structure. Frequently the membrane is vaguely darkened or the position of the sclerite is weakly indicated; I score this situation as apomorphic. I postulate that the weak form is derived rather than incipient, even though it occurs in a number of species of *Bathypogon*, one of the outgroups. My reasoning is based on the observation that the apomorphic state, which occurs primarily in the *cinctus* clade, is usually associated with a more strongly sclerotized and massive epandrium (apomorphic) where the additional strength provided by the sclerite is perhaps unnecessary.

8. Basal umbo. Two states: plesiomorphic – absent (figure 245i); apomorphic – present (+) (figure 245f). The form of the epandrium in *Bathypogon* and the

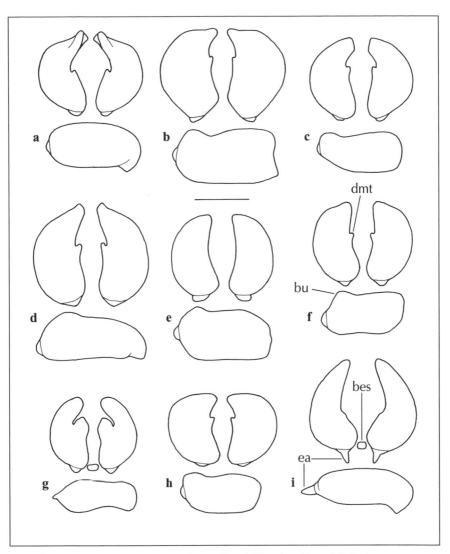

Figure 245. Forms of *Lasiopogon* epandrium: dorsal (above) and lateral (below).
a, *L. canus*; **b**, *L. bellardii*; **c**, *L. cinctus*; **d**, *L. grajus*; **e**, *L. aldrichii*; **f**, *L. testaceus*;
g, *L. prima*; **h**, *L. fumipennis*; **i**, *L. martinorum*. bes = basal epandrial sclerite,
bu = basal umbo, dmt = dorsomedial tooth, ea = epandrial apodeme. Scale line = 0.6 mm.

bivittatus group and *opaculus* section of *Lasiopogon* is the plesiomorphic state in the ingroup. Additional strengthening and grasping capability derived from heavily sclerotized basal umbos, apical flanges and teeth are apomorphic, found mostly in the *cinctus* clade and the basal lineages of the *bivittatus* clade.

9. Dorsomedial tooth. Three states: plesiomorphic – absent (figure 245e, i); apomorphic 1 – small, sclerotized tooth (+); apomorphic 2 – large, more-or-less flattened tooth-like lobe (+++). A distinctive, small, usually truncate tooth occurs at about the midpoint of the dorsomedial margin of the epandrium (figure 245f); it is homoplasious in some of the groups in which it occurs. I do not consider this homologous to the angle that sometimes develops on the basal convexity of the epandrium in the *opaculus* clade (expressed most strongly in *L. monticola*). The apomorphic 2 state is a synapomorphy of the *terricola* group; it is best developed in the pair of *L. septentrionalis* and *L. prima*, where it is disc-like (figure 245g). I am unsure of the homology between the apomorphic 1 and 2 states and so treat them as independently derived from an unmodified epandrium margin.

10. Shape of epandrium apodemes. Two states: plesiomorphic – short and broad (figure 245a-h); apomorphic – elongate (+). The *bivittatus* and *opaculus* sections are dominated by species having strong epandrium apodemes (figure 245i), although there is considerable variation in the intensity of the elongation.

11. Apical flange. Two states: plesiomorphic – absent; apomorphic – present (+). In its plesiomorphic state the apex of the epandrium half is truncate or rounded with a simple, unadorned margin (figure 245i). In various lineages the margin is toothed or otherwise modified and there is some question as to the homology of the various structures, most of which, for this reason, have not been used in the analysis. The apical flange occurs in one group of species (figure 245a).

Hypandrium and Gonocoxites

12. Shape of gonocoxal apodemes. Four states: plesiomorphic – spatulate, exposed portion usually short, apically as wide as it is long (figure 246b); apomorphic 1 – more-or-less parallel-sided and much longer than its width (+) (figure 246h); apomorphic 2 – spatulate, curved, longer than its width (+) (figure 246f); apomorphic 3 – very long and strongly spatulate, embedded in a sclerotized web that links the apodemes ventrally (+++) (figure 246c). Most other genera in the Stichopogoninae have small, shallowly lobed apodemes, and I consider the short, spatulate form to be plesiomorphic and the three apomorphic states to be independently derived. Apomorphy 3 is strongly indicative of the monophyly of the *cinctus* group. This multistate character is treated as unordered in the analysis.

Figure 246 (facing page). Forms of *Lasiopogon* hypandrium/gonocoxite complex and gonostylus: a, b, d, g, h, ventral (left) and lateral (right); c, e, f, dorsal (left) and lateral (right); and i, ventral (left) and dorsal (right).
a, *L. aldrichii*; **b**, *L. bellardii*; **c**, *L. cinctus*; **d**, *L. canus*; **e**, *L. apenninus*; **f**, *L. testaceus*; **g**, *L. prima*; **h**, *L. currani*; **i**, *L. martinorum*. gcx al = gonocoxal apical lobe, gcx ap = gonocoxal apodeme, gst = gonostylus, ml = medial lobe. Scale line = 0.3 mm.

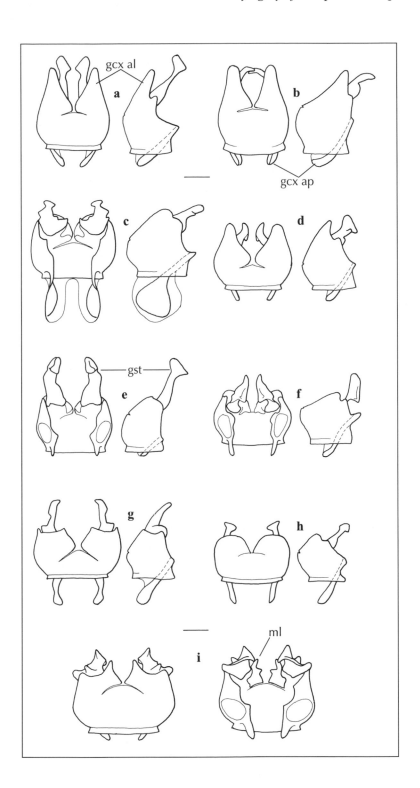

13. Length of gonocoxite apical lobes. Three states: plesiomorphic – elongate, apically acute or narrowly rounded, extending well past the articulation of the gonostylus (figure 246a, b, d); apomorphic 1 – moderately produced apically, broadly rounded and extending only a short distance past the articulation of the gonostylus (figure 246c) (++); apomorphic 2 – absent (figure 246g-i) (++). The apomorphic states are not homoplasious, but I am not confident of all my interpretations and have given them a medium weight. This multistate character is treated as unordered in the analysis. The gonocoxites in many of the putative basal lineages of the Asilidae are elongate apically, often bearing acute processes in addition to the gonostyli. This is the situation in all the Stenopogoninae and Stichopogoninae that I have examined. In *Lasiopogon*, elongate gonocoxites occur in several lineages; I consider this the plesiomorphic state. Both the plesiomorphic and apomorphic 1 states are associated with elongate gonostyli, the plesiomorphic form. Apical lobes are not associated with the medial articulation of the gonostylus and are not to be confused with the more medial lobes in the *bivittatus* group and *opaculus* section that are sometimes prominent and bear strongly sclerotized processes (see character 14). The apical lobes in these groups are absent.

14. Structure of medial lobe supporting gonostylus. Three states: plesiomorphic – simple fold on the medial margin of gonocoxite (figure 246c, e, f); apomorphic 1 – sclerotized process directed dorsally (+++) (figure 42); apomorphic 2 – a process similar to apomorphic 1 but more heavily sclerotized and strongly apically ridged, with a ventromedial tooth (+++) (figure 246i). I interpret this multistate character as ordered.

15. Setation of medial lobe supporting gonostylus. Two states: plesiomorphic – apical surface does not have a dense brush of fine hair; apomorphic – apical surface has a dense brush of fine hair (+++) (figure 26). The apomorphic state occurs in combination with the apomorphic 2 state of character 16, suggesting that the two characters are related.

16. Setal brush of gonocoxite. Three states: plesiomorphic – absent; apomorphic 1 – bristles densely arranged and of similar length, or somewhat longer medially; apomorphic 2 – lateral bristles much longer than medial ones, which are finer and more densely arranged (+++). The apomorphic state is strongly indicative of monophyly.

Subepandrial Sclerite

17. Form of sclerite. Two states: plesiomorphic – viewed from the apex, the sclerite more-or-less heart-shaped, the basal emargination more-or-less rounded, sometimes triangular (figure 247a-i); apomorphic – the sclerite more-or-less U-shaped, the halves completely separate or, more commonly, joined for a short distance near the apex, the deep basal emargination triangular or linear (++) (figure 247j-k). There is some homoplasy in this character, and the states in a few species are difficult to interpret (figure 247g). Generally in the apomorphic state, the

Figure 247. Forms of *Lasiopogon* subepandrial sclerite: ventral.
a, *L. montanus*;
b, *L. grajus*;
c, *L. canus*;
d, *L. aldrichii*;
e, *L. testaceus*;
f, *L. cinctus*;
g, *L. currani*;
h, *L. prima*;
i, *L. fumipennis*;
j, *L. martinorum*;
k, *L. californicus*.
Scale line = 0.3 mm.

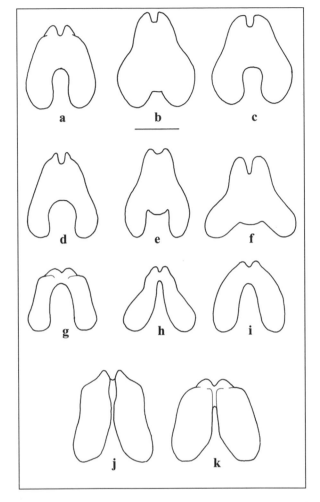

apex of the sclerite is somewhat shelf-like, more-or-less distinct from the lobes of the hypoproct (figure 247j-k).

18. Shape of sclerite spines. Two states: plesiomorphic – tapered from the base, narrowly acute and attenuate (figure 14d, e); apomorphic – more-or-less parallel-sided basally, blunt or bluntly acute (figure 14a-c)(++). These spines are a strong synapomorphy of *Lasiopogon* species. Reversal of the apomorphic state occurs in one small derived lineage.

19. Distribution of sclerite spines. Three states: plesiomorphic – absent; apomorphic 1 – unevenly distributed over the sclerite, often concentrated apically and basally (figures 36, 200, 240); apomorphic 2 – densely packed over the entire sclerotized surface (+++) (figure 28). The apomorphic state is indicative of monophyly.

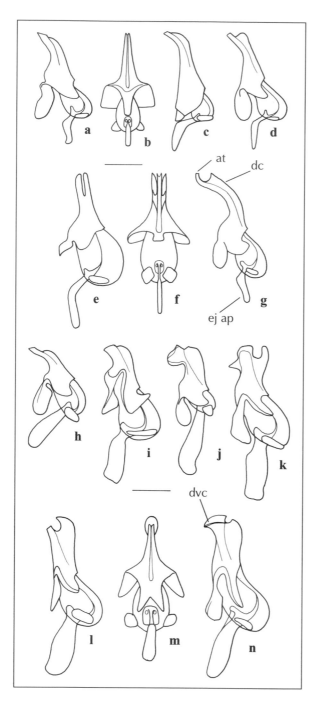

Figure 248. Forms of *Lasiopogon* phallus: lateral, except where noted. **a**, *L. montanus*; **b**, *L. montanus* (ventral); **c**, *L. apenninus*; **d**, *L. canus*; **e**, *L. cinctus*; **f**, *L. cinctus* (ventral); **g**, *L. aldrichii*; **h**, *L. testaceus*; **i**, *L. trivittatus*; **j**, *L. currani*; **k**, *L. californicus*; **l**, *L. martinorum*; **m**, *L. martinorum* (ventral); **n**, *L. coconino* at = aedeagal tube, dc = dorsal carina, dvc = dorsoventral carina, ej ap = ejaculatory apodeme. Scale line = 0.3 mm.

20. Medial unsclerotized area. Two states: plesiomorphic – medial margins of the sclerite typical in form, more-or-less evenly sclerotized (figure 13); apomorphic – medial margins have unsclerotized areas encroaching laterally (+++) (figures 94-102, 163, 176).

Gonostylus
21. Gonostylus form. Two states: plesiomorphic – longer than its width, often stalked (figure 246a-h); apomorphic – apically compressed, usually as wide as or wider than its length (+++) (figure 246i). In all genera of Stenopogoninae and Stichopogoninae examined or studied in the literature, the gonostylus is elongate and usually more-or-less linear. This is also usually the case in other asiloid families; I consider it the plesiomorphic state. In the *bivittatus* and *opaculus* sections of *Lasiopogon* the gonostylus is almost always a complex, flattened, broad structure. The exceptions are in two Chinese species where a more elongate form is secondarily derived from the apomorphic state.
22. Form of gonostylus apex. Five states: plesiomorphic – finger-like, with or without processes of various types (figure 246b-d, g); apomorphic 1 – flattened laterally into a dorsally pointed hatchet shape (+++) (figure 246a); apomorphic 2 – bird's-head shape with dorsoapical spine (+++) (figure 246f); apomorphic 3 – bird-head-shaped with a dorsoapical fold (+++) (figure 246e); apomorphic 4 – apically flattened with a medial flange and expanded lateral area (+++) (figure 246i). This multistate character is treated as unordered in the analysis.

Phallus
23. Lateral form of phallus apex. Seven states: plesiomorphic – aedeagal tube and its sheath longer than the dorsal carina, the apex of the carina rounded or smoothly tapered (figure 248a-d, i); apomorphic 1 – aedeagal tube and its sheath reflexed, dorsal carina short, falcate (+++) (figure 248g); apomorphic 2 – dorsal carina absent, the ventral margins of the paramere sheath expanded and elongated into a pair of processes, creating a trifid phallus apex (+++) (figure 248e, f); apomorphic 3 – aedeagal tube and its sheath broad, short, and bent strongly and ventrally (+++) (figure 248j); apomorphic 4 – dorsal carina produced apically, a deep gap between the carina and the aedeagal tube and its sheath (+++) (figure 248k, l); apomorphic 5 – dorsal carina only weakly produced apically, the gap between the aedeagal tube/sheath and the apex of the carina shallow (+++) (figure 248n); apomorphic 6 – neck of the phallus angled ventrally, the apex truncate, the dorsal carina more-or-less continuous with the aedeagal tube/sheath (+++) (figure 27). This multistate character is treated as unordered in the analysis.
24. Ventral shape of phallus apex. Two states: plesiomorphic – narrower than or only as wide as the basal part of the phallus neck (figures 207, 248b, f, m); apomorphic – wider than the phallus neck (++) (figure 239). This character refers to the shape of the ventral margins of the phallus apex and does not include the shape of the dorsal carina.

25. Dorsoventral carina. Three states: plesiomorphic – apex of phallus simple, without a dorsoventral carina (figure 248a-m); apomorphic 1 – apex of phallus (paramere sheath) ringed by a dorsoventral carina (++) (figure 248n); apomorphic 2 – carina distinct on ventral half of phallus only (++) (figures 117, 223). Plesiomorphically, the asilid phallus apically is a rather simple tube within a tube, the aedeagal tube inside, the paramere sheath outside; the latter may be flared or lobed at the tip. In *Lasiopogon* there is a ridge (dorsoventral carina) running from the lip-like ventral tip of the paramere sheath to the apex of the dorsal carina. In a few lineages the carina is absent; I consider this a secondarily derived condition.

26. Form of dorsal carina. Two states: plesiomorphic – linear (figures 35, 248a-b); apomorphic – apically a disc (+) (figures 27, 248l-m).

27. Lateral form of ejaculatory apodeme. Two states: plesiomorphic – broad, long and more-or-less straight (figure 248h-n); apomorphic – short, with the ventral margin strongly angled (++) (figure 248a-g). Although the apomorphic state displays no homoplasy, I have given it only a medium weight because of the difficulty of interpreting the apodeme shape in some species.

28. Apicoventral flange. Two states: plesiomorphic – absent; apomorphic – present (++) (figure 199). Ventrally, the margins of the paramere sheath do not meet along the midline. In the apomorphic state the apical parts of these margins are expanded ventrally, producing a bulge that is visible in ventral view.

Female Sternite 8

29. Midline sclerotization. Three states: plesiomorphic – full (figures 29, 53, 249a-c, e); apomorphic 1 – weak (+) (figures 66, 209); apomorphic 2 – broadly membranous (+++) (figures 45, 249d). Plesiomorphically, sternite 8 is rather stiff and boat-shaped in cross-section, the midline sclerotized; a broadly membranous midline is strongly indicative of the monophyly of the *opaculus* group, although a form of this apomorphic condition also appears in some members of the *bivittatus* group, where it is apparently independently derived. I have given the intermediate state (apomorphic 1) a low weight because the weakness of sclerotization varies considerably and interpretation of the state is often ambiguous.

30. Medial processes. Two states: plesiomorphic – absent (figure 37); apomorphic – present (+++) (figure 29). The presence of long, apically directed processes medial to the hypogynial valves in the *akaishii* group is, perhaps, the strongest evidence for the monophyly of this distinctive lineage. These processes have their lateral margins incorporated into the hypogynial valves in *L. akaishii, L. lehri* and *L. rokuroi.*

31. Lateral lobe setae of sternite 8. Two states: plesiomorphic – weak (figure 249a, b, e); apomorphic – some developed into strong bristles (+) (figures 45, 249c-d).

32. Setation of hypogynial valves. Two states: plesiomorphic – hairs (figure 39); apomorphic – strong bristles (+++) (figure 156). The apomorphic state supports the monophyly of the *opaculus* group, the only lineage in which it occurs.

Figure 249. Forms of
Lasiopogon female sternite
8: ventral view, with
setation omitted except for
lateral-lobe setae.
a, *L. bellardii*;
b, *L. fumipennis*;
c, *L. prima*;
d, *L. martinorum*;
e, *L. californicus*.
hv scl = hypogynial valve
  sclerite,
scl = secondary sclerite
  of lateral lobe.
Scale line = 0.3 mm.

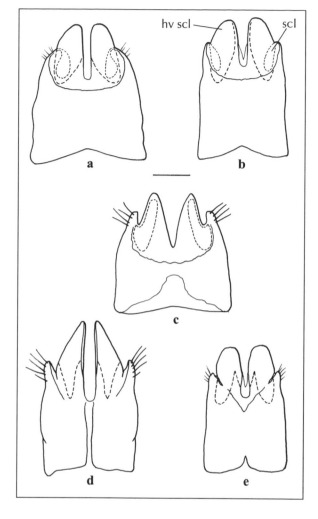

33. Hypogynial valve medial carina. Two states: plesiomorphic – absent on ventral surface; apomorphic – present (+). These ridges occur in many of the *tetragrammus* group (figure 233). They are also present on the valves of the *akaishii* group (figure 87); I am unsure of the homology of the carinae in the two groups because the structure of valve is considerably different in the two lineages.

Female Sternite 9

34. Dorsal carina. Two states: plesiomorphic – present on sternite 9 (figure 46); apomorphic – absent (+) (figure 30). The female sternite 9 (furca) takes a wide range of forms in the ingroup and outgroups and, for the most part, I have discarded the characters of this structure initially screened for analysis. For example, sternite 9 can be medially divided but a given species may often show both divided and undivided forms. Some groups show consistency of form (e.g., the *opaculus* and *akaishii* groups), but others do not. The polarity of the character states is difficult to interpret. Species of *Bathypogon* used in outgroup analysis have strong carinae, and the resemblance of sternite 9 in some of these species to those in the *opaculus* group is striking. Given the outcome of the analysis, however, the *opaculus* group is derived and the presence of a carina there is a reversal. Carinae appear in many species in the *cinctus* clade but the form of the sternite is different from that in the *opaculus* group.

35. Cross-section shape. Two states: plesiomorphic – flat; apomorphic – laterally arched (++). The plesiomorphic condition is a flat furca or one somewhat bowed basoapically. A narrow sternite arched from side to side is apomorphic. Apparently, there is no homoplasy associated with the apomorphic state; nevertheless, variation is common and the difficulty in assigning states in some specimens leads me to reduce the character's weight.

Female Sternite 10

36. Form of basal lobes. Five states: plesiomorphic – narrow but strongly sclerotized and strongly fused to the base of segment 10, apically almost meeting at the midline; apomorphic 1 – broad, strongly sclerotized, only narrowly joined to the base of segment 10, apically almost meeting at the midline (++) (figure 46); apomorphic 2 – broad, strongly sclerotized but not reaching half the distance to the midline (++) (figure 30); apomorphic 3 – short narrow lobe mostly fused to segment 10 (+); apomorphic 4 – faint or absent (++). This multistate character is treated as unordered in the analysis.

Spermathecae

37. United basal ducts. Two states: plesiomorphic – absent (figure 250c, d); apomorphic – present (+) (figure 250a, b). Plesiomorphically, the three basal ducts of the spermathecae open separately but together in a triangular pattern. In many species, there appears to be a minor amount of fusion at the base, but this is usually difficult to verify. Apomorphically, the basal ducts are significantly fused, united into a common duct, frequently for more than half their length.

38. Width of basal duct. Two states: plesiomorphic – about as wide as the striated duct (figure 250a); apomorphic – more than twice as wide as the striated duct (++) (figure 250b-d). Although the apomorphic state is not homoplasious, variation in duct width could complicate interpretation of states, and I have given the character only moderate weight.

Figure 250. Forms of
Lasiopogon spermathecae.
a, *L. montanus*;
b, *L. currani*;
c, *L. schizopygus* sp. nov.;
d, *L. quadrivittatus*.
bd = basal duct,
bulb = bulb of reservoir duct,
cd = common segment
 of basal duct,
jun = junction of basal
 duct and striated duct.
Scale line = 0.2 mm.

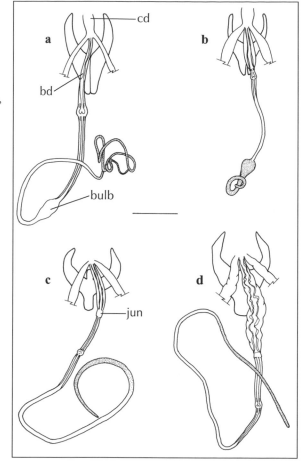

39. Length of basal duct (including the basally coalesced portion, when present). Two states: plesiomorphic – shorter than the striated duct; apomorphic – longer than the striated duct (++). The apomorphic state appears twice in the analysis.

40. Length of striated duct. Two states: plesiomorphic – more than five times longer than its width, including the valve (figure 250a, c, d); apomorphic – less than three times longer than its width (+++) (figure 250b). The striated duct is narrow in the plesiomorphic state, usually more than ten times as long as its width. Apomorphically, it is shorter, but except for the shortest and broadest form, which appears in one group only, discrete states cannot be differentiated with any confidence.

41. Basal duct margins. Two states: plesiomorphic – smooth; apomorphic – scalloped (+++). Apomorphically, the margins of the basal ducts are irregularly scalloped; this is a structural condition of the duct and is not equivalent to the surface wrinkles seen in some of the wider, sac-like ducts found mainly in some species of the *opaculus* section.

42. Striated duct width. Two states: plesiomorphic – uniform from the valve to the junction with the basal duct (figure 250c, d); apomorphic – broad at the valve, tapering to the junction with the basal duct (++) (figure 250a). I have given the apomorphic state a medium weight because it occurs twice in the ingroup; I have not seen it in the outgroups.

43. Reservoir duct bulbs. Four states: plesiomorphic – absent (figure 250c, d); apomorphic 1 – narrow, less than twice as wide as the reservoir duct (++); apomorphic 2 – broad, two or more times as wide as the reservoir duct (++) (figure 250a); apomorphic 3 – broad and incorporated into terminal reservoir (+++) (figure 250b). Spermathecae in the Asilidae take a bewildering variety of forms, but I know of no other genus that possesses bulbs or discrete swellings midway along the reservoir ducts. Apomorphic states 1 and 2 are subject to minor homoplasy in the genus. This multistate character is treated as unordered in the analysis.

44. Sclerotization of valves. Two states: plesiomorphic – weak, valves appearing clear; apomorphic – strong, valves appearing golden (++). I consider the apomorphic state to be indicative of the monophyly of the *opaculus* group even though apparently similar valves occur in two isolated species, *L. grajus* (*apenninus* group) and *L. terneicus* (*opaculus* section).

45. Differentiation of junction between basal and striated ducts. Three states: plesiomorphic – no differentiation of junction into a valve-like structure; apomorphic 1 – valve-like structure weakly sclerotized, appearing clear (+); apomorphic 2 – valve-like structure well-developed, sclertized and appearing golden (+). In the plesiomorphic state the basal ducts merge into the striated ducts without any apparent differentiation of the junction; the lobed lumen of the basal duct ends abruptly and the simple tube lining the striated duct begins. Differentiation of this junction is widespread throughout the Asilidae (Theodor 1976) with sclerotized rings or valve-like sleeves being common. Two general apomorphic states occur in *Lasiopogon*, and I have interpreted the golden sclerotized structure (apomorphic 2) as arising from a clear, more lightly sclerotized ring (apomorphic 1). There is some homoplasy associated with both forms.

46. Scales on junction between basal and striated ducts. Two states: plesiomorphic – absent (figures 39, 68); apomorphic – present (+) (figures 187, 203). In the apomorphic state, the junction between the basal and striated ducts bears small, projecting scales. Although the choice of exemplars in this analysis has fortuitously eliminated homoplasy in the cladogram, homoplasy is present when more species are considered.

47. Form of terminal reservoir. Four states: plesiomorphic – strongly coiled (figure 250a); apomorphic 1 – curled or strongly hooked (+) (figures 89, 250c); apomorphic 2 – curled, joined to bulb basally (+++) (figure 250b); apomorphic 3 – gently curved or straight (+) (figure 250d). In the ancestral state, the reservoirs are coiled; apomorphically, they are straightened and shortened to varying degrees. Apomorphic states 1 and 3 are subject to homoplasy; state 2 is strongly indicative of monophyly. This multistate character is treated as unordered in the analysis.

48. Terminal reservoir caniculi. Two states: plesiomorphic – sparse; apomorphic – abundant (++). The apomorphic state shows no homoplasy, but the assignment of states is not always clear cut, and I have given the character a medium weight only.

Phylogenetic Relationships.

Hennig86 analysis produced two equally most parsimonious trees with a length of 141 steps, a consistency index of 52 and a retention index of 86. Both trees had the same form distal to the *cinctus* clade (figure 251). The only difference between the two trees was the position of the *cinctus* group – linked to the *montanus* group or the *apenninus* group. The latter tree was chosen because the *cinctus* + *appenninus* clade is supported by a more robust, non-homoplasious character, the shape of the gonocoxite apical lobes (13). The *cinctus* + *montanus* clade is linked by a weak, homoplasious character, the colour of the halter knob (1).

I have named the two main lineages of *Lasiopogon* derived in this analysis the *cinctus* and *bivittatus* clades. For convenience, I have termed the two main lineages of the large terminal monophyletic group of the *bivittatus* clade the *bivittatus* group and the *opaculus* section (figure 251). The *opaculus* section is examined in detail beginning on page 295.

The *cinctus* clade
The *cinctus* clade is predominantly west Palaearctic in distribution and consists of the *canus, montanus, cinctus* and *apenninus* species groups. As indicated on page 275, many species belonging to this clade or related to it (or to basal lineages of the *bivittatus* clade) – that is, those having the plesiomorphic state of the phallus and gonostyli – were not assigned to any of the species groups analysed and they are excluded from the cladogram.

The monophyly of the *cinctus* clade is supported by a non-homoplasious synapomorphy, an arched sternite 9 (35). Three homoplasious synapomorphies also offer some support: the presence of a medial stripe on the scutum (6), dorsomedial teeth on the epandrium (9) and the loss of the basal sclerite (7).

The *canus* group is the sister to the rest of the clade; the apical epandrial flange (11) indicates its monophyly. The sister clade is supported by one medium-weight but uniquely derived character, the short, bent ejaculatory apodeme (27). The *montanus* group is only weakly supported by a marked halter knob (1) (with homoplasy evident in included species) and dense scutellar bristles (5). The sister clade, consisting of the *cinctus* and *grajus* groups, is defined by medium-length gonocoxite apical lobes (13:1), a character given medium weight. The monophyly of the *cinctus* group is strongly supported by three high-weight

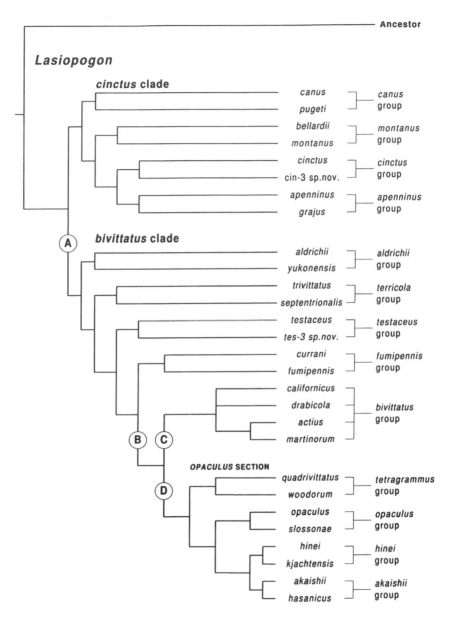

Figure 251. Overview of cladistic relationships of species groups of *Lasiopogon*. Groups are represented by exemplar species. The letters in circles refer to the parts of the cladogram detailed in figures 253 (A), 254 (B, C) and 255 (D).

synapomorphies: the enlarged gonocoxal apodemes (12:2), the trifid phallus apex (23:2) and the scalloped basal ducts (41). Other synapomorphies include the marked halter knob (1), the lack of a coxal peg (2), the absence of sternite 10 basal lobes (36:4) and, a reversal, the loss of united basal ducts (37).

The *apenninus* group, the sister of the *cinctus* group, is strongly supported as a monophyletic clade by only its unique gonostylus shape (22:3) and weakly supported by the lack of a sternite 9 carina (34).

The *bivittatus* clade

The *bivittatus* clade contains the remainder of the known species groups. Its monophyly is weakly supported; both synapomorphies are homoplasious. All species except some in the *terricola* group lack coxal pegs (2) and all groups except the *opaculus* group lack a carina on sternite 9 (34).

Basal to the rest, the *aldrichii* group is strongly supported by the uniquely derived hatchet-shaped gonostylus (22:1). The group retains the plesiomorphic elongate gonocoxite apical lobes. The sister clade is moderately supported by two

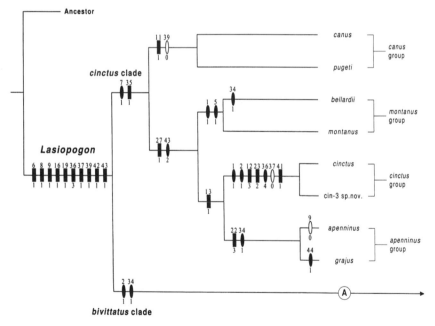

Figure 252. Cladistic relationships of species groups in the *cinctus* clade of *Lasiopogon*. Groups are represented by exemplar species. Characters are plotted on branches as rectangles (characters without homoplasy) or ovals (characters with homoplasy); open symbols are reversals. The numbers above the symbols correspond to the character numbers in the list on pages 275-89; the numbers below represent the character states described in the character list: 0 = plesiomorphic state; 1, 2, 3, etc. = apomorphic states. "A" indicates the *bivittatus* clade, the base of which is detailed in figure 253.

non-homoplasious apomorphies – gonocoxites without apical lobes (13:2) and long, narrow gonocoxal apodemes (12:1). The third apomorphy, the dorsomedial tooth on the epandrium (9:1,2), is homoplasious and is shared by only the *terricola, testaceus* and *fumipennis* groups. The *terricola* group is strongly defined by the presence of a large, flat tooth on the medial margin of the epandrium (9:2). This tooth is especially distinct in the sister species *L. prima* and *L. septentrionalis*, where it takes the form of a disc. Homoplasious characters offer some support. *L. trivittatus* and *L. septentrionalis* have a medial stripe on the scutum (6), a character that is homoplasious with respect to the *aldrichii* group and the *cinctus* clade. Others include the loss of the basal umbo of the epandrium (a reversal of 8), the presence of strong setae on the lateral lobes of sternite 9 (31), the lack of sternite 10 basal lobes (36:4) and the secondary loss of spermathecal bulbs (43). *L. terricola*, the fourth member of the group, has unusual autapomorphies, such as the lack of notal and scutellar bristles.

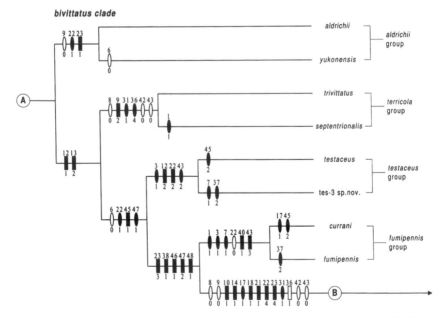

Figure 253. Cladistic relationships of species groups at the base of the *bivittatus* clade of *Lasiopogon*. Groups are represented by exemplar species. Characters are plotted on branches as rectangles (characters without homoplasy) or ovals (characters with homoplasy); open symbols are reversals. The numbers above the symbols correspond to the character numbers in the list on pages 275-89; the numbers below represent the character states described in the character list: 0 = plesiomorphic state; 1, 2, 3, etc. = apomorphic states. "A" indicates the position of these lineages in figure 251; "B" indicates the clade comprising the *bivittatus* group + *opaculus* section, the base of which is detailed in figure 254.

The monophyly of the remainder of the groups from this point is only weakly supported. These clades share differentiated junctions between the basal ducts and the striated ducts of the spermathecae (45:1,2) as well as terminal reservoirs of the spermathecae that are curled, hooked or straight rather than coiled (47:1,2,3). The *testaceus* species group is the basal clade here, strongly supported by the distinctive bird-head (with dorsoapical spine) shape of the gonostylus (22:2). Ferruginous genitalia and tibiae (3) and broad reservoir duct bulbs in the spermathecae (43:2) are other synapomorphies of the group.

Synapomorphies for the *fumipennis* group plus the *bivittatus* group and the *opaculus* section include wide basal ducts in the spermathecae (38), scales on the junction between basal and striated ducts of spermathecae (46), and abundant caniculi on terminal reservoirs of the spermathecae (48). The *fumipennis* group contains only three species, *L. fumipennis, L. currani* and *L. polensis*, which share two strong synapomorphies in the spermathecae – very short striate ducts (40) and sclerotized reservoir duct bulbs that are incorporated into the terminal reservoirs (43:3). Apomorphies subject to some homoplasy include a darkened halter knob (1) (constant in the group), ferruginous genitalia and tibiae (3), and loss of the basal sclerite (7). The presence of a finger-shaped gonostylus (22) is a reversal.

The monophyly of the clade containing the *bivittatus* group and the *opaculus* section is strongly supported by unique synapomorphies such as the sclerotized, dorsally projecting medial gonocoxal lobes (14:1,2), an apically compressed gonostylus (21) with a medial flange and expanded lateral area (22:4), and by

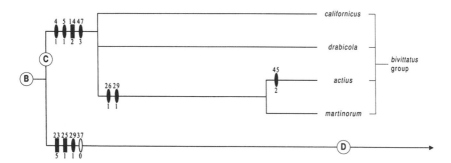

Figure 254. Cladistic relationships of the *bivittatus* group of *Lasiopogon*. The group is represented by exemplar species. Characters are plotted on branches as rectangles (characters without homoplasy) or ovals (characters with homoplasy); open symbols are reversals. The numbers above the symbols correspond to the character numbers in the list on pages 275-89; the numbers below represent the character states described in the character list: 0 = plesiomorphic state; 1, 2, 3, etc. = apomorphic states. "B" and "C" indicate the position of these lineages in figure 251; "D" indicates the *opaculus* section, which is detailed in figure 255.

several synapomorphies of the phallus apex (23:4,5,6). Other support comes from the presence of long epandrial apodemes (10), the form of the subepandrial sclerite (17) and the shape of its spines (18), and the presence of strong setae on the lateral lobes of female sternite 9 (31). Apomorphic states of several characters characterize basal clades in the genus, requiring the assumption of several reversals in this relatively derived clade; these include the loss of the basal umbo (8) and dorsomedial teeth (9) on the epandrium and the loss of reservoir duct bulbs (43) and thick striate ducts (42) in the spermathecae.

The *actius* clade is part of an unresolved trichotomy with *L. californicus* and *L. drabicola*. Because the latter species were not linked by any synapomorphies in the analysis, I have lumped all four exemplars in a *bivittatus* species group representing about 16 species. The *bivittatus* group is united on the basis of two unique synapomorphies, the strongly apically ridged medial lobes of the gonocoxites (14:2), and the strongly bifid phallus apex. Several homoplasious synapomorphies are supportive: hairs on the anepimeron (4), dense scutellar setae (5) and more-or-less straight terminal reservoirs in the spermathecae (47:3). The clade containing *L. actius* and *L. martinorum* is defined by an apically flattened dorsal carina on the phallus (26) and a medially divided sternite 8 in the female (29:2). The flattened dorsal carina is homoplasious with respect to the *akaishii* group, but I am not convinced that the structure in the two groups is homologous. The divided sternite is homoplasious with respect to the *opaculus* group.

The monophyly of the *opaculus* section is supported by the uniquely derived dorsoventral carina below the phallus apex (25) and the shape of the apex itself (23). Reduction of the midline sclerotization of female sternite 8 (29) is common across the section and the loss of a well-developed common basal duct in the spermathecae (37) occurs in all species.

The monophyly of most species groups is supported by high-weight characters. Synapomorphies of the basal *tetragrammus* group as analysed here include the expanded apex of the phallus (24), carinae on the hypogynial valves (33) and spermathecae with straight terminal reservoirs (47:3).

The remaining groups have anepimeron setae (4) and an apicoventral flange on the phallus (28). They also share strongly sclerotized golden junctions between the basal and striated ducts of the spermathecae (45). This apomorphic state also appears in three exemplars from more basal groups: *L. actius*, *L. testaceus* and *L. currani*. The synapomorphies of and the relationships among the groups of the *opaculus* section and their species are detailed beginning on page 300 and are not treated further here.

PHYLOGENY OF THE *OPACULUS* SECTION

Character Analysis

The 40 characters and their states used in the phylogenetic analysis of the species of the *opaculus* section are summarized in table 3 (next page). Each character is numbered; these numbers correspond to those on the character matrix (table 3), on the cladograms (figures 257-60) and in the discussions concerning evolutionary relationships. For each character the inferred plesiomorphic (0) and apomorphic (1 for binary characters, 2, 3,... for multistate ones) states are described. Weight values (see p. 20) assigned to character states are indicated; weights were not used in the Hennig86 analysis. Characters 6, 19, 25, 30, 36 and 38 were unordered in the analysis. Many characters in this analysis are the same as those used on pages 275-89. In order to maintain the logical order of the character list, the numbers of these shared characters were not kept equivalent. Characters used in both analyses but having different numbers are noted in the list.

Thorax and Appendages

1. Halter knob. Two states: plesiomorphic – without dark spot; apomorphic – with dark spot (++). See page 275, character 1.

2. Colour of tibiae and external genitalia. Two states: plesiomorphic – brown/black or mostly so; apomorphic – ferruginous (+). Same as character 3, page 276.

3. Anepimeron setae. Two states: plesiomorphic – absent; apomorphic – present (++). Same as character 4, page 276.

4. Scutellar bristles. Two states: plesiomorphic – arranged in a single row; apomorphic – arranged in two or more irregular rows (+). Same as character 5, page 276.

Epandrium

5. Basal sclerite. Two states: plesiomorphic – present; apomorphic – absent (+). Same as character 7, page 276.

6. Shape of dorsomedial margins. Three states: plesiomorphic – weakly or moderately concave (figures 41, 62); apomorphic 1 – strongly concave (+) (figures 78, 83); apomorphic 2 – more-or-less parallel (++) (figure 181). The *opaculus* clade is dominated by species having epandria with concave dorsomedial margins, although there is considerable variation in the intensity of this emargination. The apomorphic 1 state shows considerable homoplasy.

Hypandrium and Gonocoxites

7. Setation of medial lobe supporting gonostylus. Two states: plesiomorphic – apical surface of lobe without dense brush of fine hair; apomorphic – apical surface of lobe with dense brush of fine hair (++). Same as character 15, page 280.

Table 3. Character state matrix for species of the *opaculus* section of *Lasiopogon*. Character numbers correspond to character numbers in the text. 0 - plesiomorphic state; 1-3 - apomorphic state; ? - missing data. See text for description and weights of characters.

Characters:	12345	67891 0	11111 12345	11112 67890	22222 12345	22223 67890	33333 12345	33334 67890
Ancestor	00000	00000	00000	00000	00000	00000	00000	00000
monticola	00000	10010	00000	00041	10002	00000	10001	01000
cinereus	10110	00010	00000	00010	10000	10001	10000	01000
tetragrammus	00000	10010	00000	00011	10000	10010	00000	00000
apache	01000	00010	00000	00011	10000	00000	10000	01000
chaetosus	01000	10010	00000	00011	10000	00000	10001	00000
quadrivittatus	00000	20010	00000	00011	10001	00010	00001	01000
lavignei	00000	20010	00000	00011	10001	00010	00001	01000
coconino	00000	10010	00000	10011	10001	10010	00001	00000
oklahomensis	00000	10010	00000	10011	10000	00010	00000	10000
woodorum	00000	10010	00000	11011	10000	00010	00001	00000
chrysotus	00000	10010	00000	11011	10000	?????	?????	?????
flammeus	01000	10010	00000	11011	10000	?????	?????	?????
shermani	01000	00010	00000	00010	10100	10002	00000	01000
marshalli	00100	00010	00000	00040	10100	20102	01111	10211
piestolophus	00100	00010	00000	00040	10100	20102	01111	10211
appalachensis	00100	00010	00000	00040	10100	20102	01111	10200
schizopygus	00100	00010	00000	00040	10100	20102	01111	10200
opaculus	00100	00010	00010	00040	10100	20102	01111	10200
slossonae	00100	00010	00010	00040	00100	20102	01111	10200
akaishii	00111	11111	00101	00030	21100	01013	20111	11100
hasanicus	00111	11111	00101	00030	21100	01013	20111	11100
lehri	00111	11111	00101	00020	20100	01003	20111	11000
rokuroi	00101	11111	00101	00020	20100	01013	20101	11100
hinei	10010	11100	10202	00111	10110	00000	10111	00000
kjachtensis	11011	11100	10202	00111	10110	00000	10011	10000
leleji	10010	11100	10202	00110	10100	?????	?????	?????
phaeothysanotus	00010	00010	11203	00011	10002	?????	?????	?????
qinghaiensis	00010	00010	11203	00011	10002	00000	20111	10000
terneicus	00000	10010	00000	00040	20000	00004	11000	10000

8. Setal brush of gonocoxite. Two states: plesiomorphic – bristles densely arranged and of similar length, or somewhat longer medially; apomorphic – lateral bristles much longer than medial ones, which are finer and more densely arranged (++). Same as character 16, page 280.

Subepandrial Sclerite
9. Shape of sclerite spines. Two states: pleisiomorphic – tapered from base, narrowly acute and attenuate; apomorphic – more-or-less parallel-sided basally, blunt or bluntly acute (++). Same as character 18, page 281.

10. Distribution of sclerite spines. Two states: plesiomorphic – uneven, often concentrated apically and basally; apomorphic – densely packed over entire surface (+++). Same as character 19, page 281.

11. Medial unsclerotized area. Two states: plesiomorphic – medial margins of sclerite typical in form, more-or-less evenly sclerotized; apomorphic – medial margins with unsclerotized areas encroaching laterally (+++). Same as character 20, page 283.

12. Basodorsal angles of sclerite. Two states: plesiomorphic – rolled dorsally and hidden in ventral view; apomorphic – projecting and visible in ventral view (+++) (figures 163, 176).

Gonostylus
13. Form of dorsal flange. Three states: plesiomorphic – strong, complete, linking medial flange to lateral tooth (figure 12, top); apomorphic 1– medial only, obsolete laterally (+++) (figure 12, bottom); apomorphic 2 – lateral part an apical ridge (+++) (figure 12, middle). This multistate character is treated as unordered in the analysis.

14. Basal tooth on medial flange. Two states: plesiomorphic – absent; apomorphic – present (+++) (figure 214).

15. Form of medial flange. Four states: plesiomorphic – a more-or-less vertical ridge running dorsoventrally (figure 12, top); apomorphic 1 – ventral part expanded laterally into a flat lobe (+++) (figure 12, bottom); apomorphic 2 – flattened into a low, rounded ridge (+++) (figure 12, middle); apomorphic 3 – expanded apically and strongly narrowed (+++) (figures 161, 174).

16. Secondary medial flange. Two states: plesiomorphic – absent; apomorphic – present (++) (figure 12, top). A similar but perhaps not homologous structure appears in a few species in the *bivittatus* section.

17. Medial tooth on dorsal flange. Two states: plesiomorphic – absent; apomorphic – present (+++) (figure 238).

18. Setation of apex of medial flange. Two states: plesiomorphic – bare, or scattered setae; apomorphic – abundant long setae (+++) (figure 92).

Phallus
19. Lateral form of phallus apex. Five states: plesiomorphic – dorsal carina elongate, as long or longer than the aedeagal tube and separated from the tube by

a deep cleft (figure 248k); apomorphic 1 – apex of the dorsal carina shorter than the sheath of the aedeagal tube, pointed, separated from the sheath by a small but strong concavity (figure 35); apomorphic 2 – apex of the carina pointed, slightly longer than the sheath of the aedeagal tube and not separated from it by a concavity (+++) (figure 117); apomorphic 3 – apex of the carina rounded and smoothly continuous with the apex of the sheath of the aedeagal tube (+++) (figure 27); apomorphic 4 – apex of the dorsal carina not pointed or produced but rounded or obscure (+) (figure 43). This multistate character is treated as unordered in the analysis. Same as character 23, page 283, but the apomorphic states are modified.

20. Ventral shape of phallus apex. Two states: plesiomorphic – narrower than or only as wide as the basal part of the phallus neck; apomorphic – wider than the phallus neck (++). Same as character 24, page 283.

21. Dorsoventral carina. Three states: plesiomorphic – absent; apomorphic 1 – apex of the phallus (paramere sheath) ringed with a complete dorsoventral carina; apomorphic 2 – apex of the phallus with a dorsoventral carina on the venter only (++). Same as character 25, page 284.

22. Dorsal carina. Two states: plesiomorphic – linear; apomorphic – flattened apically (+++). Same as character 26, page 284.

23. Apicoventral flange. Two states: plesiomorphic – absent; apomorphic – present (++). Same as character 28, page 284.

24. Ventral process. Two states: plesiomorphic – basal paired projections absent; apomorphic – projections present (+++) (figure 101). The apomorphic state is indicative of the monophyly of *hinei* plus *kjachtensis*; somewhat similar structures occur in *L. albidus* (*bivittatus* group), but I do not consider them homologous.

25. Apicoventral lip of paramere sheath. Three states: plesiomorphic – moderately projecting (figure 199); apomorphic 1 – strongly projecting (+++) (figure 183); apomorphic 2 – absent (+) (figure 162). This multistate character is treated as unordered in the analysis.

Female Sternite 8
26. Midline sclerotization. Three states: plesiomorphic – full, at least in central section; apomorphic 1 –weak (+); apomorphic 2 – midline broadly membranous (+++). Same as character 29, page 284.

27. Medial processes. Two states: plesiomorphic – absent; apomorphic – present (+++). See page 284, character 30.

28. Setation of hypogynial valves. Two states: plesiomorphic – hairs; apomorphic – strong bristles (+++). See page 284, character 32.

29. Hypogynial valve medial carina. Two states: plesiomorphic – absent; apomorphic – present (+). See page 285, character 33.

Female Sternite 9
30. Shape of furca. Five states: plesiomorphic – Y-shaped (figure 38); apo-

morphic 1 – Y-shaped with arms long, narrow and divided (figure 67); apomorphic 2 – U-shaped with a dorsal carina (+++) (figure 46); apomorphic 3 – V-shaped with a slit along much of the midline (+++) (figures 30, 88); apomorphic 4 – Y-shaped with arms short, thick and divided (figure 226). The shape of the furca (sternite 9) is more stable in the species groups of the *opaculus* section than it is in most of the other groups of the genus and can be used more confidently here than in the analysis of species groups (p. 275). The apomorphic 1 and 4 states are autapomorphies of *L. cinereus* and *L. terneicus* respectively; I have included these for the sake of completeness. This multistate character is treated as unordered in the analysis.

Female Sternite 10
31. Form of basal lobes. Three states: plesiomorphic – strongly sclerotized and apically meeting or almost meeting at the midline (figure 46); apomorphic 1 – posterior edge reaching half to three-quarters the distance to the midline (+) (figure 54); apomorphic 2 – posterior edge not reaching half the distance to the midline (++) (figure 30). Same as character 36, page 286.

Spermathecae
32. Sclerotization of valves. Two states: plesiomorphic – weak, appearing clear; apomorphic – strong, appearing golden (++). Same as character 44, page 288.

33. Striated duct setulae. Two states: plesiomorphic – absent (figure 39); apomorphic – present (+) (figure 203).

34. Differentiation of junction between basal and striated ducts. Two states: plesiomorphic – junction lightly sclerotized, appearing clear; apomorphic – valve-like structure well-sclerotized and appearing golden (++). Same as character 45, page 288.

35. Scales on junction between basal and striated ducts. Two states: plesiomorphic – absent (figure 68); apomorphic – present (+) (figures 187, 203). Same as character 46, page 288.

36. Form of terminal reservoir. Two states: plesiomorphic – gently curved or straight (+) (figures 39, 97); apomorphic – curled or strongly hooked (+) (figure 47). Same as character 47, page 288.

37. Terminal reservoir warts. Two states: plesiomorphic – absent (figure 55); apomorphic – present (+) (figure 31).

38. Reservoir duct setulae. Three states: plesiomorphic – absent; apomorphic 1 – present, directed apically (++) (figure 89); apomorphic 2 – present, directed basally (++) (figure 203). Although there apparently is no homoplasy associated with either of the apomorphic states, and although I consider these states indicative of monophyly, I have given them medium weights because in some species they are difficult to assess.

39. Reservoir duct sclerotization. Two states: plesiomorphic – base striate (figure 31); apomorphic – base thickly sclerotized and without striations (+++)

(figure 134). The apomorphic state is strongly indicative of monophyly; it is associated with the apomorphic condition of character 40.

40. Reservoir duct caniculi. Two states: plesiomorphic – few to moderate amounts on the base just apical to the striated part (figures 47, 121); apomorphic – dense on the base just apical to the striated part (+++) (figure 134).

Phylogenetic Relationships

Hennig86 analysis produced six equally most parsimonious trees with a length of 109 steps, a consistency index of 50 and a retention index of 82. There is a discrepancy between the number of species groups proposed in the initial species-group analysis (figure 255) and the number determined by the analysis of all the species in the *opaculus* section (figure 256). The original *tetragrammus* group, as defined by the exemplars *L. quadrivittatus* and *L. woodorum* (figure 255), was assumed to contain *L. apache, L. chaetosus, L. monticola, L. shermani* and *L. terneicus,* but these species were split off into other groups in the analysis of the *opaculus* section (figure 256). No synapomorphy linked them to the species of the *tetragrammus* group.

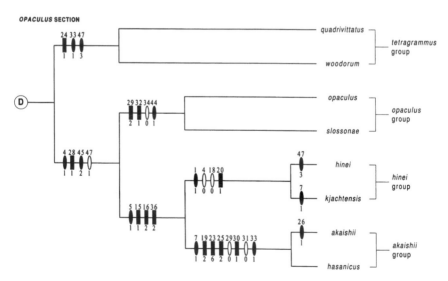

Figure 255. Cladistic relationships of the species groups in the *opaculus* section of *Lasiopogon*. Groups are represented by exemplar species. Characters are plotted on branches as rectangles (characters without homoplasy) or ovals (characters with homoplasy); open symbols are reversals. The numbers above the symbols correspond to the character numbers in the list on pages 275-89; the numbers below represent the character states described in the character list: 0 = plesiomorphic state; 1, 2, 3, etc. = apomorphic states. "D" indicates the position of this lineage in figure 251.

All trees had the same topology from *L. monticola* to *L. kjachtensis* (figure 256). The composition of the *tetragrammus* species group was constant, but variation occurred in its relationship to *L. cinereus, L. shermani, L. apache* and *L. chaetosus*. The relationships among these four species also varied. Figures 256-60 illustrate one of these trees. In general, I found few robust characters in the plesiomorphic species basal to the *opaculus* group. Except for the internal structure of the *tetragrammus* group itself, which is reasonably supported, the paucity

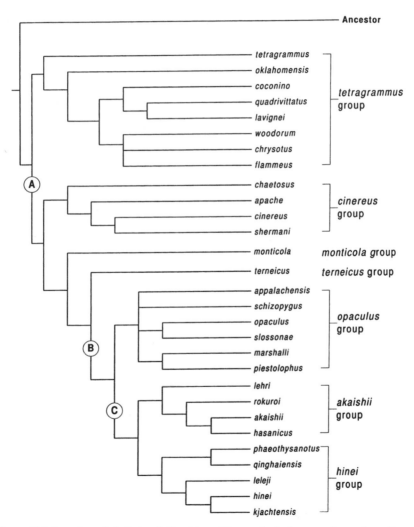

Figure 256. Overview of cladistic relationships of the species in the *opaculus* section of *Lasiopogon*. The letters in circles refer to the parts of the cladogram detailed in figures 258 (A), 259 (B) and 260 (C).

of strong synapomorphies gives me little confidence in the resulting weakly supported basal clades. On the other hand, the derived clades are mostly well supported.

The *tetragrammus* species group

The eight species of the *tetragrammus* group are linked by the presence of a carina on the hypogynial valves (29). All species but the basal *L. tetragrammus* have a secondary flange on the medial flange of the gonostylus (16), an apomorphy that is homoplasious in one lineage of the outgroups. The subsequent dichotomy is supported by a single weak character – scales on the junction between the basal and striated ducts in the spermathecae (35).

Species in one of the resulting clades (*L. coconino, L. quadrivittatus* and *L. lavignei*) share a single highly weighted character – a strong apicoventral lip on the phallus paramere sheath (25:1). A sister species relationship between *L. quadrivittatus* and *L. lavignei* is defined by a moderately weighted character – parallel medial margins of the epandrium (6) – and by the loss of the secondary gonostylus flange (16) and the presence of warts on the terminal reservoirs of the

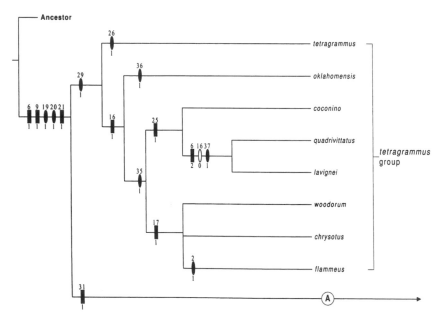

Figure 257. Cladistic relationships of the *tetragrammus* group of *Lasiopogon*. Characters are plotted on branches as rectangles (characters without homoplasy) or ovals (characters with homoplasy); open symbols are reversals. The numbers above the symbols correspond to the character numbers in the list on pages 295-300; the numbers below represent the character states described in the character list: 0 = plesiomorphic state; 1, 2, 3, etc. = apomorphic states. "A" indicates the clade comprising the *cinereus* group and others, which are detailed in figures 258-260.

spermathecae (37). The sister group is an unresolved trichotomy among *L. woodorum*, *L. chrysotus* and *L. flammeus*; monophyly is supported by one uniquely derived character – a small, medial tooth on the dorsal flange of the gonostylus. I suspect that the former two species are sisters; the trichotomy of this clade might be resolved with the discovery of the females of *L. chrysotus* and *L. flammeus*.

The *cinereus* species group

The large clade sister to the *tetragrammus* group is weakly supported by one synapomorphy – reduced basal lobes on sternite 10 (31).

The *cinereus* species group is basal to the rest of the clade. It is only weakly defined by a highly homoplasious apomorphy, ferruginous genitalia and tibiae (2) (*L. cinereus* itself is plesiomorphic). The relationships among the constituent species are weak also, and represent the least stable parts of the six most parsimonius trees. In a number of characters (e.g., 6, 23, 30), *L. shermani* is more similar to members of the *opaculus* group than to others of the *cinereus* group.

The *monticola* species group

The *monticola* group comprises one species, *L. monticola*, weakly linked to the remaining taxa of the clade on the basis of apomorphies of the phallus (19: 1, 2, 3) and scales on the spermathecal junctions (35). Autapomorphies include the

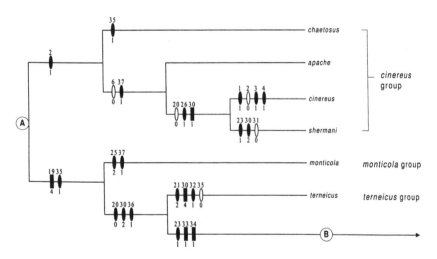

Figure 258. Cladistic relationships of the *cinereus, monticola* and *terneicus* groups of *Lasiopogon*. Characters are plotted on branches as rectangles (characters without homoplasy) or ovals (characters with homoplasy); open symbols are reversals. The numbers above the symbols correspond to the character numbers in the list on pages 295-300; the numbers below represent the character states described in the character list: 0 = plesiomorphic state; 1, 2, 3, etc. = apomorphic states. "A" indicates the position of these lineages in figures 256 and 257. "B" indicates the clade comprising the *opaculus, hinei* and *akaishii* groups; the former is detailed in figure 259.

absence of an apicoventral lip on the phallus sheath (homoplasious with respect to *L. phaeothysanotus* and *L. qinghaiensis*) and the presence of warts on the terminal reservoirs of the spermathecae (37).

The *terneicus* species group

The characteristics of *L. terneicus* are difficult to interpret. The species has an unusual number of striking autapomorphies that are not included in the analysis (form of basal duct, gonostylus, medial lobe of gonocoxite and phallus) that make its placement in the genus problematic. The form of the phallus (figure 223) puts *L. terneicus* into the *opaculus* section. The presence of a complete dorsal flange (13) and medial flange (15) (plesiomorphic in the *opaculus* section), even though they are strikingly different from those occurring in other species, logically place the species basal to the *akaishii* group. Although the sister-group relationship of *L. terneicus* with the clade containing the *opaculus, akaishii* and *hinei* groups is weakly supported, it will have to do until more characters are studied. Synapomorphies defining this relationship are a narrowed phallus apex (20) (a homoplasious character reversal), apomorphic states of sternite 9 form (30: 2, 3, 4), and curled or hooked terminal reservoirs in the spermathecae (36).

The *opaculus* species group

The *opaculus* group is basal to the Asian *akaishii* and *hinei* species groups. Two moderately weighted synapomorphies support the monophyly of this clade – the presence of an apicoventral flange (23) (known elsewhere only in *L. shermani*), and strongly sclerotized, golden junctions between the basal and striated ducts of the spermathecae (34). The presence of setulae on the striated ducts of the spermathecae (33) is a third synapomorphy. The latter two characters do not exhibit homoplasy.

The monophyly of the *opaculus* group itself is strongly supported by synapomorphies such as the presence of spines on the hypogynial valves (28), a broadly membranous midline on female sternite 8 (26:2), strongly sclerotized, golden spermathecal valves (32), and a U-shaped female sternite 9 (30:2). Other synapomorphies include anepimeron setae (3) and basally directed setulae on the reservoir ducts (38:2). Two character reversals occur – epandrial halves with dorsomedial margins are moderately or weakly concave (6) and the basal lobes of female sternite 10 are strongly produced medially (31).

The relationships among the six species of the group are not completely resolved in the cladogram. *L. opaculus* and *L. slossonae* are united by the presence of a small basal tooth on the dorsal margin of the medial flange of the gonostylus (14). The monophyly of *L. marshalli* and *L. piestolophus* is strongly indicated by two high-weight synapomorphies, the thickly sclerotized base of the reservoir ducts (39) and the ring of dense caniculi apical to the striated part of the reservoir ducts (40). The other two species, *L. appalachensis* and *L. schizopygus,* form an unresolved polychotomy of four branches with the above two species pairs.

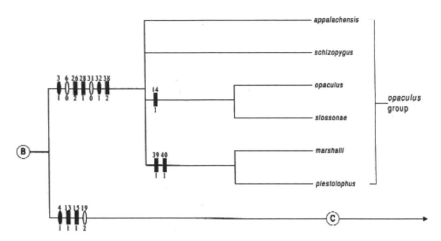

Figure 259. Cladistic relationships of the *opaculus* group of *Lasiopogon*. Characters are plotted on branches as rectangles (characters without homoplasy) or ovals (characters with homoplasy); open symbols are reversals. The numbers above the symbols correspond to the character numbers in the list on pages 295-300; the numbers below represent the character states described in the character list: 0 = plesiomorphic state; 1, 2, 3, etc. = apomorphic states. "B" indicates the position of these lineages in figures 256 and 258; "C" indicates the clade comprising the hinei and akaishii groups, which is detailed in figure 260.

The *akaishii* species group

The apical two species groups, *akaishii* and *hinei*, share densely arranged scutellar bristles (4) and modifications to the dorsal (13:1,2) and medial (15:1-3) flanges. The *akaishii* group is one of the most distinctive groups in the section; the monophyly of the four species is strongly supported by several high-weight synapomorphies – densely distributed spines on the subepandrial sclerite (10), medial processes on female sternite 8 (27) and female sternite 9 with a medial slit (30:3). Shared medium-weight characters (all homoplasious) include anepimeron setae (3), dense setae on medial lobes of the gonocoxites (7), long lateral setae in the gonocoxite setal brush (8), the dorsoventral carina present on the ventral side of phallus only (21) and strongly reduced basal lobes on female sternite 10 (31:2).

All species except the basal *L. lehri* have carinae on the hypogynial valves (29) and apically directed setulae on the striated portion of the reservoir ducts (38:1). *L. rokuroi* is the sister to the species pair of *L. akaishii* and *L. hasanicus*, which is defined by the smoothly rounded phallus apex (19:2) and the apically flattened dorsal carina of the phallus (22).

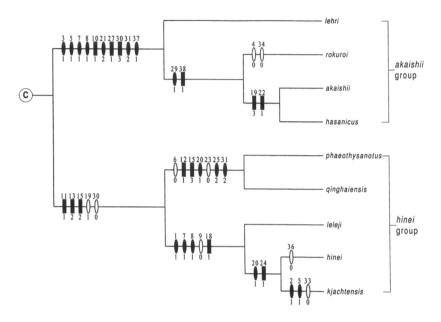

Figure 260. Cladistic relationships of the *akaishii* and *hinei* groups of *Lasiopogon*. Supporting character states are plotted on branches as rectangles (characters without homoplasy) or ovals (characters with homoplasy); open symbols are reversals. The numbers above the symbols correspond to the character numbers in the character list on pages 295-300; the numbers below represent the character states described in the charater list: 0 = plesiomorphic states; 1, 2, 3, etc. = apomorphic states. "C" indicates the position of this lineage in figures 256 and 259.

The *hinei* species group

High-weight synapomorphies indicating the monophyly of the *hinei* group include a subepandrial sclerite with medial unsclerotized areas encroaching laterally (11) and the gonostylus with the lateral part of the dorsal flange raised into an apical ridge (13:2). Each subgroup has a separate synapomorphy for the form of the medial flange. Two character reversals occur, one in the structure of the phallus apex (19), the other in the shape of female sternite 9 (30).

The group consists of two distinctive subgroups. The species pair of *L. phaeothysanotus* and *L. qinghaiensis* is united by several synapomorphies. The two most convincing are the expanded basal angles of the subepandrial sclerites (12) and the apically expanded and strongly narrowed medial flange of the gonostylus (15:3). Other synapomorphies include the apically widened phallus (20) (shows considerable homoplasy), the absence of an apicoventral lip on the phallus (25) (homoplasious with respect to *L. monticola*) and strongly reduced basal lobes on female sternite 10 (31:2). Two reversals occur – the medial margin of the epandrium is weakly concave (6) and the apicoventral flange of the phallus is complete (23).

The second subgroup contains three species (*L. leleji, L. hinei* and *L. kjachtensis*), their monophyly defined by abundant setae on the apex of the medial flange of the gonostylus (18), a dark spot on the halter knob (1), dense setae on the medial lobes of the gonocoxites (7) and long lateral bristles in the gonocoxite setal brush (8). The latter two apomorphies are shared with the *akaishii* group. There is a single character reversal – the spines of the subepandrial sclerite take on the plesiomorphic attenuate shape (9).

Within this clade, *L. hinei* and *L. kjachtensis* are sister species linked by one highly weighted character – paired projections at the base of the ventral process of the phallus (24). A second synapomorphy is the expanded phallus apex (20).

BIOGEOGRAPHY OF *LASIOPOGON*

Except for the *opaculus* section, the relationships among the species of *Lasiopogon* are unresolved. For the most part, discussions linking patterns of phylogeny with those of distribution in the majority of *Lasiopogon* species are not possible except at the species group level in Nearctic clades basal to the *opaculus* section. Many West Palaearctic, and a few East Palaearctic species have not been placed in species groups and any detailed discussion of distribution patterns in the *cinctus* clade is impossible even at the group level.

DISTRIBUTIONAL PATTERNS

The *cinctus* clade and other West Palaearctic species

Lasiopogon is strictly Holarctic in distribution (see p. 47, and map 1 on p. 5) with about equal numbers of species in the Nearctic and Palaearctic regions. West Palaearctic species all show what I consider plesiomorphic states of gonostylus and phallus form in the male and sternite 8 form in the female. Related species occur in Asia in diminishing concentration from Turkey through the Caucasus (*L. avetianae* Richter) to the East Palaearctic – Kazakhstan (*L. zaitzevi* Lehr), the Altai (*L. tuvensis* Richter), central China (*L. eichingeri* Hradský and *L.* unc-4 sp. nov.) and the mountains of Taiwan (*L. solox* Enderlein).

Some of the paucity of species known from western and central Asia reflects a lack of collecting, but active collectors and systematists, notably P.A. Lehr (1984a, 1996 and many others) and V. Richter (1962, 1968, 1972, 1975, 1977), have studied many of the areas in the vast southern borderlands of the former U.S.S.R. and have found only three species of *Lasiopogon* among a myriad of other new asilid species and even genera. The gradient of species numbers descending from west to east is probably real, although perhaps not as striking as it appears.

The few European species used in the analysis of species groups all are contained in the *cinctus* clade along with the three species of the *canus* group (*L.* can-1 sp. nov., *L. canus* Cole and Wilcox, *L. pugeti* Cole and Wilcox), which live in the northwestern Nearctic. *L. canus* has an East Beringian distribution, ranging from Alaska east to the Tuktoyaktuk Peninsula just east of the mouth of the Mackenzie River and south to the southern Yukon (Cannings 1997). *L.* can-1 sp. nov. is from central Alberta. *L. pugeti* is a species of the margins of coastal streams ranging from southern British Columbia to northern Oregon.

The presence of a West Nearctic group as the basal clade to the groups containing many (if not all) of the western Palaearctic fauna is unexpected. The distribution of *L. canus* today is undoubtedly the result of a Beringian history (Cannings 1997), suggesting that the relationship with Eurasian clades is a transpacific rather than a transatlantic one.

The distinctive *cinctus* group is European. The widespread *L. cinctus* (Fabricius) ranges from Britain and Scandinavia east to the forests of European Russia, mostly north of the Alps and the Carpathian ranges. In southern Europe this species is replaced by several species of more restricted distribution, for example, *L. macquarti* (Perris) in southwestern France and *L. peusi* (Hradský) in the Peloponnesos of Greece. Eastward, *L. intermedius* Oldenberg is known from Romania, *L.* cin-6 sp. nov. from western Turkey and *L. novus* Lehr from the Ukraine. Several undescribed species live in southeastern France, Spain and Greece.

The *apenninus* group is the sister group to the *cinctus* group, at least in the limited analysis undertaken here. These are montane species from southwestern Europe: *L. apenninus* Bezzi from the Apennines of central Italy, *L. grajus* Bezzi from the western Alps of Switzerland and Italy, and *L.* app-1 sp. nov. from both the Alps and Pyrenees.

The *montanus* group is also mainly western European in distribution. *L. bellardii* Jaennicke, ranges from the Alps of Italy and Switzerland to the Croatian coast; the range of *L. montanus* Schiner extends east to Romania and Greece.

There are about 30 additional species of *Lasiopogon* recorded from Europe and western Turkey, half of which are undescribed. Most of the undescribed species are known from only one or two localities. Similarly, almost all of the named and better known species have restricted ranges, although the amount of material available from southern France, Spain, Greece and the Balkans is surprisingly small and the true distributions of many species are unknown. Concentrations of species occur in the Pyrenees and central Spain, the western Alps, northern Italy, the central Danube basin, northern Greece and the southern Balkans, and western Turkey.

The basic pattern found in the West Palaearctic fauna is one of few widespread species and many species with restricted distributions in various mountain ranges, river valleys and coastal areas. As noted above, the number of species drops quickly east of Turkey. Except for the Siberian *L. hinei*, there are no *opaculus* section species in the Old World between the Atlantic Ocean and western China. This suggests early disjunction of the Palaearctic and Nearctic faunas and extensive subsequent speciation in a European landmass isolated from North America.

The *bivittatus* clade exclusive of the *opaculus* section

The *bivittatus* clade is mainly Nearctic with the most derived clades being East Palaearctic in origin. Apparently, there is no close connection of Nearctic and West Palaearctic faunas within this clade, although the basal groups share both plesiomorphic and apomorphic states with the *cinctus* clade.

The basal group of the *bivittatus* clade, the *aldrichii* group, includes five species from northwestern north America. *L. aldrichii* Melander is a typically distributed Cordilleran species (Scudder 1979) ranging in the Cascades and Rockies from southern British Columbia and Alberta south to New Mexico. The other species show more restricted distributions. Known only from southern Yukon, *L. yukonensis* Cole and Wilcox is an East Beringian species (Cannings 1997). *L. pacificus* Cole and Wilcox lives along coastal rivers and beaches from the Fraser River valley of British Columbia to southern Oregon; *L.* ald-1 sp. nov., replaces it on the banks of coastal rivers in northern California. Both species are part of the Pacific Coastal element of Scudder (1979) (the Pacific Maritime element of Van Dyke (1919)). *L.* ald-2 sp. nov. is an inland (Great Basin element (Scudder 1979)) equivalent in Idaho and eastern Washington.

The four species of the *terricola* group show distinct differences in their distributions. *L. terricola* (Johnson) ranges from the northeast coast of the United States northwest across the Great Plains to the Rocky Mountains of Alberta. *L. trivittatus* Melander belongs to the Great Plains element (Scudder 1979) and is known only from the high plains of western Alberta and Montana. The sister species, *L. prima* Adisoemarto and *L. septentrionalis* Lehr share a Beringian geographical relationship. The former is an East Beringian species ranging from Alaska and the Yukon south to central Alberta; the latter is recorded west of the Bering Strait from the Kolyma Basin to the Lena River and thus is a typical West Beringian taxon.

The *testaceus* group includes six Cordilleran species restricted to a rather narrow band of mountain ranges arcing from central Oregon south to central Arizona. *L.* tes-3 sp. nov. lives on the eastern slopes of the Cascade Range and adjacent plateaus in central Oregon. To the south, *L. testaceus* Cole and Wilcox, *L.* tes-1 sp. nov., and *L.* tes-2 sp. nov. range in the Sierra Nevada. *L.* tes-4 sp. nov. is known only from the San Bernardino Mountains of southern California and *L.* tes-5 sp. nov. from the mountains of central Arizona.

In contrast, the *fumipennis* group has a disjunct distribution, with *L. currani* Cole and Wilcox, a species of the Appalachian/Atlantic Coastal element (Scudder 1979) ranging from Georgia north to eastern Ontario and the others in the Cordilleran element from the western mountains. *L. fumipennis* Melander ranges from southwestern British Columbia south in the Cascades and Sierra Nevada to central California and in the Rocky Mountains to Colorado. *L. polensis* Lavigne is known only from southeastern Wyoming.

Two species remain unplaced, but probably belong among the basal clades of the *bivittatus* clade. *L. delicatulus* Melander is most similar to species of the *ter*-

ricola group. It is a Cordilleran species distributed in the Cascade Mountains from Washington to northern California. *L.* unc-7 sp. nov. belongs to the Great Basin element and lives in the arid plateaus of western Nevada.

The *bivittatus* group is restricted to the western Nearctic. Most of the species belong to the Pacific Coastal element and live along Pacific beaches or coastal rivers and streams. A few occur east of the Cascade and Coast ranges and belong to the Great Basin element; all of these, except *L. albidus* Cole and Wilcox, which inhabits arid shrub-steppe, live along streams. *L. martinorum* Cole and Wilcox and *L. ripicola* Cole and Wilcox, from Idaho and eastern Washington, are examples. Denizens of northern ocean beaches include *L. actius* Melander (ranging from the Queen Charlotte Islands to Oregon), *L. willametti* Cole and Wilcox (ranging from Vancouver Island to northern California, and also living along coastal rivers as well as streams east of the Cascade Range), *L. dimicki* Cole and Wilcox (Pacific beaches in Oregon), and *L. bivittatus* Loew (beaches from Washington south to central California). *L. arenicola* (Osten Sacken) and *L. littoris* Cole are restricted to coastal beaches in central California.

The remainder of the species are mostly from the coastal hills and western valleys of California – *L. californicus* Cole and Wilcox, *L. drabicola* Cole and Wilcox, *L. gabrieli* Cole and Wilcox, and *L. zonatus* Cole and Wilcox. There is much variation in some of these species, and some specimens from northern California, Idaho and eastern Oregon may represent undescribed species. In the south, *L.* biv-1 sp. nov. and *L.* biv-3 sp. nov. are known from northern Baja Mexico, the farthest south the genus ranges in western North America: the latter also occurs in southern California. *L.* biv-2 sp. nov. is a Cordilleran species of the central Sierra Nevada.

Some patterns in the basal groups of the *bivittatus* clade are worth noting. The five clades basal to the *opaculus* section are dominated by western (west of 100° W longitude) Nearctic species; three of the five groups and 33 of 36 species are exclusively western. *L. terricola* is almost transcontinental in the *terricola* group, but does not penetrate the Cordillera. Another member, *L. septentrionalis*, is Asian. *L. currani* is an eastern member of the *fumipennis* group. Thus, east-west disjunctions in North America are rare in these clades, occurring only twice, if *L. terricola* is considered eastern in origin.

A Beringian connection plays a role. Like the *canus* group in the *cinctus* clade, which has an East Beringian endemic (*L. canus*) and a Pacific coastal species (*L. pugeti*), the *aldrichii* group contains a Yukon endemic (*L. yukonensis*) and two south coastal species (*L. pacificus* and *L.* ald-1). This pattern of disjunction likely indicates an early, more widespread distribution of the clade separated in the Pleistocene by glaciation and survival of species in the Beringian refugium (Scudder 1979; Danks et al.1997) and the West Coast refugium (Scudder 1979; Van Dyke 1919) in the south. A different Beringian pattern, that of sister species disjunct across the Bering Strait, occurs in *L. prima* and *L. septentrionalis*, as noted above.

The two largest clades, the *bivittatus* and *testaceus* groups, show relatively restricted ranges when compared to the other groups. In the *bivittatus* group the main patterns are disjunctions along different stretches of coastline, among different river basins, or between coastal habitats and transmountain areas. In the *testaceus* group the main disjunctions are predominantly north-south among different ranges of the western mountains.

The *opaculus* section

The *tetragrammus* species group
The *tetragrammus* group is Nearctic with both eastern and western members. The basal species, *L. tetragrammus*, is northeastern, ranging from rivers in the boreal forests of James Bay to the Great Lakes and New England (map 11, p. 258). Its sister clade has *L. oklahomensis* of the southern Great Plains (map 9, p. 244) basal to two groups of three species each, one group restricted to the West, the other to the East. *L. oklahomensis* and *L. quadrivittatus* are the only representatives of the Great Plains element (Scudder 1979) in the section. *L. quadrivittatus* is widespread along rivers on the western plains from northeastern British Columbia south to New Mexico, extending west into the Rockies in the southern parts of the range (map 8, p. 216). Its sister species, *L. lavignei*, is known only from one locality in the Rocky Mountains of north central Wyoming (map 12, p. 266). The sister to this pair, *L. coconino*, inhabits the mountains of central and eastern Arizona and western New Mexico (map 11).

The sister group to these western species is Appalachian and contains three species of unresolved relationship. *L. woodorum* is the most widespread and the only lowland species, ranging from Indiana east to Virginia and Delaware (map 12). *L. flammeus* and *L. chrysotus* are mountain denizens known from only three and two males respectively, the former from North Carolina (map 12), the latter from Tennessee (map 11).

The general pattern in the *tetragrammus* group is one of disjunctions between clades showing modest diversity in the western plains and Rocky Mountains and in the eastern mountains and lowlands.

The *cinereus* species group
Three of the four species in the cinereus group are from the western mountains and plateaus – *L. cinereus* and *L. apache* are Cordilleran, *L. chaetosus* is from the Great Basin element. The fourth species, *L. shermani*, is restricted to the southern Appalachians (map 2, p. 110).

L. chaetosus, basal to the rest of the group, inhabits dry grassland and shrub-steppe from eastern Washington and central Idaho to Nevada and Utah (map 7, p. 192). *L. apache* is known only from the White and San Francisco mountains of eastern Arizona (map 2).

L. cinereus is the most widespread of the group, and is one of the most often encountered Nearctic *Lasiopogon* species. It ranges from northern British

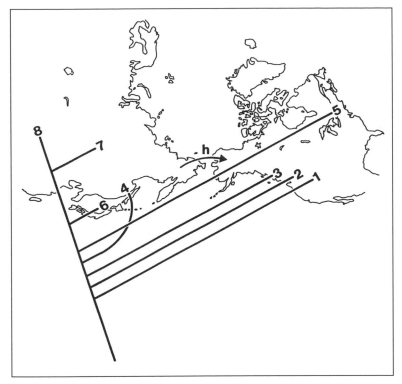

Map 13. Biogeography of the species groups in the *opaculus* section of *Lasiopogon*. Hypothesized phylogeny of the species groups superimposed on the geographical areas that they inhabit. 1, *tetragrammus* group; 2, *cinereus* group; 3, *monticola* group; 4, *terneicus* group; 5, *opaculus* group; 6, *akaishii* group; 7, *hinei* subgroup of *hinei* group; 8, *phaeothysanotus* subgroup of *hinei* group. The arrow "h" indicates the populations of *L. hinei* isolated in North America in the Pleistocene.

Columbia and the Rocky Mountains of Alberta south on the coast and in the Cascades and Sierra Nevada to central California; in the Rocky Mountains it reaches southern Utah and Colorado (map 2).

In summary, the *cinereus* group is western montane in origin with the derived species pair showing a Rocky Mountain/Appalachian disjunction.

The *monticola* species group

L. monticola is an abundant species of the Cordilleran element, ranging from the mountains of southern British Columbia south in the Cascades and Sierra Nevada to central California and in the western Rockies to central Utah (map 6, p. 178). The putative sister group is the remainder of the section, which includes only eastern Nearctic and eastern Palaearctic clades.

The *terneicus* species group

L. terneicus is the sole species in the group and is known from only a few localities in Primorskiy Kray (Far-eastern Russia) (map 10, p. 251).

The *opaculus* species group

The *opaculus* group is predominantly of the Appalachian element (Scudder 1979) in eastern North America, with species living along streams in the mountains or, frequently, in lowland areas distant from high elevations. The basal polychotomy (four branches) in the group makes it difficult to associate distribution with phylogeny.

L. opaculus is the most widely ranging species, recorded from such diverse places (among others) as the plains of central Iowa, the beaches of Lake Huron, the southern foothills of the Appalachian Mountains in Mississippi, and the mountains themselves in Georgia and the Carolinas (map 7, p. 192). Its sister species, *L. slossonae*, is widespread in the northeast, but not so widely ranging; in the western part of its range in Michigan and Ohio it is sympatric with *L. opaculus*. Eastward, it is common along streams from West Virginia to New England. There is one record from Quebec (map 9, p. 244).

The other resolved clade, *L. marshalli* plus *L. piestolophus*, is more southern in distribution. The former is recorded from the New River of Virginia (map 8, p. 216), the latter from the lowlands of southern Alabama (map 9).

L. appalachensis is restricted to the Appalachians of Tennessee, Kentucky and West Virginia (map 6, p. 178). The final species in the group, *L. shizopygus*, is mainly known from the mountains and piedmont of Georgia and the Carolinas, with an outlier in northern Mississippi (map 8).

Speciation in the distinctive *opaculus* group and subsequent dispersal of species evidently has occurred only within the general area of the group's origin, the Appalachian Mountains and surrounding lowlands. To find the group's closest relatives, however, one must look to eastern Asia.

The *akaishii* species group

The four known species of the *akaishii* group live in the Japanese archipelago and the adjacent mainland of Russia. The basal species, *L. lehri*, is rather common along streams in the southern half of Primorskiy Kray (map 5, p. 155). The sister clade of three species consists of two Japanese species and one from Russia. *L. rokuroi*, an inhabitant of Honshu and Kyushu, is the sister of the species pair of *L. akaishii* from Hokkaido and the Kurile Islands, and *L. hasanicus* from mainland Primorskiy Kray (map 3, p. 127)

The basic pattern is one of vicariance across the Sea of Japan or even the Korea Strait. The presence of members of the *akaishii* group on the Korean peninsula and in Manchuria could be expected. I have not seen Korean material, and even the number of specimens from Japan is minimal. The cladistic pattern requires two vicariance events, one separating *L. lehri* on the mainland from the ancestor of the remainder of the group in Japan, the other splitting *L. hasanicus* from the Japanese taxa (map 14, below). Dispersal could also be invoked to help explain these relationships.

The *hinei* species group

The *hinei* group consists of two main clades, a species pair from western China and a widely distributed clade of three species, one of which has recolonized the Nearctic.

L. phaeothysanotus and *L. qinghaiensis* are represented by only four specimens from the high cold plateaus of western Qinghai Province in China (map 5, p. 155, and 10, p. 251), more than 600 km north of Lhasa. The ratio of numbers of *Lasiopogon* specimens to numbers of species (9:4) that I have seen from China (from the Academia Sinica and European museums – all specimens are from Qinghai Province) hints that many species remain to be discovered in that huge region. This, in turn, implies that any explanations of the relationships within the *hinei* group are preliminary at best.

L. leleji is known from a single male from Primorskiy Kray (map 10). The Russian *L. kjachtensis* is recorded from the Buryatskayan steppes between Lake Baikal and the Mongolian border (map 5). Its sister species, *L. hinei*, has the largest range of any *Lasiopogon*, living from the tundra of Arkhangelskaya Oblast in European Russia east to central Alberta in northwestern North America (map 4, p. 137). *L. hinei* has a typical Palaearctic-East Beringian distribution resulting from the stranding of the East Beringian population in Alaska and Yukon

Map 14. Biogeography of the *akaishii* group of *Lasiopogon*. Hypothesized phylogeny of the *akaishii* group superimposed on the geographical areas of eastern Asia inhabited by the species: 1, *L. lehri*; 2, *L. rokuroi*; 3, *L. hasanicus*; 4, *L. akaishii*.

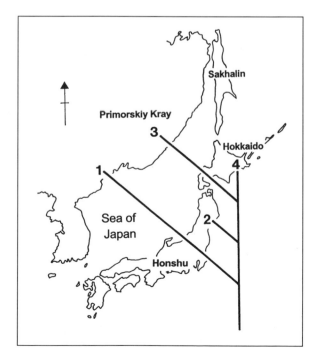

when the Bering land bridge disappeared at the close of the Pleistocene (Cannings 1997) (map 13, p. 313).

THEORETICAL PREAMBLE

The cladistic biogeographic method (vicariance biogeography) is an integration of geological and geographical data with hypotheses of phylogenetic history of organisms. In theory, the historical relationships of geographical areas should be reflected by the phylogenetic relationships of taxa that occur in these areas (Brundin 1972, Ball 1976, Platnick and Nelson 1978). Thus, the goal of biogeography is the search for congruence between phylogenetic patterns and earth history (Nelson and Platnick 1981, Humphries and Parenti 1986, Wiley 1988); any symmetry adds credence to the patterns discovered in both disciplines (Yeates and Irwin 1996).

Both vicariance and dispersal are valid mechanisms used to explain the patterns found in the distribution of organisms. Vicariance is the fragmentation of populations or larger groups, or even whole biotas through major environmental changes, mainly the creation of barriers to movement and genetic interchange. The tectonic movement of terrains, orogeny, sea-level changes, glaciation and major shifts in climate are critical vicariant events that shape the evolution and distribution of species. Dispersal is the movement of organisms over the earth, either passively or actively. It is often a haphazard phenomenon mediated by wind or water or stimulated by behavioural factors. In general, dispersal involves individuals or small groups.

Cladistic biogeography incorporates hypotheses of vicariance as much as possible and dispersal as little as possible. Its proponents advocate it as a parsimony method because the number of hypothesized dispersal events is minimized (Maddison 1993). More importantly, vicariance hypotheses are testable by comparison with physical historical events (Wiley 1988), but past dispersals are mostly impossible to document and explanations invoking them are immune to falsification (Rosen 1978).

The explanations that I offer for the geographical patterns in *Lasiopogon* involve a mixture of vicariance and dispersal, with stress placed on the former. The explanations are, of course, testable hypotheses, attempts to correlate the phylogenies outlined in the cladograms and geophysical events such as those described by Matthews (1979, 1980), Noonan (1986, 1988a) and Hallam (1981).

GEOLOGICAL AND CLIMATIC HISTORY

Hallam (1981), Noonan (1986, 1988a, b), Raven and Axelrod (1974), and Yeates and Irwin (1996) described the evolution of the continental landmasses relevant to this study. By the end of the Jurassic, Pangaea was divided into northern Laurasia and southern Gondwanaland by the Tethys Sea. Epicontinental seas then separated Laurasia into two land masses: Asiamerica (Asia plus western North America joined at high latitudes) and Euramerica (Europe plus eastern North America). Europe and Asia were separated by the Turgai Sea from the late Jurassic (163-169 mybp) to the late Eocene (about 30 mybp). The Mid-continental Seaway divided eastern and western North America until the early Tertiary (60-70 mybp), when transcontinental interchange of biotas was again possible.

The opening of Atlantic Ocean began from the south in mid-Cretaceous times (90-95 mybp), but evidently, several land connections between Europe and eastern North America persisted in the north (Matthews 1979, 1980), the latest one, the Thulean route, until the Miocene, about 20 mybp (Noonan 1988a). But cooling in the Eocene and Oligocene about 30-35 mybp probably acted as a climatic barrier to warm temperate species in these regions.

Asia was connected to North America through the early Tertiary; movement of forest faunas and floras was not fully disrupted until the Pliocene, about 3 mybp (Matthews and Telka 1997), when the Bering Strait and Bering Sea initially developed a modern configuration. The Beringian land connection was present intermittently when sea levels fell during periods of cold climate and finally disappeared at the end of the Wisconsinan glaciation.

Warm, azonal climates characterized the northern hemisphere during early Tertiary and tropical or subtropical forests occurred near the Bering land bridge from the Paleocene to the mid Eocene (Matthews 1980). The northern hemisphere did not cool dramatically until the Eocene-Oligiocene boundary (35 mybp).

Mixed forests grew in the Canadian arctic in the mid Miocene (about 16 mybp) and apparently also near the land bridge in East and West Beringia. Diverse coniferous forests were present near the bridge until the Pliocene breach (Matthews and Telka 1997). There is no definite evidence of tundra in North America at the Miocene-Pliocene boundary, and probably tundra did not form in the Beringian region until as late as 2.5 mybp, after the bridge had disappeared (Matthews and Telka 1997).

Climatic cooling around the Eocene-Oligocene boundary (Askevold 1991) and the uplift of the western Cordillera from the late Oligocene through the Pliocene (Stanley 1989) initiated striking modifications to the environments of North America. With the subsequent cooling and drying of the mid continent, the once widespread broadleaved forests shifted eastward and major disjunctions between eastern and western biotas occurred. Most mesophytic temperate floras were isolated in eastern and western North America. In the West, an arid-adapted flora, including grasslands, developed in the lowlands in the Miocene and mesic conifer forests withdrew to higher elevations (Cronquist 1978, Lafontaine 1982, Axelrod and Raven 1985). In the Great Basin, the flora and fauna had a modern aspect by the Miocene; subsequent changes in the biota "mainly reflect evolution and migration at the level of species and, to a lesser extent, to genera, in response to regional conditions and repeated fluctuations in climate" (Cronquist 1978).

Widespread glaciations in the Pleistocene radically affected the northern areas of the Holarctic Region and have further complicated distributional patterns of plants and animals. Present patterns of species in previously glaciated regions of North America result mainly from post-glacial dispersal from ice-free areas to the south or from unglaciated refugia. The largest and best documented of these, the Beringian refugium (Danks *et al.* 1997), plays an important role in the history of some *Lasiopogon* taxa.

Glaciation also affected unglaciated environments. In western North America, glacial episodes produced relatively lower temperatures and higher humidity and rainfall; mesic-adapted plants such as conifers, many with northern affinities, expanded at the expense of xeric-adapted species (Cronquist 1978), although even during glacial periods mesic forests were not widespread in lowland areas (Lafontaine 1982). In the warmer, drier interglacial periods, xeric-adapted species expanded their ranges. Such fluctuations created unstable islands of habitat out of mountains and valleys. During the Wisconsinan glacial maximum, montane vegetation spread to lower elevations in the Rocky Mountains, eliminating the connections of arid habitat between the Great Basin and the Great Plains to the east.

The northern half of the Great Plains south of the ice front during the last glacial maximum was apparently covered by a spruce-dominated forest. Pine savannah occurred in warmer, drier parts of the southern plains such as the Nebraska Sandhills; these areas presumably acted as refugia for xeric-adapted species (Hubbard 1973, Lafontaine 1982)

In Asia, there are several major geological events that might bear on the history of *Lasiopogon*. India, rifted from Gondwanaland, collided with the Asian mainland in the mid Eocene, about 45 mybp. The resulting upthrust of the Himalayas occurred through the Miocene and reached its strongest phase in the Pliocene-Pleistocene (McKenzie and Sclater 1973).

The Japanese Basin and the Sea of Japan developed through seafloor spreading rather rapidly in the early part of the mid Miocene (about 15 mybp) creating the Japanese archipelago (Honza 1995, Tsuchi 1997). Since the Miocene, the history of the islands has been complex, with numerous land connections to the mainland via the Ryukyu Islands, the Korean Strait and Hokkaido-Sakhalin (La Perouse and Tatarskiy straits) (Minato et al. 1965, Taira 1990). The earliest and broadest connections were with the Korean peninsula on and off between the Miocene and Pleistocene. In the Pleistocene, there were major links among the islands and with the mainland, including via Sakhalin; the islands were not completely severed from the mainland until the Holocene (Minato et al. 1965, Taira 1990).

HISTORICAL BIOGEOGRAPHY OF *LASIOPOGON*

Fossil Asilidae

The earliest putative fossil asilid, *Asilus ignotus* Westwood, is reported from the early Jurassic of England (187-208 mybp), but Evenhuis (1994), in documenting this record, doubted its accuracy. The oldest known specimen definitely assigned to the Asilidae is from the Santana Formation of the early Cretaceous of Brazil, which dates to about 110 mybp (Grimaldi 1990). *Araripogon axelrodi* Grimaldi has plesiomorphic asilid wing venation, a strong mystax and leg bristles and from the description, appears similar to some modern members of the Stenopogoninae. Unfortunately, the genitalia are not preserved.

The Green River shales of the Utah Eocene (36.6-57.8 mybp; or more accurately, from the later part of the early Eocene to the later part of the mid Eocene, 47-52mybp) (Wilson 1978) contain specimens described as *Asilus* (Cockerell 1921) from the Asilinae, considered by most, if not all, authors to be the youngest subfamily of robber flies. *Proctacanthus*, also in the Asilinae, is reported as early as the Eocene-Oligocene from England (Evenhuis 1994).

The Florissant shales of Colorado (Oligocene, 23.7-36.6 mybp) present a wide array of asilids (23 species in 15 genera), many of derived status (James 1939, Grimaldi 1990, Evenhuis 1994). Almost all the specimens have been assigned to extant genera, including: *Leptogaster* and *Tipulogaster* (Leptogastrinae); *Cophura, Lestomyia* and *Nicocles* (Dasypogoninae); *Ceraturgus, Dioctria* and *Holopogon* (Stenopogoninae); and *Asilus, Machimus* and *Philonicus* (Asilinae). *Senoprosopis* is an example of a genus recorded that is today restricted to the neotropics. The Florissant fossil site was a subtropical savannah woodland (Cronquist 1978).

Other asilids come from North American and European fossil beds and amber sources from the Oligocene to the Pliocene. No fossils of *Lasiopogon* are known. However, given the mainly plesiomorphic nature of the genus its origin likely predates the Eocene shales, which, as outlined above, contain examples of other more derived lineages.

The Jurassic and Pangaea

Grimaldi (1990) put the probable origin of the Asiloidea near the Jurassic-Cretaceous boundary (about 144 mybp) and that of the Brachycera in the Middle Jurassic. He noted that "The influence of continental tectonics on distributions must be regarded, at least at the outset of biogeography studies on flies, as anything but trivial". In slightly later studies, Yeates and Grimaldi (1993) and Yeates and Irwin (1996) postulated that the Asiloidea arose in Pangaean times, that is, before the break-up of Pangaea in the late Jurassic. The latter work demonstrates that the distribution of Apioceridae and plesiomorphic Mydidae exhibits complete congruence with the geological vicariance of the continents where they live today, giving strong support to the hypothesis that the ancestral apiocerids and mydids spread from Laurasia to Gondwanaland before the fragmentation of Pangaea in the Jurassic. The late Jurassic (about 160-144 mybp) was the latest the vicariance between Laurasia and Gondwanaland taxa could have occurred (Yeates and Irwin 1996).

Yeates (1994) gave strong support to a clade containing the Asilidae, Apioceridae and Mydidae, and suggested that an Apioceridae + Mydidae clade is possibly the sister group to the Asilidae, a hypothesis strengthened by Yeates and Irwin (1996). This suggests that the Asilidae is at least as old as the apiocerids and mydids, and a Pangaean origin for the family is reasonable, even though the earliest known fossil asilid is from the early Cretaceous. The extant stenopogonines *Tillobroma* (southern South America) and *Hypenetes* (southern Africa) possibly diverged after the separation of southern Africa and South America. If so, the populations have been distinct for well over 100 million years, given the estimates of continental separation (Raven and Axelrod 1974). Despite this, the two genera are so close morphologically that they were long considered congeneric.

The lack of any phylogenetic hypothesis dealing with plesiomorphic asilid lineages such as the genera now contained in the Stenopogoninae is a major problem in estimating the age of *Lasiopogon*. Much more needs to be learned about the relationships among *Lasiopogon* and its relatives. Nevertheless, in the absence of contrary evidence, the fact that Holarctic *Lasiopogon* and Australasian *Bathypogon* share apparently unique and significant apomorphies in the structure of the epandrium suggests that they may be sister groups resulting from the vicariance of Laurasia and Gondwanaland.

My best explanation for the phylogeny of extant *Lasiopogon* taxa uses generalized geological events beginning in the Tertiary. Major Jurassic/Cretaceous geographical phenomena, such as the Mid-continental Seaway dividing North America, do not readily help explain the formation of *Lasiopogon* lineages. The lack of evidence for any evolutionary divergence in *Lasiopogon* during approximately 120 million years between the late Jurassic and early Tertiary might be considered a drawback to a hypothesized Jurassic origin of the genus.

Early Vicariance and the *cinctus* clade

The original *Lasiopogon* fauna, sharing the plesiomorphic states of phallus and gonostylus, was probably widespread in Eocene times after the disappearance of the Mid-continental Seaway in central North America. The nature of the vicariance event that might have produced the *cinctus* and *bivittatus* clades is unclear, but because *cinctus* is primarily western Palaearctic and *bivittatus* is Nearctic, the breaking of the land connection between Europe and North America at the

Eocene-Oligocene boundary is a candidate. Continental configurations and climates suggest an approximate age of 30 to 35 mybp for vicariance of Euramerican insects (Noonan 1988a). Subsequently, the two main clades radiated especially in the western Palaearctic and western Nearctic.

Many species from the western Palaearctic have not been analysed in my study and the biogeography of the fauna in that region is difficult to assess. Environmental change resulting from Oligocene cooling and the uplift of complex mountain systems such as the Alps, Apennines, Pyrenees from the Miocene to the Pleistocene undoubtedly stimulated the production of a number of lineages. For example, one of these may have been the vicariance event that divided the lowland *cinctus* group from the montane *apenninus* group. Pleistocene glaciations and the concomitant fluctuations of temperatures, sea levels and vegetation zones must have also been important in determining present-day species distributions and even some speciation events.

The western Nearctic distribution of the *canus* group, which is basal to the *cinctus* clade, suggests a Beringian origin, but there are no clues as to the age of the lineage. Ancestors of the *canus* group may have inhabited the eastern Palaearctic in the past, although no close relatives apparently occur there now.

L. canus is restricted to Alaska, the Yukon and extreme northwestern Northwest Territories, and is the most northerly distributed asilid in the Nearctic. It has a range typical of a species that spent the Wisconsinan glaciation in eastern Beringia, then dispersing only short distances south and east after the ice melted (Cannings 1997). *L. canus* is clearly adapted to cold environments and can live north of the tree line in shrub-tundra habitats as well as in taiga. Its undescribed sister species, *L.* can-1 sp. nov., is known only from central Alberta. Probably the ice sheets divided the original ancestral population and allowed differentiation of the two extant species. *L.* can-1 sp. nov. would then have spent the glacial maximum south of the ice and dispersed northwards following the retreat of the glaciers.

The sister to this pair is *L. pugeti*, which lives along Pacific coastal rivers from Puget Sound to the Columbia River. It lives in warmer, moister habitats than its relatives and has probably lived well south of Beringia since before the Pleistocene. The most logical explanation for the pattern in the *canus* group has an ancestral species occupying stream edges in the forests and open areas of Beringia during the Miocene or Pliocene before the inundation of the land bridge about 3 mybp (Matthews and Telka 1997). The cooling of the region and the reduction of the mesic conifer forests to taiga or even tundra, the latter appearing about 2.5 mybp (Matthews and Telka 1997), might have acted as the vicariance event that split the lineage in two. *L. pugeti* remained in warm coastal forests to the south, the ancestor of *L. canus* and *L.* can-1 sp. nov in taiga and shrub tundra to the north.

The *bivittatus* clade

General Trends
The bi*vittatus* clade appears to have originated in the Nearctic. The ages of the basal species groups are a matter of speculation, as are the vicariance events that produced them. If the clade arose at the onset of the Oligocene, as suggested above, much of the later radiation of the taxa could be explained by the environmental upheavals resulting from the mountain uplift, drastic cooling and other climatic fluctuations that characterized the Oligocene, Miocene and more recent times.

I discuss the *opaculus* section later (pp. 324-28), but a point should be made here about the age of the dichotomies between the North American and Asian clades (the *terneicus* group and the rest of the section; the *opaculus* group and the clade containing the *hinei* plus *akaishii* groups). The latest age of these splits can be estimated; this helps in the dating of events that affected the more basal parts of the *bivittatus* clade.

The clades are largely inhabitants of mesic mixed forests in eastern North America and eastern Asia. Presumably, these are remnants of a much wider ranging biome (but likely not a uniform one in time or place (Matthews 1979)) that stretched across the northern Holarctic in the Tertiary (Allen 1983). Cooling in the Oligocene and Miocene resulted in the disruption of this forest as a conduit for warm-adapted forest insects by late Miocene (Matthews 1980). Thus, the Asian clades of the *opaculus* section must have inhabited eastern Asia before the late Miocene, and perhaps earlier.

If these vicariance events occurred in the mid Miocene, the earlier branching of the species group lineages in the *bivittatus* clade were likely Oligocene to mid Miocene in age. Judging from present distributions, most of this early evolution occurred in, or adjacent to, the western mountains, which were forming at the time. Axelrod and Raven (1985) believed that the Cordilleran flora had become essentially modern by the mid Oligocene (27 mybp) and many representative temperate genera of trees were even present by the early Eocene. By the mid Oligocene many near modern species of pines grew in the Cordillera (Axelrod 1986). In western montane habitats, *Lasiopogon* is primarily a genus of mid-elevation montane or subalpine coniferous forests. The existence of the major tree constituents of the modern flora (*Abies, Pinus, Picea*) in the region at the time suggests that associated modern insect groups might also be present (Askevold 1991).

Basal Groups
Few of the phylogenetic relationships within the basal species groups of the *bivittatus* clade are known and little can be said about the biogeography of the smaller clades within them. The basal *aldrichii* group may have split from the lineage leading to the rest of the clade as result of the orogenies of the late Tertiary. Except for the widespread *L. aldrichii* of the central Cordillera, the group is dis-

tinctly northwestern, more than any other basal group. It has montane and Pacific coastal components separated by the Cascade Mountains, which reached maximum uplift in the Pliocene and Pleistocene (Lafontaine 1982). One of the species, *L. yukonensis*, is restricted to the southern and central Yukon and distributionally qualifies as an East Beringian species. It appears to be most closely related to the Pacific coastal species in the group, *L. pacificus* and *L.* ald-1 sp. nov. It may represent a relict population of a once more widespread southern species, derived in a similar manner to the southern steppe Noctuidae documented in Beringia by Lafontaine and Wood (1988).

The *terricola* group's relationships are mostly unresolved. *L. terricola* and *L. trivittatus* are both found east of the Rocky Mountains, but the former is western and the latter is primarily eastern; details of their phylogenetic relationship are unknown. *L. terricola* has probably invaded the Great Plains from the East since the Pleistocene, mainly along river drainages; this is a common pattern seen especially in many aquatic insect groups (Lehmkuhl 1980).

L. prima is East Beringian and *L. septentrionalis* is West Beringian; this is the only American/Eurasian species pair known in *Lasiopogon*. Their distribution to the northwest, across the Rocky Mountains from the rest of the group, suggests mid Tertiary uplift of the northern Rockies was responsible for the vicariance of their ancestor from the rest of the lineage. The species are primarily inhabitants of taiga, and *L. prima*, at least, has followed treed river valleys to the Arctic coast. It also has expanded its range southward along the mountains into western Alberta since the last glaciation. The species pair is likely at least three million years old, and I assume that its ancestor was a Nearctic species. Sister species from the taiga in East and West Beringia represent a vicariance pattern that results from Pliocene separation of forest habitats in Beringia after climatic cooling (Lafontaine and Wood 1988) or by the initial Pliocene formation of the Bering Strait itself (Matthews and Telka 1997). This is seen in a number of taxa, but is perhaps best documented in the noctuid moths (Lafontaine and Wood 1988).

The split between the ancestor of the *testaceus* group and the rest of the *bivittatus* clade must also be a product of mountain uplift. The species are confined to the narrow bands and islands of mountain ranges stretching from the southern Cascades through the Sierra Nevada and San Bernardinos to central Arizona. Even though the phylogenetic pattern suggests that the origin of the group is rather old, some of the modern species may be much younger, dating from the time of maximum uplift of the Sierras and Cascades in the Pliocene and Pleistocene (Lafontaine 1982).

Kavanaugh (1988) documented examples of *Nebria* (Carabidae) in a Cascade/Sierran montane element that show vicariance patterns in these mountains. He postulated that subspecies pairs of *Nebria* developed through isolation in unglaciated regions of the Sierras during the last glacial period. Orogeny and glaciation were probably the main factors of vicariance acting on *Lasiopogon* in these western ranges. Clarification of the vicariance patterns must wait a phylogenetic analysis of the interesting *testaceus* group.

The *Lasiopogon* species of the *testaceus* group or *bivittatus* group (e.g., *L. biv*-2 sp. nov.) that evolved in the Sierras are more-or-less restricted to these ranges. Harper et al. (1978) noted that Sierran taxa evolved in less continental environments (moister and less thermally variable) than those of the Rocky Mountains and Great Basin ranges. This factor may be important in confining the more western mountain species to the Sierras; Johnson (1977) considered this true for birds.

The *fumipennis* group has a sister relationship to the rest of the *bivittatus* clade, and the origins of both groups were probably in the western mountains. Two of the three species in the group, *L. fumipennis* and *L. polensis*, are Cordilleran; *L. currani* is an Appalachian/Atlantic coastal species. Although no phylogeny has been developed for the group, phallus morphology suggests that *L. fumipennis* and *L. currani* are sister species. Speciation may have occurred early as a result of vicariance across the drying mid-continent in the Miocene; alternatively, ancestral populations may have been severed much later by the environmental repercussions of Pleistocene ice advance.

In the species group cladogram (figure 251), the *bivittatus* group is basal to the *opaculus* section. The group is Pacific coastal in origin, likely initially isolated there in the early Miocene orogenies that began the uplift of the Sierra Nevada and Cascades. The *actius* subgroup shows strong disjunct distributions across the Cascades, with some species living on coastal beaches and others along rivers on the interior plateaus. Kavanaugh (1988) showed a similar pattern in the *eschscholtzii* species subgroup of the carabid genus *Nebria*. In *Lasiopogon*, these are probably the result of later vicariance events associated with Pliocene or Pleistocene orogeny or the effects of glaciation on habitats. Species related to *L. californicus* range mostly in lowland California or along the western foothills of the Sierras and mountains of southern California. None of the interrelationships among the species has been clarified.

The *opaculus* section

The *opaculus* section, as analysed in the cladogram of species (figure 256), comprises the most derived clades of *Lasiopogon*. The general phylogenetic pattern has western Nearctic clades basal to branches of eastern Nearctic and eastern Asian taxa (map 13, p. 313). Equivalent patterns are seen in other groups of insects, such as the caddisflies. In *Rhyacophila*, the western North American *divaricata* branch is basal to the eastern North American *torva* and eastern Palaearctic and Oriental *nigrocephala* branch (Hamilton and Morse 1990).

The *tetragrammus* group and the remainder of the clade arising along with it probably originated about the early Miocene. The initial dichotomy is east-west, perhaps produced as a result of vicariance across a drying mid-continent. One vicar is *L. tetragrammus*, a common northeastern species now inhabiting lands once covered by the Wisconsinan ice sheets, and the sole eastern species to penetrate the boreal forest (map 11, p. 258); the other is the lineage leading to the remainder of the group.

L. oklahomensis is basal to this lineage; it is a prairie (probably riparian) species apparently restricted to the Oklahoma plains (map 9, p. 244). Such endemics are rare, for much of the plains fauna originated in surrounding biomes (Lehmkuhl 1980). Perhaps it once was more widespread in the grasslands or savannah of the plains and may now be a relict from one of the Wisconsinan dryland refugia mentioned by Hoffman and Jones (1970) and Lafontaine (1982). The rest of the lineage divides into two subgroups; these groups form an east-west pair separated by the Great Plains. Such disjunct patterns are common, the result of orogeny and periodic barriers resulting from alternating climatic conditions from the Eocene right through to the post-Pleistocene (Noonan 1988b).

L. coconino is the sister to the *L. quadrivittatus/L. lavignei* species pair. This disjunction is comparable to Noonan's (1988b) Zone 10 barrier, which separates sister taxa in the southern Rockies from those of the middle Rockies. *L. coconino* is isolated on the central plateaus and mountains of Arizona and New Mexico (map 11). *L. quadrivittatus* is a wide-ranging species of the western Prairies and eastern Rockies (map 8, p. 216); *L. lavignei* is known from one location in the Rockies of Wyoming (map 12, p. 266).

The vicariance of *L. coconino* and the species pair cannot be dated. It may have resulted from mountain uplift or from late Pleistocene glacial activity. The biogeography of the insect fauna of the American Southwest in the Pleistocene is a giant puzzle to be solved; climatic changes over the complex physiography resulted in a myriad of speciation events and shifts in species distributions (Elias 1994, pp. 102-105). Ball (1965) postulated Pleistocene speciation in *Scaphinotus* carabid beetles in the American Southwest. Species are restricted to coniferous forests (as are *Lasiopogon* species in the region) with desert regions isolating them. During cooler, moister glacial periods, the insects could disperse through lowland habitats but during the warmer interglacials they were isolated in the mountains. The separation of *L. quadrivittatus* and *L. lavignei* might have happened in the same manner.

In the mountains of northern New Mexico, populations of *L. quadrivittatus* are morphologically distinct from populations farther north. They are darker and less setose, and were originally described by Cole and Wilcox (1938) as a separate species, *L. aridus*. These may represent relictual populations displaced southward in the Pleistocene. Askevold (1991) considered this the case for several species of *Plateumaris* (donaciine chrysomelid beetles) occurring in the same mountain ranges, such as the Jemez Mountains of New Mexico.

The eastern clade is a trichotomy of *L. woodorum, L. chrysotus* and *L. flammeus*. According to locality records, *L. woodorum* is separated into two populations, one on either side of the Appalachians (map 12). This may be an artifact of collecting. But if it is a real pattern, it may reflect the effect of glaciation; severe conditions in the mountains may have divided a once more widely distributed population. *L. chrysotus* lives in the central Appalachians (map 12), *L. flammeus* in the south (map 11, p. 258). Nothing much can be said of their biogeography without a more resolved phylogeny.

The second and main lineage of the section is of western Cordilleran origin. The *cinereus* group is basal to the remaining clades. As it stands, it is mostly a western montane group with the derived *cinereus/shermani* pair exhibiting a Cordilleran/Appalachian disjunction. The basal species, *L. chaetosus* lives in arid Great Basin shrub-steppe (map 7, p. 192) and was probably derived from a western ancestor that also gave rise to the more montane lineage of *L. apache* and the *L. cinereus/L. shermani* pair. *L. apache* is evidently confined to the White Mountains of Eastern Arizona (map 2, p. 110), probably originating through montane vicariance with the ancestor of the *L. cinereus/L. shermani* pair. The divergence of these sister species appears to be a typical east-west vicariance, but may have been a rather early one because the two species have many morphological differences. *L. shermani* is a southern Appalachian/Piedmont Plateau species (map 2). *L. cinereus* is a wide-ranging Cordilleran species (map 2) and one of the most variable *Lasiopogon* species. The populations living along streams in the coastal hills from Oregon to northern California are especially different from the rest (originally described as *L. atripennis* Cole and Wilcox) and probably spent at least the Wisconsinan glaciation in the Pacific coastal refugium, separated from other populations.

L. monticola and *L. terneicus* are sequential basal taxa to the rest of the section. *L. monticola* is a widespread and common montane species in the Cordillera (map 6, p. 178). Its position in the phylogeny is consistent with the early radiation of western taxa in the changing environments of the West. But *L. terneicus* is an eastern Palaeartic species from far-eastern Russia (map 10, p. 251) and its position basal to the *opaculus* group is problematic in that it requires two successive America/Asia vicariance events in the shrinking mixed forests of the north before the mid Miocene.

The *opaculus* group is Appalachian in origin, and likely arose as a vicar of the Asian clade, as noted above, by the shrinking of the Holarctic mesic mixed forests before the mid Miocene. There are many examples of insects showing this pattern of eastern North America/eastern Asia disjunction, and many of them are evidently very old. The closest relatives of an Appalachian *Panorpa* (Mecoptera) species group live in China (Byers 1988). Sinclair and Saigusa (1997) documented a sister-species pair of the seepage fly genus *Trichothaumalea* from Japan and the Appalachians. The subgenus *Calorhamphomyia* of the Empididae shows the same pattern (Saigusa 1997).

Explanations of the pattern of evolution in the *opaculus* group are hindered by the poor cladistic resolution of the taxa. Nevertheless, if the ranges documented are more-or-less accurate and major gaps are not the result of a lack of collecting, a few points can be made. The sister species *L. marshalli* and *L. piestolophus* show disjunction across the length of the southern Appalachians; the former is known from the New River in the uplands of southwestern Virginia (map 8, p. 216), the latter from riparian habitats in the Alabama lowlands (map 9, p. 244). The two species are very close morphologically and difficult to tell apart. They may be young species, the result of vicariance during Pleistocene glaciation.

The *L. opaculus/L. slossonae* species pair is much more widespread and clearly there has been considerable dispersal since the Wisconsin glaciation. Although the two species are sympatric over wide areas, especially in the Midwest, they are allopatric in others. *L. slossonae* (map 9) is common in the northern Appalachians (New England), but *L. opaculus* (map 7, p. 192) is absent from the region; *L. opaculus* is common at high elevations in the Carolinas and Georgia, but *L. slossonae* is absent. These populations may represent the remnants of another north-south vicariance event. About all that can be said is that the complex history of mountain uplift, climatic change, glaciation and shifting habitats has produced at least six closely related species in the *opaculus* group.

The Asian clade of the *opaculus* section evidently diverged into two lineages in eastern Asia sometime before the mid Miocene. There are probably some unknown species in the *hinei* group in China that might provide clues to the history of these events, but as yet there is little to go by. The vicariance might have occurred in what is now Far-eastern Russia; both the *akaishii* and *hinei* groups have members in Primorskiy Kray. The stimulus for this vicariance is unknown, although all members of the *hinei* group except *L. leleji* are distributed to the north or east in regions influenced by much more continental climatic regimes. Miocene cooling perhaps split the ancestral population into warm-adapted and cool-adapted taxa, ancestral to the *akaishii* and *hinei* groups respectively.

The *akaishii* group inhabits mesic mixed forest and adjacent habitats in Primorskiy Kray and Japan, although *L. akaishii* itself ranges into the southern Kurile Islands (map 3, p. 127) where a cool maritime climate prevails (Tatewaki 1957).

The Japanese Basin and the Sea of Japan developed through sea-floor spreading rather rapidly in the early part of the mid Miocene, about 15 mybp, creating the Japanese archipelago (Honza 1995, Tsuchi 1997). This suggests that the disjunction of the basal *L. lehri* (mainland, map 5, p. 155) and the rest of the group is the result of this island building. Map 14 (p. 315) illustrates the phylogeny of the group in relation to the geographical distribution of the species.

L. rokuroi is more widely distributed on the larger islands (map 3) except Hokkaido, where *L. akaishii* occurs. The distribution and phylogenetic patterns indicate vicariance or dispersal of the ancestor of the *L. akaishii/L. hasanicus* species pair across the narrow Tsugaru Strait between Honshu and Hokkaido. Pleistocene sea-level fluctuations bridged and broke this waterway several times during the Pleistocene (Minato *et al.* 1965). During this period there were extensive connections with the Russian mainland through Sakhalin, which would have allowed the dispersal of fly populations off the islands. Subsequent separation of the islands from the Russian mainland in the Holocene would then have produced the vicariance required for the speciation of *L. akaishii* and *L. hasanicus;* the latter lives on the mainland today (map 3).

The *hinei* group has two subgroups: a pair of species on the high cold plateau of Qinghai in western China, and the widespread clade containing the mainly Siberian *L. hinei* (map 4, p. 137), *L. kjachtensis* of the Trans-Baikal steppes (map

5, p. 155) and *L. leleji* (map 10, p. 251) of the Primorskiy forests. Separation of the two subgroups probably occurred with the orogeny of the Himalayan massif. India, rifted from Gondwanaland, collided with the Asian mainland in the mid Eocene, about 45 mybp. The resulting upthrust of the Himalayas occurred through the Miocene and reached its strongest phase in the Pliocene-Pleistocene (McKenzie and Sclater 1973). Many montane taxa evolved during this uplift; Cumming (1989) postulated a vicariance of certain species of *Symmorphus*, a genus of Eumenine wasps.

By interpolation, the origin of the *L. leleji* lineage and the *L. hinei/L. kjachtensis* pair is probably near Pliocene in age, but the potential cause of the split is unknown. The *L. hinei/L. kjachtensis* speciation event probably resulted from habitat divergence in the Lake Baikal region about the early Quaternary; *L. hinei* is primarily a forest/riparian species, *L. kjachtensis* a grassland one.

Populations of *L. hinei* have since dispersed west almost to Scandinavia and east to northwestern North America. Its distribution in the Nearctic is consistent with a Wisconsinan residence in the Beringian refugium (Cannings 1997). The species is quite capable of thriving in cold taiga at the tree line. Since the disappearance of the continental ice sheets, *L. hinei* has dispersed only slightly east and south to central Alberta. As far as is known, *L. hinei* is the only *Lasiopogon* living in both the Old and New worlds. Its isolation east of the Bering Strait has evidently been too short to result in any obvious morphological change. In eastern Asia it colonized Sakhalin, probably during the Pleistocene when that island was connected to the mainland. Those populations are darker and more setose than those of the adjacent mainland.

In summary, the radiation of *Lasiopogon* species appears to encompass the Cenozoic and its origin may well lie deeper in the Mezozoic. Evolution and dispersal must produce a broad range of possible and often conflicting patterns (Noonan 1988b); the geographic histories presented here can only tentatively be associated with hypothesized events. Despite this, both the histories and the phylogenies upon which they are based can be tested when new data become available.

CONCLUDING REMARKS

I consider this study a first step in the organization and analysis of information on the genus *Lasiopogon*. The work contains a number of contributions to the systematics of *Lasiopogon*, but much research remains before a clear understanding can be reached about the phylogenetic and biogeographical relationships in the genus as a whole.

Lasiopogon (along with most other asilid taxa) has never before been studied using phylogenetic techniques or biogeographical hypotheses. The examination of the Palaearctic and Nearctic *Lasiopogon* faunas as a single entity is a major contribution, although the main drawback to the study's conclusions is the lack of phylogenetic detail on most of the western Palaearctic species. The incorporation of abundant East Asian material in a North American-based project such as this is not unique, but it is not always possible. Access to the important collection amassed by P.A. Lehr in Vladivostok produced some of the more useful insights into the history of the genus.

Detailed characterization of the genitalia of the species studied was a major goal of the work, and perhaps remains its main contribution. These structures had never before been examined in detail in *Lasiopogon*, and I consider them critical for any phylogenetic analysis of the genus. Indeed, the genitalic morphology examined herein may aid future phylogenetic work in the Asilidae generally. A number of morphological structures, such as the basal sclerite and the form and setation of the subepandrial sclerite have not previously been recognized or described.

Although solving the mystery of the taxonomic position of *Lasiopogon* within the family was not a priority of the project, some progress was made in this search. A cursory analysis of the relationship of *Lasiopogon* to the Stichopogoninae and some related genera in the Stenopogoninae suggests areas where future research might prove fruitful.

At the outset of the study, the genus appeared to be well known taxonomically. But an examination of material from 85 collections revealed 43 undescribed species; the 14 new species in the *opaculus* section are fully described in this work. An additional 6 undescribed species are outlined in an unpublished manuscript by Kovár and Hradský (1996).

Future research needs are legion. The highest priority is a more thorough investigation of the Palaearctic fauna outside the *opaculus* section. These species are mostly European. Preliminary work has been done, but these species must be incorporated in a phylogenetic analysis of *Lasiopogon* as a whole. Such a project would confirm or refute the hypotheses presented here and would strengthen our understanding of the patterns of evolution in the genus.

I urge entomologists to continue the search for more characters that can better define monophyletic groups. Examination of microstructure could prove fruitful, and molecular characters may help elucidate relationships in clades such as the *opaculus* group. The immature stages of *Lasiopogon* are difficult to find, but some intrepid entomologist, following in Melin's (1923) footsteps, may be successful in collecting and rearing the larvae and pupae of different species. These stages will provide important characters that will augment those from adult flies. Characters from ecological, behavioural and life-history studies may also be useful in unravelling the phylogeny of the genus.

Although it is a daunting task, a comprehensive phylogenetic overview of the subfamilies and tribes of the Asilidae is vital for the success of future generic revisions in the family. The production of a satisfactory phylogeny would help confirm a ground plan for the robber flies and would significantly improve understanding of character states in the family. A more solid understanding of outgroup relationships would benefit future research on *Lasiopogon* and many other related genera.

The collection of more *Lasiopogon* specimens is still a priority, especially in poorly studied Palaearctic regions where diversity appears to be high: Turkey and the Palaearctic Middle East, and the steppes and mountains of Central Asia, especially China. The fauna in North America is better known, but four of the seven species from the Appalachian Mountains and adjacent areas are new, and females are unknown in two of theses species. The mountains of the southwest United States might yield a few more undescribed species.

Finally, I make a plea for biological information about *Lasiopogon*. We know very little about the habitat requirements, larval habitats and behaviour, prey selection, reproductive behaviour and life histories of these lively little flies.

ACKNOWLEDGEMENTS

Steve Marshall, my thesis advisor, helped and encouraged me throughout the study. The curators, collection managers and private collectors who loaned the specimens vital for the study are listed in the Materials and Methods section; I thank them all. I am especially grateful to those who helped me personally at their institutions: P. Arnaud, Jr (CASC), D. Grimaldi (AMNH), C. Thompson (USNM), J.M. Cumming (CNCI), L. Munari and E. Ratti (MCNV), B. Hübl and E. Krasser (NMBA), R. Contreras-Lichtenberg (NHMW), C. Leonardi (MSNM), P.A. Lehr and A. Lelej (IBPV).

I am grateful to those entomological friends who invited me into their homes or took me on collecting trips in faraway places: Paul Arnaud, Eric Fisher, Milan Hradský, Steve Marshall, Hannah Nadel, Riley Nelson, Geoff Scudder, Jeff Skevington, John Swann and Monty Wood. My friends in Russia, Pavel Lehr, Arkady Lelej and Nicolai Kurzenko in Vladivostok, and Sergei Bukhalo and Evgeniy Tikhmenev in Magadan made my Russian research a delight.

My friends Monty and Grace Wood, as well as being generous, encouraging and hospitable over the years, have taught me much about natural history. Eric Fisher has been a constant and enthusiastic source of help and information about robber flies and their biology.

The Royal British Columbia Museum and the Government of British Columbia provided educational leave and significant financial support. Travel to Far-eastern Russia was supported by research travel grants from the Friends of the Royal B.C. Museum and the International Education Services (University of Guelph). The University of Guelph also offered a Taffy Davison Memorial Research Travel Grant through the Ontario Agricultural College and a Department of Environmental Biology graduate fellowship. The Entomological Society of Canada provided a Graduate Research Travel Grant to defray the costs of a trip to museums in Europe. Additional support came from an NSERC operating grant to Steve Marshall. The Russell Munn Foundation, managed by Lorna Klohn, helped fund my tuition.

Other entomological friends and colleagues offered advice (both personal and scientific), translations and help with literature over the years – Robb Bennett, Jeff Cumming, Jim O'Hara, Brad Sinclair, John Swann, Richard Vockeroth and

Terry Wheeler. Victor Golini translated important Italian literature. Rolof Idema drew the *Lasiopogon* portrait in figure 1.

My compatriots at the Royal B.C. Museum supported me in innumerable ways; I cannot thank them all individually here, but special thanks go to Dave Blades, who expertly made insect labels, ran computer phylogenetic programs and produced the cladogram figures. Adolf Ceska translated obscure Russian papers. Tracy Coyle and Tara Steigenberger electronically produced the figures from my original illustrations. Bill Barkley, Ted Miller, Grant Hughes and Peter Newroth offered administrative support; Gerry Truscott guided the manuscript through to publication.

Steve Marshall, Eric Fisher and Monty Wood reviewed the manuscript.

My family, especially my wife, Joan Kerik, helped immeasurably.

APPENDIX
Checklist of the species
and species groups of *Lasiopogon*

This checklist contains the 118 species of *Lasiopogon*: 69 represent valid names and 49 are new species. These include 6 from a manuscript by Kovár and Hradský (1996) – their manuscript names are not given here, but the species are listed by number. Other new species not described in this thesis are designated by an abbreviation of the relevant species group and a number. Evidently, there are more undescribed species in the material at hand, but the specimens require more study. The groups are arranged in phylogenetic order; the species within them are in alphabetical order.

The *Cinctus* Clade

The *canus* species group
1. *Lasiopogon canus* Cole and Wilcox (Western Nearctic: Alaska and Yukon to Tuktoyaktuk Peninsula)
2. *L. pugeti* Cole and Wilcox (Western Nearctic: coastal rivers in Washington and northern Oregon)
3. *L.* can-1 sp. nov. (Western Nearctic: stream sides in central Alberta)

The *montanus* species group
1. *L. bellardii* Jaennicke (Western Palaearctic: montane meadows, Italy to Croatia)
2. *L. montanus* Schiner (Western Palaearctic: montane meadows, Italy to Romania and Greece)

The *apenninus* species group
1. *L. apenninus* Bezzi (Western Palaearctic: montane meadows, Apennine Mountains of central Italy)
2. *L. grajus* Bezzi (Western Palaearctic: the montane meadows of the western Alps)
3. *L.* app-1 sp. nov. (Western Palaearctic: the montane meadows of the French Alps and Pyrenees)

The *cinctus* species group
1. *L. cinctus* (Fabricius) (Western Palaearctic: northern Europe)
2. *L.* K/H-1 sp. nov. (Kovár and Hradský ms) (Western Palaearctic: Greece, Macedonia)
3. *L. intermedius* Oldenberg (Western Palaearctic: Romania)
4. *L. macquarti* (Perris) (Western Palaearctic: France, Landes)
5. *L. novus* Lehr (Western Palaearctic: Ukraine)
6. *L. peusi* Hradský (Western Palaearctic: Greece: Peloponnesos)
7. *L.* cin-1 sp. nov. (Western Palaearctic: Spain, Aragon, Tereul)
8. *L.* cin-2 sp. nov. (Western Palaearctic: Spain, Huesca)
9. *L.* cin-3 sp. nov. (Western Palaearctic: France: Provence)
10. *L.* cin-4 sp. nov. (Western Palaearctic: France: Alps-de-Haute-Provence)
11. *L.* cin-5 sp. nov. (Western Palaearctic: Greece: Macedonia)
12. *L.* cin-6 sp. nov. (Western Palaearctic: Turkey)

The *Bivittatus* Clade

The *aldrichii* species group
1. *L. aldrichii* Melander (Western Nearctic: the mountains of southern British Columbia and Alberta to Utah and New Mexico)
2. *L. pacificus* Cole and Wilcox (Western Nearctic: sea beaches and coastal streams from the Fraser River valley of British Columbia to Oregon)
3. *L. yukonensis* Cole and Wilcox (Western Nearctic: southern Yukon)
4. *L.* ald-1 sp. nov. (Western Nearctic: coastal stream sides in northern California)
5. *L.* ald-2 sp. nov. (Western Nearctic: stream sides in Idaho and eastern Washington)

The *terricola* species group
1. *L. prima* Adisoemarto (Western Nearctic: Alaska and Yukon to Alberta)
2. *L. septentrionalis* Lehr (Eastern Palaearctic: Russia east of the Lena River)
3. *L. terricola* (Johnson) (Nearctic: northeastern U.S.A. to Alberta)
4. *L. trivittatus* Melander (Western Nearctic: stream sides in Alberta and Montana)

The *testaceus* species group
1. *L. testaceus* Cole and Wilcox (Western Nearctic: the Sierra Nevada of California and Nevada)
2. *L.* tes-1 sp. nov. (Western Nearctic: the Sierra Nevada of California and Nevada)
3. *L.* tes-2 sp. nov. (Western Nearctic: the Sierra Nevada (White Mountains) of California)

4. *L.* tes-3 sp. nov. (Western Nearctic: the eastern slopes of the Cascade Range in central Oregon)
5. *L.* tes-4 sp. nov. (Western Nearctic: the San Bernardino Mountains of southern California)
6. *L.* tes-5 sp. nov. (Western Nearctic: the mountains of central Arizona)

The *fumipennis* species group
1. *L. currani* Cole and Wilcox (Eastern Nearctic: Ontario and eastern U.S.A.)
2. *L. fumipennis* Melander (Western Nearctic: British Columbia to California)
3. *L. polensis* Lavigne (Western Nearctic: Wyoming)

The *bivittatus* species group
1. *L. actius* Melander (Western Nearctic: the sea beaches of British Columbia to Oregon)
2. *L. albidus* Cole and Wilcox (Western Nearctic: the grasslands and shrub-steppe of eastern Washington and Oregon)
3. *L. arenicola* (Osten Sacken) (Western Nearctic: the sea beaches of central California)
4. *L. bivittatus* Loew (Western Nearctic: the sea beaches of Washington to central California)
5. *L. californicus* Cole and Wilcox (Western Nearctic: stream sides in western California)
6. *L. dimicki* Cole and Wilcox (Western Nearctic: the sea beaches of Oregon)
7. *L. drabicola* Cole and Wilcox (Western Nearctic: southern California lowlands)
8. *L. gabrieli* Cole and Wilcox (Western Nearctic: southern California lowlands)
9. *L. littoris* Cole (Western Nearctic: the sea beaches of central California)
10. *L. martinorum* Cole and Wilcox (Western Nearctic: stream sides in eastern Washington)
11. *L. ripicola* Melander (Western Nearctic: stream sides in eastern Washington and western Idaho)
12. *L. willametti* Cole and Wilcox (Western Nearctic: stream sides, mostly coastal, and sea beaches from southwestern British Columbia to northern California)
13. *L. zonatus* Cole and Wilcox (Western Nearctic: stream sides in California)
14. *L.* biv-1 sp. nov. (Western Nearctic: northern Baja California)
15. *L.* biv-2 sp. nov. (Western Nearctic: the Sierra Nevada of central California and Nevada)
16. *L.* biv-3 sp. nov. (Western Nearctic: Southern California and northern Baja California)

The *opaculus* section

The *tetragrammus* species group
1. *L. chrysotus* sp. nov. (Eastern Nearctic: Tennessee)
2. *L. coconino* sp. nov. (Western Nearctic: Arizona and New Mexico)
3. *L. flammeus* sp. nov. (Eastern Nearctic: North Carolina)
4. *L. lavignei* sp. nov. (Western Nearctic: Wyoming)
5. *L. oklahomensis* Cole and Wilcox (Western Nearctic: Oklahoma)
6. *L. quadrivittatus* Jones (Western Nearctic: Great Plains and eastern Rocky Mountains)
7. *L. tetragrammus* Loew (Eastern Nearctic: James Bay to Michigan and Connecticut)
8. *L. woodorum* sp. nov. (Eastern Nearctic: Indiana east to Delaware)

The *cinereus* species group
1. *L. apache* sp. nov. (Western Nearctic: Arizona)
2. *L. chaetosus* Cole and Wilcox (Western Nearctic: Washington to Nevada and Utah)
3. *L. cinereus* Cole (Western Nearctic: British Columbia and Alberta to California and Colorado)
4. *L. shermani* Cole and Wilcox (Eastern Nearctic: North Carolina to Georgia)

The *monticola* species group
1. *L. monticola* Melander (Western Nearctic: southern British Columbia to California and Utah)

The *terneicus* species group
1. *L. terneicus* Lehr (Eastern Palaearctic: Russia, Primorskiy Kray)

The *opaculus* species group
1. *L. appalachensis* sp. nov. (Eastern Nearctic: West Virginia to Tennessee)
2. *L. marshalli* sp. nov. (Eastern Nearctic: Virginia)
3. *L. opaculus* Loew (Eastern Nearctic: Ontario and Iowa to Mississippi and Georgia)
4. *L. piestolophus* sp. nov. (Eastern Nearctic: Alabama)
5. *L. schizopygus* sp. nov. (Eastern Nearctic: Mississippi to North Carolina)
6. *L. slossonae* Cole and Wilcox (Eastern Nearctic: Quebec and Michigan to West Virginia)

The *akaishii* species group
1. *L. akaishii* Lehr (Eastern Palaearctic: Japan, Hokkaido; Russia, Kurile Islands)
2. *L. hasanicus* Lehr (Eastern Palaearctic: Russia, Primorskiy Kray)

3. *L. lehri* sp. nov. (Eastern Palaearctic: Russia, Primorskiy Kray)
4. *L. rokuroi* Hradský (Eastern Palaearctic: Japan, Honshu and Kyushu)

The *hinei* species group
1. *L. hinei* Cole and Wilcox (Palaearctic and Western Nearctic: Siberia and Far-eastern Russia; Alaska and Yukon to the Mackenzie River delta)
2. *L. kjachtensis* Lehr (Eastern Palaearctic: Russia, Buryatskaya)
3. *L. leleji* sp. nov. (Eastern Palaearctic: Russia, Primorskiy Kray)
4. *L. qinghaiensis* sp. nov. (Eastern Palaearctic: China, Qinghai)
5. *L. phaeothysanotus* sp. nov. (Eastern Palaearctic: China, Qinghai)

Uncertain placement

1. *L. avetianae* Richter (Western Palaearctic: Caucasus Mountains)
2. *L. bezzii* Engel (Western Palaearctic: lowland Italy)
3. *L. K/H-2* sp. nov. (Kovár and Hradský ms) (Western Palaearctic: lowland central Europe)
4. *L. delicatulus* Melander (Western Nearctic: Cascade Mountains)
5. *L. delphinensis* Bezzi (Western Palaearctic: Western Alps)
6. *L. eichingeri* Hradský (Eastern Palaearctic: China, Qinghai)
7. *L. fourcatensis* Timon-David (Western Palaearctic: Pyrenees)
8. *L. K/H-3* sp. nov. (Kovár and Hradský ms) (Western Palaearctic: no data available)
9. *L. immaculatus* Strobl (Western Palaearctic: Alps)
10. *L. K/H-4* sp. nov. (Kovár and Hradský ms) (Western Palaearctic: Austria)
11. *L. lichtwardti* Oldenberg (Western Palaearctic: lowland central Europe)
12. *L. K/H-5* sp. nov. (Kovár and Hradský ms) (Western Palaearctic: Greece, Peloponnesos)
13. *L. nanus* Oldenberg (Western Palaearctic: coastal Croatia)
14. *L. nitidicauda* Bezzi (Western Palaearctic: Italian Alps)
15. *L. K/H-6* sp. nov. (Kovár and Hradský ms) (Western Palaearctic: the mountains of Bulgaria)
16. *L. pilosellus* Loew (Western Palaearctic: Turkey)
17. *L. pusillus* Bezzi (Western Palaearctic: lowland central Europe)
18. *L. soffneri* Hradský and Moucha (Western Palaearctic: Bulgaria)
19. *L. solox* Enderlein (Eastern Palaearctic: Taiwan)
20. *L. spinisquama* Bezzi (Western Palaearctic: lowland northern Italy)
21. *L. tarsalis* Loew (Western Palaearctic: Turkey)
22. *L. tuvensis* Richter (Eastern Palaearctic: Russia, Tuvinskaya)
23. *L. velutinus* Bezzi (Western Palaearctic: Portugal)
24. *L. zaitzevi* Lehr (Eastern Palaearctic: Kazakhstan)
25. *L. unc-1* sp. nov. (Western Palaearctic: Germany, Bavaria)

26. *L.* unc-2 sp. nov. (Western Palaearctic: lowland southern Germany)
27. *L.* unc-3 sp. nov. (Western Palaearctic: Turkey)
28. *L.* unc-4 sp. nov. (Eastern Palaearctic: China, Qinghai)
29. *L.* unc-5 sp. nov. (Western Palaearctic: southeastern France)
30. *L.* unc-6 sp. nov. (Western Palaearctic: Switzerland, Valais)
31. *L.* unc-7 sp. nov. (Western Nearctic: Great Basin)
32. *L.* unc-8 sp. nov. (Western Palaearctic: French Pyrenees)
33. *L.* unc-9 sp. nov. (Western Palaearctic: Spanish Pyrenees)
34. *L.* unc-10 sp. nov. (Western Palaearctic: Greece, Macedonia)
35. *L.* unc-11 sp. nov. (Western Palaearctic: Spain)

REFERENCES

Adisoemarto, S. 1967. The Asilidae (Diptera) of Alberta. *Quaestiones Entomologicae* 3: 3-90.

Adisoemarto, S., and D.M. Wood. 1975. The Nearctic species of *Dioctria* and six related genera (Diptera: Asilidae). *Quaestiones Entomologicae* 11: 505-76.

Allen, R.T. 1983. Distribution patterns among arthropods of the North Temperate deciduous forest biota. *Annals of the Missouri Botanical Garden* 70: 616-28.

Aoki, A. 1949. On some alpine Asilidae from Formosa and Japan. *Kontyu* 17: 59-60 (in Japanese).

Arnett, R.H. Jr, G.A. Samuelson and G.M. Nishida. 1993. *The Insect and Spider Collections of the World*, 2nd edition. Gainesville, Florida: Sandhill Crane Press.

Artigas, J.N., and N. Papavero. 1990. The American genera of Asilidae (Diptera): keys for identification with an atlas of female spermathecae and other morphological details. V. Subfamily Stichopogoninae G.H. Hardy, with descriptions of new genera and species and new synonymies. *Boletin de la Sociedad de Biologia de Concepcion, Chile* 61: 39-47.

————. 1991a. The American genera of Asilidae (Diptera): keys for identification with an atlas of female spermathecae and other morphological details. VII.2. Subfamily Stenopogoninae Hull – Tribes Acronychini, Bathypogonini (with description of a new genus) and Ceraturgini, and a catalogue of the Neotropical species. *Boletin de la Sociedad de Biologia de Concepcion, Chile* 55: 247-55.

————. 1991b. The American genera of Asilidae (Diptera): keys for identification with an atlas of female spermathecae and other morphological details. VII.7. Subfamily Stenopogoninae Hull – Tribe Cyrtopogonini, with descriptions of four new genera and one new species and a catalogue of the Neotropical species. *Boletin de la Sociedad de Biologia de Concepcion, Chile* 61: 55-81.

————. 1993. The American genera of Asilidae (Diptera): keys for identification with an atlas of female spermathecae and other morphological details. VII.6. Subfamily Stenopogoninae Hull - Tribes Phellini, Plesiommatini, Stenopogonini and Willistoninini. *Gayana Zoologia* 57: 309-21.

339

Askevold, I.S. 1991. Classification, reconstructed phylogeny, and geographic history of the New World members of *Plateumaris* Thomson, 1859 (Coleoptera: Chrysomelidae: Donaciinae). *Memoirs of the Entomological Society of Canada* 157: 1-175.

Axelrod, D.I. 1986. Cenozoic history of some western American pines. *Annals of the Missouri Botanical Garden* 73: 565-641.

Axelrod, D.I., and P.H. Raven. 1985. Origins of the Cordilleran flora. *Journal of Biogeography* 12: 21-47.

Back, E.A. 1909. The Robber-flies of America north of Mexico belonging to the subfamilies Leptogastrinae and Dasypogoninae. *Transactions of the American Entomological Society* 35: 137-400.

Baker, N.T., and R.L. Fischer. 1975. A taxonomic and ecologic study of the Asilidae of Michigan. *Great Lakes Entomologist* 8: 31-91.

Ball, G.E. 1965. The genus *Scaphinotus* (Coleoptera: Carabidae) and the Pleistocene epoch in southwestern United States. *Proceedings of the XII International Congress of Entomology, London 1964*: 460-61.

Ball, I.R. 1976. Nature and formulation of biogeographical hypotheses. *Systematic Zoology* 24: 407-30.

Bezzi, M. 1921. Il genere *Lasiopogon* Loew. *Bollettino del Laboratorio di Zoologia Generale e Agraria della R. Scuola Superiore d'Agricoltura in Portici* 11: 250-81.

Bromley, S.W. 1934. The robber flies of Texas (Diptera: Asilidae). *Annals of the Entomological Society of America* 27: 74-113.

———. 1946. Guide to the insects of Connecticut, Part VI. The Diptera or True Flies of Connecticut. Asilidae. *Connecticut State Geological and Natural History Survey Bulletin* 69.

Brundin, L. 1972. Phylogenetics and biogeography. *Systematic Zoology* 21: 69-79.

Byers, G.W. 1988. Geographic affinities of the North American Mecoptera. *Memoirs of the Entomological Society of Canada* 144: 25-30.

Cannings, R.A. 1996. Taxonomy and distribution of *Lasiopogon montanus* Schiner and *L. bellardii* Jaennicke (Diptera: Asilidae), two common robber flies from the mountains of Western and Central Europe. *Entomologica Scandinavica* 27: 347-59.

———. 1997. Robber Flies (Diptera: Asilidae) of the Yukon. In *Insects of the Yukon*, edited by H.V. Danks and J.A. Downes. Ottawa: Biological Survey of Canada (Terrestrial Arthropods).

Castellani, O., and G. Crivaro. 1966. Contributo alla conoscenza degli Asilidi Palaearctici. *Bollettino Associazione Romana Entomologia* 20: 50-60.

Clements, A.N. 1951. The use of the prosternum in classifying Asilidae (Diptera). *Proceedings of the Royal Entomological Society of London* (B) 20: 10-14.

Cockerell, T.D.A. 1921. Some Eocene insects from Colorado and Wyoming. *Proceedings of the U.S. National Museum* 59 (2358): 29-39.

Cole, F.R. 1916. New species of Asilidae from southern California. *Psyche* 23: 63-69.

————. 1924. Notes on the dipterous family Asilidae, with descriptions of new species. *Pan-Pacific Entomologist* 1: 7-13.

————. 1969. *The Flies of Western North America*. Berkeley: University of California Press.

Cole, F.R., and A.L. Lovett. 1919. New Oregon Diptera. *Proceedings of the California Academy of Sciences*, Series 4, 9: 221-55.

Cole, F.R., and J. Wilcox, 1938. The genera *Lasiopogon* Loew and *Alexiopogon* Curran in North America (Diptera-Asilidae). *Entomologica Americana* 43: 1-90.

Coquillett, D.W. 1910. The type-species of the North American genera of Diptera. *Proceedings of the U.S. National Museum* 37: 499-647 (no. 1719).

Crampton, G.C. 1942. Guide to the insects of Connecticut, Part IV. The Diptera or True Flies of Connecticut. The external morphology of the Diptera. *Connecticut State Geological and Natural History Survey Bulletin* 64.

Cronquist, A. 1978. The biota of the intermountain region in geohistorical context. Intermountain Biogeography: A symposium. *Great Basin Naturalist Memoirs* 2: 3-15.

Cumming, J.M. 1989. Classification and evolution of the Eumenine wasp genus *Symmorphus* Wesmael (Hymenoptera: Vespidae). *Memoirs of the Entomological Society of Canada* 148: 1-168.

Cumming, J.M., B.J. Sinclair and D.M. Wood. 1995. Homology and phylogenetic implications of male genitalia in Diptera-Eremoneura. *Entomologica Scandinavica* 26: 120-51.

Curran, C.H.1934. *The Families and Genera of North American Diptera*. New York: Ballon Press.

Danks, H.V., J.A. Downes, D.J. Larson and G.G.E. Scudder. 1997. Insects of the Yukon: characteristics and history. In *Insects of the Yukon*, edited by H.V. Danks and J.A. Downes. Ottawa: Biological Survey of Canada.

Elias, S.A. 1994. *Quaternary Insects and Their Environments*. Washington: Smithsonian Institution Press.

Engel, E.O. 1930. Asilidae (Part 24). In *Die Fliegen der Palaearktischen Region*, vol. 4, edited by E. Lindner. Stuttgart: Schweizerbart'sche.

Evenhuis, N.L., 1994. *Catalogue of the Fossil Flies of the World (Insecta: Diptera)*. Leiden: Backhuys Publishers.

Fabricius, J.C. 1781. *Species Insectorum*, vol. 2. Hamburg: C.E. Bohn.

————. 1805. *Systema Antliatorum*. Reprinted 1970. Vaals: Asher and Co.

Fallén, C.R. 1814. *Asilici Sveciae*. Lund: Litteris Berlingianis.

Farris, J.S. 1988. Hennig86, version 1.5. Computer software. Privately published, Port Jefferson, New York.

Fisher, E.M. 1977. A review of the North American genera of Laphystiini with a revision of the genus *Zabrops* Hull (Insecta: Diptera: Asilidae). *Proceedings of the California Academy of Sciences*, Series 4, 41(5): 183-213.

Fisher, E.M. 1986. A reclassification of the Robber Fly tribe Andrenosomini, with a revision of the genus *Dasyllis* Loew (Diptera: Asilidae). Unpublished PhD thesis, University of California, Riverside.

Fisher, E.M., and J. Wilcox. 1997. Catalogue of the robber flies of the Nearctic Region. Unpublished draft.

Geller-Grimm, F. 2000. Robber Flies (Asilidae). Internet site at http://www.geller-grimm.de/asilidae.htm

Grimaldi, D. 1990. Diptera. In *Insects from the Santana Formation, Lower Cretaceous, of Brazil*, edited by D. Grimaldi. *Bulletin of the American Museum of Natural History* 195: 1-191.

Hallam, A. 1981. Relative importance of plate movements, eustasy, and climate in controlling major biogeographical changes since the early Mesozoic. In *Vicariance Biogeography: A Critique*, edited by G. Nelson and D.E. Rosen. New York: Columbia University Press.

Hamilton, S.W., and J.C. Morse. 1990. Southeastern caddisfly fauna: origins and affinities. *Florida Entomologist* 73: 587-600.

Hardy, G.H. 1930. Fifth contribution towards a new classification of Australian Asilidae (Diptera). *Proceedings of the Linnean Society of New South Wales* 55(3): 249-60.

―――. 1934-35. The Asilidae of Australia, Parts 1-4. *Ann. Mag. nat. Hist* 13: 498-526; 14: 1-35; 16: 161-87, 405-26.

―――. 1948. On classifying Asilidae. *The Entomologist's Monthly Magazine* 84: 116-19.

Harper, K.T., D.C. Freeman, W.K. Ostler and L.G. Klikoff 1978. The flora of Great Basin mountain ranges: diversity, sources and dispersal ecology. Intermountain Biogeography: a symposium. *Great Basin Naturalist Memoirs* 2: 81-103.

Hennig, W. 1965. Phylogenetic systematics. *Annual Review of Entomology* 10: 97-116.

―――. 1966. *Phylogenetic Systematics*. Urbana: University of Illinois Press.

Hine, J.S. 1909. Robberflies of the genus *Asilus*. *Annals of the Entomological Society of America* 2: 136-72.

Hobby, B.M. 1931. The British species of Asilidae (Diptera) and their prey. *Transactions of the Entomological Society of the South of England* 6 (1930): 1-42.

Hoffmann, R.S., and J.K. Jones Jr. 1970. Influence of late-glacial and post-glacial events on distribution of recent mammals on the northern Great Plains. In *Pleistocene and Recent Environments of the Central Great Plains*, edited by W. Dort Jr and J.K. Jones Jr. Special Publication 3. Lawrence:University of Kansas.

Honza, E. 1995. Spreading mode of backarc basins in the western Pacific. *Tectonophysics* 251: 139-52.

Hradský, M. 1981. Drei neue ostpalaearktische *Lasiopogon*-Arten (Diptera, Asilidae, Stichopogonini). *Traveaux du Museum d'Histoire Naturelle*

"Grigore Antipa" 23: 177-82.

Hradský, M. 1982. Zur Kentniss der Asilidae der Balkanhalbinsel, insbesondere Griechenlands. *Faunistische Abhandlungen Museum für Tierkunde Dresden* 9: 179-84.

Hradský, M., and J. Moucha. 1964. Raubfliegen (Diptera, Asilidae) Bulgariens. *Acta Faunistica Entomologica Musei Nationalis Pragae* 10: 23-30.

Hubbard, J.P. 1973. Avian evolution in the aridlands of North America. *The Living Bird* 12: 155-96.

Hull, F.M. 1962. Robber Flies of the world: the genera of the family Asilidae. *Smithsonian Institution Bulletin* 224 (Parts 1 and 2): 1-907.

Humphries, C.J., and L.R. Parenti. 1986. *Cladistic Biogeography.* Oxford Monographs on Biogeography No. 2. Oxford: Clarendon Press.

Ionescu, M.A., and M. Weinberg. 1971. *Fauna Republicii Socialiste Romania.* Insecta, vol. 11, Fasc. 11. Diptera-Asilidae.

Irwin, M.E. 1976. Morphology of the terminalia and known ovipositing behaviour of female Therevidae (Diptera: Asiloidea), with an account of correlated adaptations and comments on phylogenetic relationships. *Annals of the Natal Museum* 22: 913-35.

Irwin, M.E., and L. Lyneborg. 1981. The genera of Nearctic Therevidae. *Illinois Natural History Survey Bulletin* 32 (1980): 188-277.

Jaennicke, F. 1867. Beiträge zur Kenntniss der europäischen Bombyliden, Acroceriden, Scenopiniden, Thereviden und Asiliden. *Berliner entomologische Zeitschrift* 11: 63-94.

James, M.T. 1939. A preliminary review of certain families of Diptera from the Florissant Miocene beds. II. *Journal of Paleontology* 13: 42-48.

———. 1941. The robber flies of Colorado (Diptera: Asilidae). *Journal of the Kansas Entomological Society* 14: 27-53.

Johnson, C.W. 1900. Notes and descriptions of seven new species and one new genus of Diptera. *Entomological News* 11: 323-28.

Johnson, N.K. 1977. Intermountain biogeography: a symposium. *Great Basin Naturalist Memoirs* 2: 137-59.

Jones, P.R. 1907. A preliminary list of the Asilidae of Nebraska, with a description of a new species. *Transactions of the American Entomological Society* 33: 273-86.

Karl, E. 1959. Vergleichend-morphologische Untersuchungen der männlichen Kopulationsorgane bei Asiliden (Diptera). *Beiträge Entomologie* 9: 619-80.

Kavanaugh, D.H. 1988. The insect fauna of the Pacific northwest coast of North America: present patterns and affinities and their origins. In *Origins of the North American Insect Fauna*, edited by J.A. Downes and D.H. Kavanaugh. *Memoirs of the Entomological Society of Canada* 144.

Knutson, L.V. 1972. Pupa of *Neomochtherus angustipennis* (Hine), with notes on feeding habits of robber flies and a review of publications on morphology of immature stages. (Diptera: Asilidae). *Proceedings of the Biological Society of Washington* 85: 163-78.

Kovár, I., and M. Hradský. 1996. Contribution to European species of the genus *Lasiopogon* Loew (Diptera, Asilidae). Unpublished manuscript.

Lafontaine, J.D. 1982. Biogeography of the genus *Euxoa* (Lepidoptera: Noctuidae) in North America. *The Canadian Entomologist* 114:1-53.

Lafontaine, J.D., and D.M. Wood. 1988. A zoogeographic analysis of the Noctuidae (Lepidoptera) of Beringia, and some inferences about past Beringian habitats. In *Origins of the North American Insect Fauna*, edited by J.A. Downes and D.H. Kavanaugh. *Memoirs of the Entomological Society of Canada* 144.

Lavigne, R.J. 1969. A new species of *Lasiopogon* from Wyoming (Diptera: Asilidae). *Journal of the Kansas Entomological Society* 42: 363-65.

—————. 1972. *Asilidae of the Pawnee National Grasslands in Northeastern Colorado*. Science Monograph 25. Laramie: University of Wyoming Agricultural Experiment Station.

—————. 1999. *Bibliography Update for the Asilidae (Insecta: Diptera)*. Science Monograph 55. Laramie: University of Wyoming Agricultural Experiment Station.

Lavigne, R.J., and F. Holland. 1969. *Comparative Behavior of Eleven Species of Wyoming Robber Flies (Diptera, Asilidae)*. Science Monograph 18. Laramie: University of Wyoming Agricultural Experiment Station.

Lavigne, R.J., D.S. Dennis and J.A. Gowen. 1978. *Asilid Literature Update 1956-1976, Including a Brief Review of Robber Fly Biology*. Science Monograph 36. Laramie: University of Wyoming Agricultural Experiment Station.

Lavigne, R.J., and S.W. Bullington. 1981. New species of *Cyrtopogon* from Wyoming. *Annals of the Entomological Society of America* 74: 414-18.

Lehmkuhl, D.M. 1980. Temporal and spatial changes in the Canadian insect fauna: patterns and explanation. The Prairies. *The Canadian Entomologist* 112: 1145-59.

Lehr, P.A. 1962. Some aspects of the evolution of robber flies. *Inst. Zasch. Rast. Kazakhstan, Alma-Ata* 7: 347-82.

—————. 1984a. Assassin flies of the tribe Stichopogonini (Diptera: Asilidae) from the fauna of the U.S.S.R. 1. *Zoologicheskii Zhurnal* 63: 696-706 (in Russian).

—————. 1984b. Assassin flies of the tribe Stichopogonini (Diptera: Asilidae) from the fauna of the U.S.S.R. 2. *Zoologicheskii Zhurnal* 63: 859-64 (in Russian).

—————. 1988. Family Asilidae. In *Catalogue of Palaearctic Diptera*, vol. 5: *Athericidae-Asilidae*, edited by A. Soos and L. Papp. Amsterdam: Elsevier.

—————. 1996. *Robber Flies of Subfamily Asilinae (Diptera, Asilidae) of Palaearctic: Ecological and Morphological Analysis, Taxonomy and Evolution*. Vladivostok: Dalnauka (in Russian).

Lindner, E. 1966. Zur Kenntnis der Asiliden-Gattung *Lasiopogon* Loew in der Alpen (Diptera). *Stuttgarter Beiträge zur Naturkunde* 164: 1-3.

Loew, H. 1847. Über die europäischen Raubfliegen (Diptera Asilica). *Linnaea Entomologica* 2: 384-568.

———. 1866. Diptera Americae septentrionalis indigena. *Berliner Entomologische Zeitschrift* 10: 1-54.

———. 1874. Neue nordamerikanische Dasypogonina. *Berliner Entomologische Zeitschrift* 18: 353-77.

Londt, J.G.H. 1985. Afrotropical Asilidae (Diptera) 10. The genus *Hypenetes* Loew. *Annals of the Natal Museum* 26(2): 377-405.

———. 1993. *Psilinus* Wulp, 1899 and *Spanurus* Loew, 1858 – new synonyms of *Rhabdogaster* Loew, 1858 (Diptera: Asilidae: Stenopogoninae). *Journal of African Zoology* 107: 383-92.

———. 1994. Afrotropical Asilidae (Diptera) 25. A key to the genera of the subfamily Stenopogoninae with new synonymy and descriptions of six new genera. *Annals of the Natal Museum* 35: 71-96.

Lundbeck, W. 1908. *Diptera Danica. Genera and Species of Flies Hitherto Found in Denmark*, part 2: *Asilidae, Bombyliidae, Therevidae, Scenopinidae*. Copenhagen: G.E.C. Gadd.

Macquart, J. 1838. Diptères exotiques nouveaux ou peu connus. *Mem. Soc. Sci. Agric. et Arts, Lille* 1(2): 5-207.

Maddison, D.R. 1993. Systematics of the Holarctic beetle subgenus *Bracteon* and related *Bembidion* (Coleoptera: Carabidae). *Bulletin of the Museum of Comparative Zoology* 153: 143-299.

Maddison, W.P., M.J. Donoghue and D.R. Maddison. 1984. Outgroup analysis and parsimony. *Systematic Zoology* 33: 83-103.

Majer, J.M. 1997. European Asilidae. In *Contributions to a Manual of Palaearctic Diptera*, vol 2: *Nematocera and Lower Brachycera*, edited by L. Papp and B. Darvas. Budapest: Science Herald.

Marshall, S.A. 1985. A revision of the New World species of *Minilimosina* Rohácek (Diptera: Sphaeroceridae). *Proceedings of the Entomological Society of Ontario* 116: 1-60.

———. 1987. Systematics of *Bitheca*, a new genus of New World Sphaeroceridae (Diptera). *Systematic Entomology* 12: 355-80.

Martin, C.H. 1968. The new family Leptogastridae (the grass flies) compared with the Asilidae (robber flies)(Diptera). *Journal of the Kansas Entomological Society* 41:70-100.

———. 1975. The Generic and Specific Characters of Four Old and Six New Asilini Genera in the Western United States, Mexico and Central America (Diptera: Asilidae). *Occasional Papers of the California Academy of Sciences* 119.

Martin, C.H., and J. Wilcox. 1965. Family Asilidae. In *A Catalogue of the Diptera of America North of Mexico*, edited by A. Stone, C.W. Sabrosky, W.W. Wirth, R.H. Foote and J.R. Coulson. Washington: U.S. Department of Agriculture.

Matthews, J.V. Jr. 1979. Tertiary and Quaternary environments: historical background for an analysis of the Canadian insect fauna. In *Canada and its Insect Fauna*, edited by H.V. Danks. *Memoirs of the Entomological Society of Canada* 108.

Matthews, J.V. Jr. 1980. Tertiary land bridges and their climate: backdrop for development of the present Canadian insect fauna. *The Canadian Entomologist* 112: 1089-1103.

Matthews, J.V. Jr, and A. Telka. 1997. Insect fossils from the Yukon. In *Insects of the Yukon*, edited by H.V. Danks and J.A. Downes. Ottawa: Biological Survey of Canada (Terrestrial Arthropods).

McAlpine, J.F. 1981. Morphology: terminology – adults. In *Manual of Nearctic Diptera*, edited by J.F. McAlpine, B.V. Peterson, G.E. Shewell, H.J. Teskey, J.R. Vockeroth and D.M. Wood, vol. 1. Monograph 27. Ottawa: Agriculture Canada.

McKenzie, D.P., and J.G. Sclater. 1973. The evolution of the Indian Ocean. *Scientific American* 228: 62-72.

Meigen, J.W. 1851 (reissue of 1820 volume). *Systematische Beschreibung der Bekannten Europäischen Zweiflügeligen Insecten*. Halle: Schmidt Verlag.

Melander, A.L. 1923. The genus *Lasiopogon* (Diptera, Asilidae). *Psyche* 30: 130-45.

Melin, D. 1923. Contributions to the knowledge of the biology, metamorphosis and distribution of the Swedish asilids. *Zoologiska Bidrag Fron Uppsala* 8: 1-317.

Metz, C.W., and J.F. Nonidez. 1924. The behavior of the nucleus and chromosomes during spermatogenesis in the robber fly, *Lasiopogon bivittatus*. *Biological Bulletin* 46: 153-62.

Minato, M., M. Gorai and M. Hunahashi, eds. 1965. *The Geologic Developments of the Japanese Islands*. Tokyo: Tsukiji Shokan Co.

Nelson, G., and N. Platnick. 1981. *Systematics and Biogeography / cladistics and vicariance*. New York: Columbia University Press.

Nelson, C.R. 1987. Robber flies of Utah (Diptera: Asilidae). *Great Basin Naturalist* 47: 38-90.

Nixon, K.C. 1992. CLADOS, version 1.2 Computer software. L.H. Bailey Hortorium, Cornell University, Ithaca, New York.

Noonan, G.E. 1986. Distribution of insects in the northern hemisphere: continental drift and epicontinental seas. *Bulletin of the Entomological Society of America* 32: 80-84.

———. 1988a. Faunal relationships between eastern North America and Europe as shown by insects. In *Origins of the North American Insect Fauna*, edited by J.A. Downes and D.H. Kavanaugh. *Memoirs of the Entomological Society of Canada* 144.

———. 1988b. Biogeography of North American and Mexican insects, and a critique of vicariance biogeography. *Systematic Zoology* 37: 366-84.

O'Hara, J.E. 1983. Classification, phylogeny and zoogeography of the North American species of *Siphona* Meigen (Diptera: Tachinidae). *Quaestiones*

Entomologicae 18:261-380.

Oldenberg, L. 1924. Zur Kenntnis der Asiliden-Gattung *Lasiopogon* (Diptera). *Deutsche Entomologische Zeitschrift* 5: 441-448.

Oldroyd, H. 1969a. The family Leptogastridae (Diptera). *Proceedings of the Entomological Society of London*, Series B, 38: 27-31.

———. 1969b. *Diptera Brachycera, Section A — Tabanoidea and Asiloidea.* Handbooks for the Identification of British Insects, vol. 9, part 4. London: Royal Entomological Society of London.

———. 1970a. *Diptera, I — Introduction and Key to Families.* Handbooks for the Identification of British Insects, vol. 9, part 1. London: Royal Entomological Society of London.

———. 1970b. Studies of African Asilidae (Diptera). I. Asilidae of the Congo Basin. *Bulletin of the British Museum (Natural History), Entomology* 24: 207-334.

———. 1974. An introduction to the robber flies (Diptera: Asilidae) of South Africa. *Annals of the Natal Museum* 22: 1-172.

Osten Sacken, C.R. 1877. Western Diptera: descriptions of new genera and species of Diptera from the region west of the Mississippi, and especially from California. *Bulletin of the U.S. Geological and Geographical Survey of the Territories* III(2) 1877: 189-354.

———. 1878. *Catalogue of the Described Diptera of North America*, 2nd edition. Smithsonian Miscellaneous Collection, vol. 16. Washington: Smithsonian Institution.

Papavero, N. 1973a. Studies of Asilidae (Diptera) systematics and evolution. I. A preliminary classification in subfamilies. *Arquivos de Zoologia, Sao Paulo* 23: 217-74.

Perris, E. 1852. Seconde excursion dans les Grandes-Landes. *Annales de la Societé Linnéenne de Lyon* 1852: 145-216.

Platnick, N.I., and G. Nelson. 1978. A method of analysis for historical biogeography. *Systematic Zoology* 27: 1-16.

Poulton, E.B. 1906. Predaceous insects and their prey. Part 1: Predaceous Diptera, Neuroptera, Hemiptera, Orthoptera and Coleoptera. *Transactions of the Entomological Society of London* 1906: 323-409.

Pritchard, A.E. 1943. Revision of the genus *Cophura* Osten Sacken. *Annals of the Entomological Society of America* 36: 281-309.

Raven, P.H., and D.I. Axelrod. 1974. Angiosperm biogeography and past continental movements. *Annals of the Missouri Botanical Garden* 61: 539-673.

Reichardt, H. 1929. Untersuchungen über den Genitalapparat der Asiliden. *Zeitschrift für Wissenschaftliche Zoologie* 135: 17-301.

Richter, V.A. 1962. New and little known species of robber-flies (Diptera, Asilidae) in the Armenian S.S.R. *Doklady Akademia Nauk Armyanskoi S.S.R.* 34: 37-41.

———. 1968. The predaceous robber flies (Diptera, Asilidae) of the Caucasus. *Opredeliteli po faune U.S.S.R.* 97: 1-284.

Richter, V.A. 1972. On the fauna of robber flies of the subfamilies Dasypogoninae and Laphriinae (Diptera: Asilidae) of Mongolia. *Insects of Mongolia* 1: 785-90.

————. 1975. Asilidae, Subfamily Dasypogoninae (Diptera). Ergebnisse der zoologischen Forschungen von Dr Z. Kasab in der Mongolei. *Folia Entomologica Hungarica* 38: 337-340.

————. 1977. A new species of robber-fly of the genus *Lasiopogon* Loew (Diptera, Asilidae) from the southern Tuva. *Insects of Mongolia* 5: 685-87.

Rondani, C. 1856. Asilidae. *Dipterologiae Italicae Prodromus* 1: 1-226.

Rosen, D.E. 1978. Vicariant patterns and historical explanation in biogeography. *Systematic Zoology* 27: 159-88.

Saigusa, T. 1997. The geographical distribution of the genus *Rhamphomyia* (Diptera: Empidiae) in eastern Asia. In *Current situation and Prospects of Dipterology in Asia* (conference proceedings, 8 March 1997). Seoul: Korea University.

Schiner, J.R. 1862. *Fauna Austriaca. Die Fliegen (Diptera)*. I Theil. IX. Family Asilidae. Vienna: Carl Gerold's Sohn.

Scudder, G.G.E. 1979. Present patterns in the fauna and flora of Canada. In *Canada and Its Insect Fauna*, edited by H.V. Danks. *Memoirs of the Entomological Society of Canada* 108.

Séguy, E. 1927. *Faune de France*. 17: *Diptères (Brachycères)*. Paris: Fédération Francaise des Sociétés de Sciences Naturelles.

Schaeffer, C.F.A. 1916. New Diptera of the family Asilidae with notes on known species. *Journal of the New York Entomological Society* 24: 65-69.

Sinclair, B.J., J.M. Cumming and D.M. Wood. 1994. Homology and phylogenetic implications of male genitalia in Diptera - Lower Brachycera. *Entomologica Scandinavica* 24: 407-32.

Sinclair, B.J., and T. Saigusa. 1997. Review of the seepage-fly genus *Trichothaumalea* Edwards (Diptera: Culicomorpha: Thaumaleidae). Abstract. The Third Asia-Pacific Conference of Entomology, Taichung, Taiwan.

Stanley, S.M. 1989. *Earth and Life Through Time*, 2nd edition. New York: W.H. Freeman and Co.

Taira, A. 1990. *The Origins of Japan*. Tokyo: Iwanami-Shoten Co. (in Japanese).

Tatewaki, M. 1957. Geobotanical studies on the Kurile Islands. *Acta Horti Gotoburgensis* 21: 43-72.

Theodor, O. 1976. *On the Structure of the Spermathecae and Aedeagus in the Asilidae and Their Importance in the Systematics of the Family*. Jerusalem: Israel Academy of Sciences and Humanities.

————. 1980. *Fauna Palestina. Diptera: Asilidae*. Jerusalem: Israel Academy of Sciences and Humanities.

Timon-David, J. 1950. Dipterès des Pyrénées Ariégeoises: notes écologiques et biogéographiques. *Bulletin de la Société d'Histoire Naturelle de Toulouse* 85: 11-25.

Tsuchi, R. 1997. Marine climatic responses to neogene tectonics of the Pacific Ocean seaways. *Tectonophysics* 281: 113-24.

Van Dyke, E.C. 1919. The distribution of insects in western North America. *Annals of the Entomological Society of America* 12: 1-12.

Watrous, L.E., and Q.D. Wheeler. 1981. The out-group comparison method of character analysis. *Systematic Zoology* 30: 1-11.

Weinberg, M. 1978. Contribution to the knowledge of the morphology and biology of the species *Lasiopogon montanus* Schin. (Diptera, Asilidae) from the southern Carpathian mountains. *Traveaux du Museum d'Histoire Naturelle "Grigore Antipa"* 19: 293-96.

Weinberg, M., and G. Bächli. 1995. *Diptera: Asilidae.* Insecta Helvetica (Fauna), vol. 11. Geneva: Schweizerischen Entomologischen Gesellschaft.

Wheeler, Q.D. 1986. Character weighting and cladstic analysis. *Systematic Zoology* 35: 102-109.

Wheeler, T.A. 1991. Systematics of the New World *Rachispoda* Lioy (Diptera: Sphaeroceridae). Unpublished PhD thesis, University of Guelph, Canada.

Whitfield, F.G.S. 1925. The relationship between the feeding habits and structure of the mouthparts in the Asilidae. *Proceedings of the Zoological Society of London* 1925: 599-638.

Wilcox, J. 1966. *Efferia* Coquillett in America north of Mexico (Diptera: Asilidae). *Proceedings of the California Academy of Sciences* 34: 85-234.

Wilcox, J., and C.H. Martin. 1936. A review of the genus *Cyrtopogon* Loew in North America (Diptera- Asilidae). *Entomologica Americana* 16 (new series): 1-95.

Wiley, E.O. 1981. *Phylogenetics: The Theory and Practise of Phylogenetic Systematics.* New York: John Wiley.

————. 1988. Vicariance Biogeography. *Annual Review of Ecology and Systematics* 19: 513-42.

Williston, S.W. 1908. *Manual of North American Diptera*, 3rd edition. New Haven, Conn.: James T. Hathaway.

Wilson, M.V.H. 1978. Paleogene insect faunas of western North America. *Quaestiones Entomologicae* 14: 13-34.

Wood, D.M. 1990. Ground plan of the male genitalia of Brachycera (Diptera). In Abstract Volume (3rd Supplement), Second International Congress of Dipterology (Bratislava, Czechoslovakia).

————. 1991. Homology and phylogenetic implications of male genitalia in Diptera. The ground plan. In *Proceedings of the Second International Congress of Dipterology*, edited by L. Weismann, I. Országh and A.C. Pont. The Hague: SPB Academic Publishing.

Wood, G.C. 1981. Asilidae. In *Manual of Nearctic Diptera*, vol. 1, edited by J.F. McAlpine, B.V. Peterson, G.E. Shewell, H.J. Teskey, J.R. Vockeroth and D.M. Wood. Agriculture Canada Monograph 27. Ottawa: Agriculture Canada.

Woodley, N.E. 1989. Phylogeny and classification of the "Orthorrhaphous"

Brachycera. Chapter 15. In *Manual of Nearctic Diptera*, vol. 3, edited by J.F. McAlpine and D.M. Wood. Monograph 32. Ottawa: Agriculture Canada.

Yeates, D.K. 1994. The cladistics and classification of the Bombyliidae (Diptera: Asiloidea). *Bulletin of the American Museum of Natural History* 219: 1-191.

Yeates, D.K., and D. Grimaldi. 1993. A new *Metatrichia* window fly (Diptera: Scenopinidae) in Dominican amber, with a review of the systematics and biogeography of the genus. *American Museum Novitates* 3078: 1-8.

Yeates, D.K., and M.E. Irwin. 1996. Apioceridae (Insecta: Diptera): cladistic reappraisal and biogeography. *Zoological Journal of the Linnean Society* 166: 247-301.

INDEX

Page numbers in **bold** locate a species description, in ***bold italics*** a species distribution map, and in *italics* a figure, table or another kind of map.